Springer Proceedings in Physics 135

Springer Proceedings in Physics

Please view available titles in *Springer Proceedings in Physics* on series homepage http://www.springer.com/series/361/

Editors

Joung Hwan Lee · Habin Lee

Jung-Sik Kim

EKC 2009 Proceedings of EU-Korea Conference on Science and Technology

Springer

Editors

Joung Hwan Lee
The University of Sheffield
The Adv. Manufacturing
Research Centre with Boeing
Wallis Way, Catcliffe
S60 5TZ Rotherham
United Kingdom
E-mail: jhlee@kseauk.org

Habin Lee
Brunel University
Kingston Lane
UB8 3PH Uxbridge
United Kingdom
Email: habin.lee@kseauk.org

Jung-Sik Kim
Loughborough University
Aeronautical And Automotive
Engineering
LE11 3TU Loughborough
United Kingdom
E-mail: j.kim@kseauk.org

ISSN 0930-8989 e-ISSN 1867-4941

ISBN: 978-3-642-13623-8 e-ISBN: 978-3-642-13624-5

DOI 10.1007/978-3-642-13624-5

Springer Heidelberg Dordrecht London New York

Library of Congress Control Number: 2010928127

Cover Design: SPi Publisher Services

Printed on acid-free paper

Springer is part of Springer Science+Business Media (www.springer.com)

Preface

It gives us great pleasure to welcome you to this issue of the *Springer Book Series*. This issue is dedicated to the publication of selected papers written by researchers in Europe and Korea from the *EU- Korea Conference on Science and Technology* (http://www.ekc2009.org, 2009).

The year of 2009 bears a special significance to us, the Korean Scientists' and Engineers' Association in the UK (KSEAUK), as we reach this year, the 35th anniversary of the association's establishment in the UK, 35, the age of no skepticism but self-reliance. The KSEAUK laid its foundation for the last 35 years through a mountain of trials and errors and, also we have been significantly improved in terms of quality and quantity for the last 7 years. We now dare to have the confidence to say that we have contributed to the development of science and technology for our homeland.

Following the successful hosting of last year's EU-Korea Conference (EKC2008) in Heidelberg, Germany, the KSEAUK is very pleased to continue this tradition in 2009 in Reading, UK, with the hosting of EKC2009 at Wokefield Park near London. We are truly excited to be able to provide the opportunity and the venue where Europe's Korean scientists and engineers and also those scientists and engineers in Korea can meet in one place so as to exchange information and network with one another.

The KSEAUK, has prepared this year's EKC conference, under the title of "Science and Technology Closer to Humanity-Greenness" with the focus on "Green Science and Technology". Currently, it appears that the world is debilitated from the global economic downturn and in efforts to overcome this crisis, new technologies are constantly sought after. However, many of these efforts seem have reached the end of their line regarding new technologies in mono-disciplinary researches alone and the solution to this issues seems to spring out of interdisciplinary research approaches. Amongst these, the key Issue appears to be the interdisciplinary technologies combining both the issues of environment and energy solutions which involve a vast number of projects carried out by various groups of scholars from all over the world. I believe that South Korea is no exception here.

In line with this trend, we prepared an opportunity where those of you who work in the areas of advanced sciences and technologies in Europe and those of you who work in the same areas in Korea and those interested in the related environmental and energy policies in Korea, can exchange ideas.

During the event, parallel with this theme of EKC2009, the Korean Federation of Science and Technology Societies (KOFST) hosted a symposium entitled "Ultra-Programme", focussing on the area of environmentally friendly technology and energy in depth. Moreover a variety of expert presentations ranging from nanotechnology to aerospace engineering delivered at EKC2009, giving you a plenty of opportunities to information exchange and awareness of the issues.

We would like to thank the EKC2009 organising committee members, especially, Ki-Jun Lee, President of KOFST and Soo-Sam Kim, Vice-president of KOFST, Man-Wook Han, President of KOSEAA, Chang-Hoon Jun, President of ASCoF. Joon-Weon Seok, President of VeKNI, volunteers and sponsors who have made this conference possible.

Finally, the EKC2009 would not have been successful without the huge amount of efforts from the local organising committee members including Soo Hwan Kim, Namshik Han, Sung Eun Bae, Sung Joo Lee, Sungwook Park, Minsoo Kim, Jawon Song, Seongjin Baek, Young Duck Kang, Joseph Lee, Chang-Ho Choi, Anne Kim, Rira Kim, Ik Soo Kim and last but not least Jaeyoun Oh who helped a lot for preparing this book.

Thank you very much and please enjoy reading.

March 2010

Joung Hwan Lee
(EKC2009 Chair, President of KSEAUK)
Habin Lee
(Program Chair, SIG-T Leader of KSEAUK)
Jung-Sik Kim
(Vice President, KSEAUK)
Volume Editors

Table of Contents

Past, Present and Future Contribution on Research and Development of Aero-Propulsion Systems in Korea

Changduk Kong[1]

Abstract. Korean government has recently allocated and invested large research and development fund in both engineering and science related projects including, information technology, marine, aerospace and high speed train transportation etc.

Among them, aerospace engineering is an importance area for promoting the next generation economic and strategic growth. This paper hence aims to reflect past, present and future contribution on the research and development of air-breathing propulsion systems that has been carried out by several major aero propulsion related research institutes, companies, universities and societies in Korea.

Research institutions, companies and universities related to R&D of aerospace engineering in the Republic of Korea are structured in a more or less departmental manner where upcoming research projects are designated to be carried out by certain specialized research centers to handle the given project. However it is rational to classify them according to the research areas, activities performed and according to their long-term future anticipated projects.

1 Introduction

We can not underestimate the importance of aero power plants in producing the thrust force for flying vehicle. This makes aero propulsion engines a fundamental consideration in aircraft development. Early aircraft widely used reciprocating engines. However, since the development of the first experimental Whittle gas turbine engine in 1937, aero gas turbine engine became the major power plant for aero vehicles that are now more diverged depending on their applications. The aero-propulsion can be greatly divided into air-breathing engine and the engine related system. The air-breathing engine can be classified into turbojet, turbofan, turboshaft, turboprop, propfan, unducted fan, ramjet, scramjet, hybrid air-breathing engines, and each engine has the proper operation flight envelop depending on its flight velocity and altitude[1]. The air breathing engine is composed of some major components such as compressor, turbine, burner, nozzle, intake, fan, propeller, reduction gear, shaft, case, rotor blade and propfan, and some sub-systems such as engine controller, lubrication, fuel, cooling, starting and ignition

[1] Department of Aerospace Engineering, Chosun University,
375 Seosuk-dong Gwangju South Korea
cdgong@chosun.ac.kr

J.H. Lee et al. (Eds.): EU-Korea Conference on Science and Technology, SPPHY 135, pp. 1–16.
springerlink.com

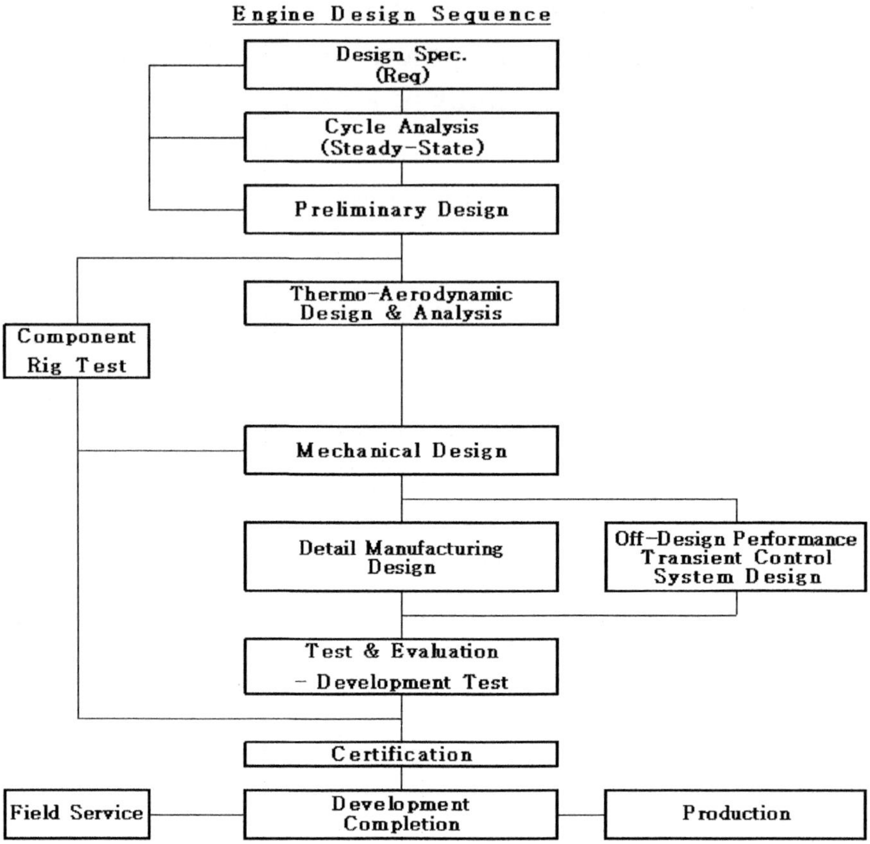

Fig. 1. Typical aero-gas turbine design procedure

system. Recent advanced technologies have been developed for more energy efficient such as low fuel consumption, more environment friendly such as low noise and low emission, more reliable and long life gas turbine engines in worldwide. Typical development procedure used in the development of engines is as shown in Fig. 1. [1]

The propulsion system integration also is an important stage towards the development of aero vehicles. Figure 2 shows various activities of the propulsion system integration and design steps such as engine selection, inlet sizing, nozzle sizing, aerodynamic analysis and test of inlet and nozzle, propulsion system performance analysis and test, structural design and analysis, etc. [2]

Korea has a short history in air breathing propulsion system that dates back in the 1970s. In order to obtain the gas turbine development technology, some major companies and research institutes have invested development cost of more than 0.1 billion US dollars since 1991. [3] The Agency for Defense Development (ADD) has contributed greatly to research and development of air breathing

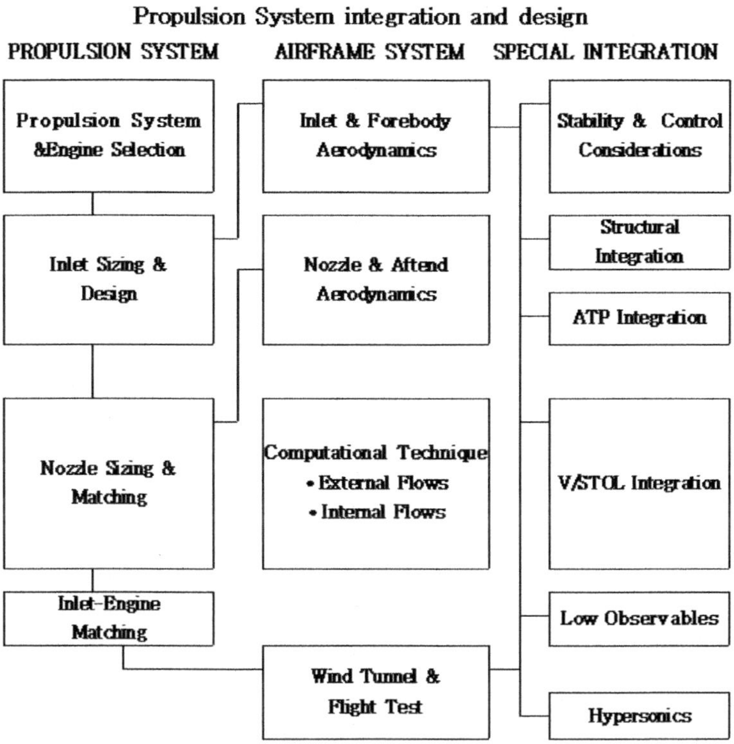

Fig. 2. Propulsion System integration procedure

engine in Korea. At the start of 1980, the 1334N thrust class turbojet engine was developed for a decoy mission RPV (see Fig. 3). This marked the beginning of turbojet engine development in Korea. By the end of 1999, the 4448N thrust class turbo jet engine (SS-760K) was developed for a special flight vehicle. Projects supervised and developed under the watch of the author as a head of Aero-propulsion Division of ADD at the time.

Currently Korea has the capability to develop a 1000HP class gas turbine with an additional 0.1 billion US dollars invested for 5 year started in 2006. The

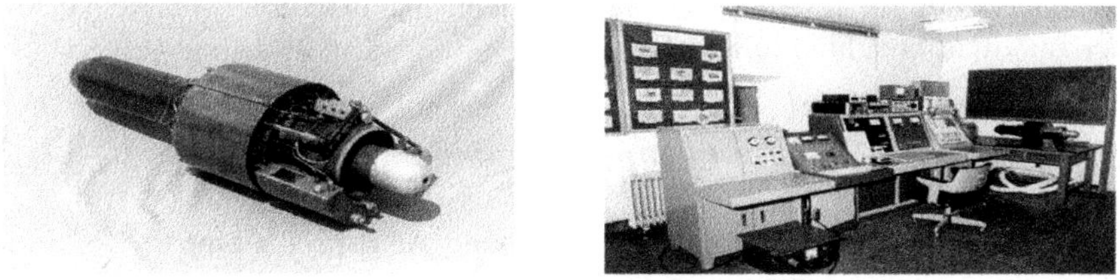

Fig. 3. Expendable 1334N class turbojet engine and engine control room

Korean aero engine has an annual market size about 0.4 billion US dollars including 51% for military and 49% for civil. Korea has a mid-term technological aero engine development plan for the domestic development of a high altitude UAV's gas turbine engine, international co-development of the KHP's turbo shaft engine, the domestic development of the KHP's APU, the international co-development of a commercial turbofan engine, etc. Korean research institutions and university are not only inland based but also work in collaboration with other overseas countries like USA, Russia, Japan, Israel and UK among others serving as a good guide to evaluate Korean aerospace potentials and to understand the general global trend in aero-engine development. Figure 4 shows 2007 technology road map for development of aero power plant parts which was generated by Korean government. [3]

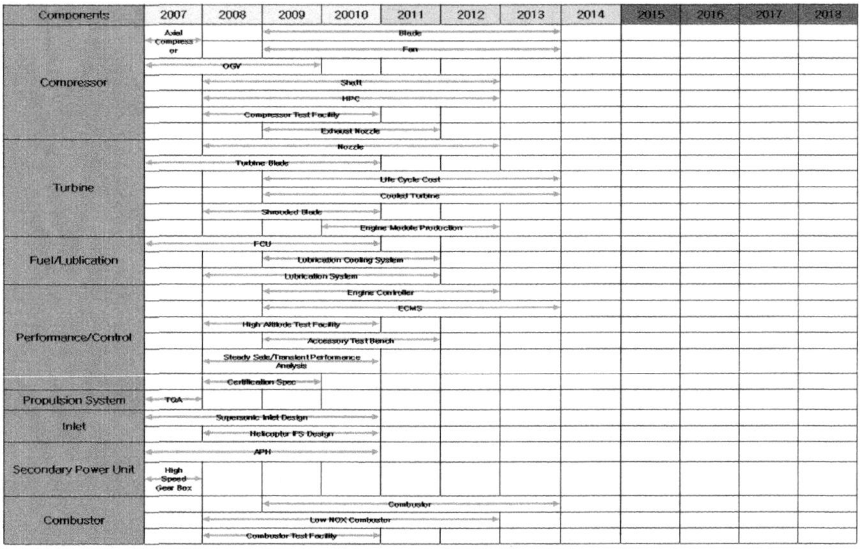

Fig. 4. 2007 Technology road map for development of aero power plant parts

2 Current Status of Korean Air breathing Engines

The following article briefly introduces some major research institutes, industries and universities for research and development, manufacturing, maintenance and overhaul of air breathing in Korea.

2.1 Samsung Techwin Company

Samsung Techwin Company (Former Samsung Aerospace Company) is a major air breathing engine company in Korea. This company started operation in 1970 by offering overhauls and licensed production of military and civil aircraft engines for example T-56, T-53, T-700 for UH-60 helicopter, LM2500, A250, J-79 for F-4 fighter, J-85 (1980) for F-5A,B fighter, F-100(1994) for F-16 fighter, F404(2002) for T-50 supersonic advanced trainer and A-50 attack, F110(2005) for F-15K fighter etc., and they have also performed co-works with domestic and international institutes and companies for research, development and production of K77 APU, F100 turbo charger(1996), turbo master(1997), turbo compressor SM1000 (micro scale) and SM6000(large scale)(2001), Turbo green 1200(1997), 100kW APU(2002), T700 turbo shaft engine modification for KHP helicopter(2007) , APU for KHP helicopter (2007), SS760K turbojet engine for the anti-ship missile, etc. (see Fig.. 5) [5]

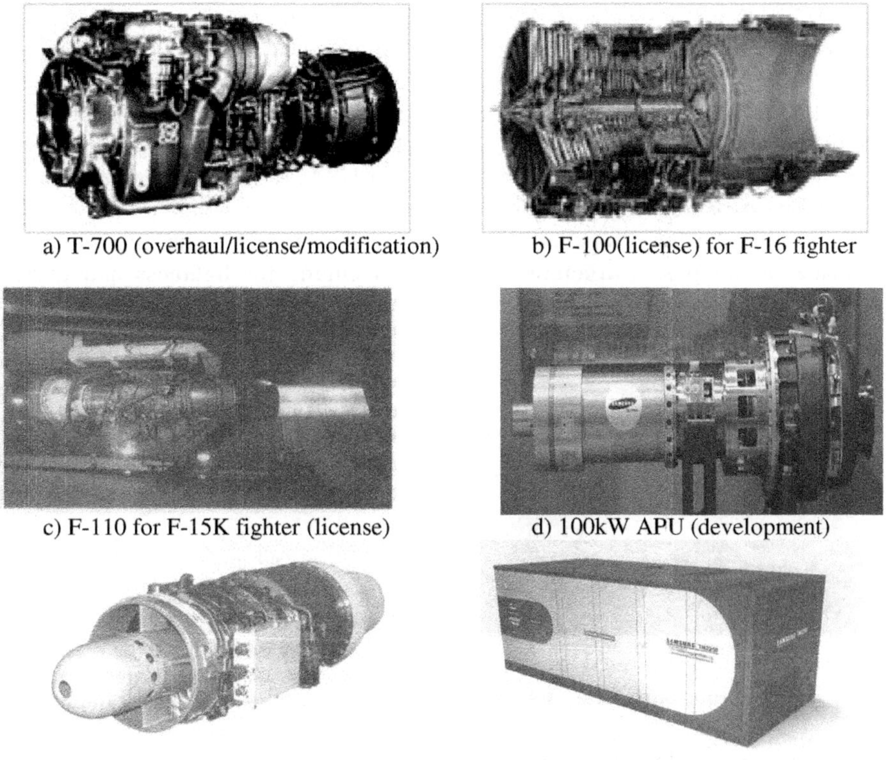

a) T-700 (overhaul/license/modification) b) F-100(license) for F-16 fighter

c) F-110 for F-15K fighter (license) d) 100kW APU (development)

e) SS-760K Turbojet (development) f) SM6000 Turbo compressor

Fig. 5. Aero engines overhauled, licensed and developed by Samsung Techwin

The SS-760K turbojet engine was especially co-developed for a special flight vehicle by the Agency for Defense Development and Samsung Techwin Company between 1990 to 2002. This engine was the first successful production jet engine in Korea, and the author prepared its development plan and was initially responsible for development of this engine. This engine is now produced for flight vehicles and UAVs, and has several derivatives such as turbofans and scaled turbojet engines.

2.2 *Korea Aerospace Industry*

Korea Aerospace Industry (KAI) began operation in 1999 and is mainly an aircraft manufacturing company. Since establishment, KAI has produced several military use aircrafts such as the basic trainer KT-1, the advanced supersonic trainer T-50 (see Fig. 6), the attack A-50 and the reconnaissance UAV which were developed by ADD and KAI.[7] The company hence has a department for development of the propulsion system integration for those developed aircrafts. Recently this company is co-working with ADD and KARI for developing the next generation Korean fighter KFX and the next generation Korean helicopter project KHP for military use (see Fig. 7) and the 10~20 seat business jet for civil use. The preliminary design for the propulsion system including air induction system, exhaust system, fuel supply system, APU system, rotor, and transmission system were completed in July, 2007, with the detailed design set for completion by the end of 2007. Recently, in order to establish BASA (Bilateral Aviation Safety Agreement) system with FAA (Federal Aviation Agency) of USA, Korean government started to develop a small aircraft (4 seat class) with a reciprocating engine that is called KPP (Korean Poston Propeller) aircraft (see Fig. 8). This aircraft will fully be made of composite structure and a diesel engine for lightness and energy efficiency.

a) Supersonic Advanced Trainer T-50

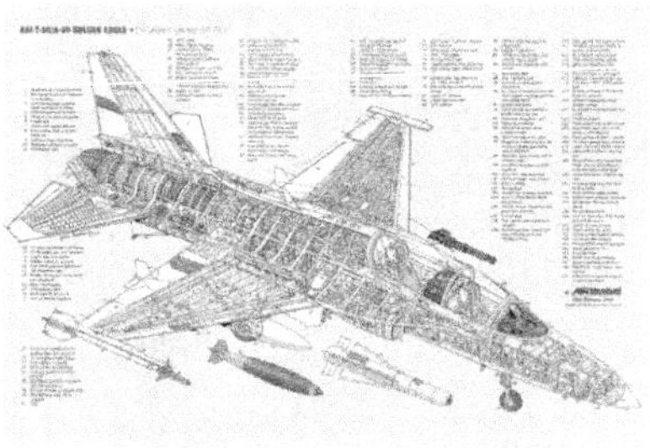

b) Schematic view of T-50

c) F404 Low bypass turbofan engine

Fig. 6. Advanced supersonic trainer T-50 with turbofan engine F404 developed by KAI

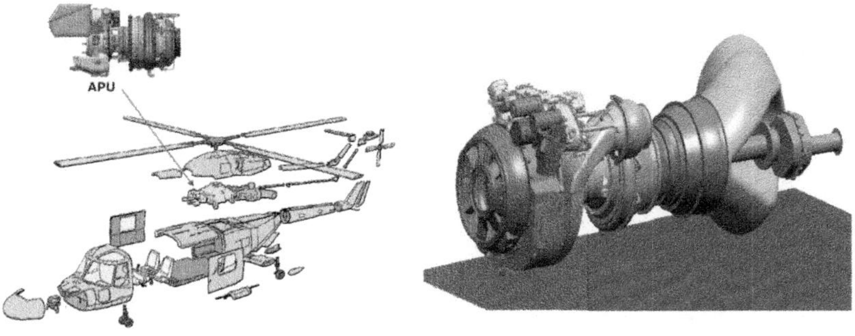

a) Propulsion system with engine, APU, induction, exhaust, transmission, fuel and rotor systems (Fully developed)
b) T-700 Turbo shaft engine (Partially modified)

Fig. 7. KHP development project

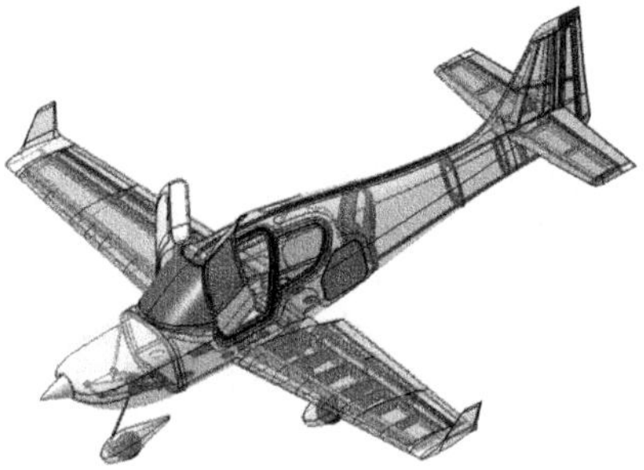

Fig. 8. 4 seat class all composite aircraft using diesel engine for establishing BASA with FAA

2.3 Korean Air Line

Korean Air Line (KAL) is the biggest air transportation service company in Korea. This company has a maintenance and overhaul facility for its civil fleets and engines and a research institute to develop light aircrafts, components and parts for aircrafts and engines, etc. In 1975, KAL developed a light aircraft named Chang-Gong 91 with a reciprocating engine and has manufactured 4 prototype aircrafts. Since that time, several aircrafts, for examples the 8 seat twin propeller engine composite aircraft in 1990s, Bandi composite aircraft and its modification aircraft in 2000s, have been developed by Korean Aerospace Research Institute.

Recently in order to establish BASA system with USA, Korean government started to develop a small aircraft (4 seat class) with a reciprocating engine for CFR-23 class. (see Fig. 9)[8]

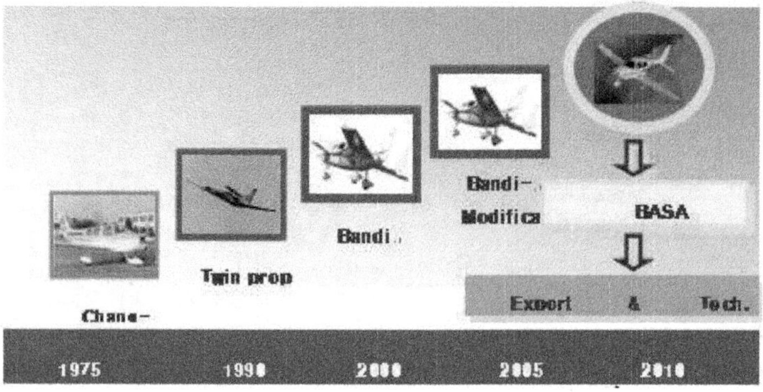

Fig. 9. R&D activities of civil small aircrafts in Korea

2.4 Agency for Defense Development

Agency for Defense Development (ADD) was founded in 1970 to carry out research and development of military weapon system including tank, gun, missile, ship, aircraft, etc. The propulsion department of this organization has developed small turbojets such as previously mentioned the 1334N thrust class turbojet and the 4448N thrust class turbo jets(SS-760K) and turbofan engines and ramjet engines for special purpose UAV vehicles. Figure 10 shows the 667N thrust micro turbojet engine for a small scale UAV which has been developed by ADD since 2007. This engine has a centrifugal compressor, a reversible flow combustor, a stage axial turbine, rotary type fuel nozzle system and a DEEC engine controller. The aircraft development group has developed the propulsion system integrations for the basic military trainer KT-1(see Fig. 11), the advanced supersonic trainer KTX-2 (current T-50) and the next generation Korean fighter KFX (see Fig. 12). Figure 11 shows the KT-1 with 950 SHP turboprop engine (PT6A-62), the propulsion system integration test view of KTX-1 (1st version of KT-1) with 550 SHP turboprop engine (PT6A-25A). To date ADD has started to develop the propulsion system for a long duration UAV in about 13500m altitude. A 1200HP class turboprop engine will be used for this purpose. However the engine and the related propulsion system will be slightly modified to meet the mission requirements. [6]

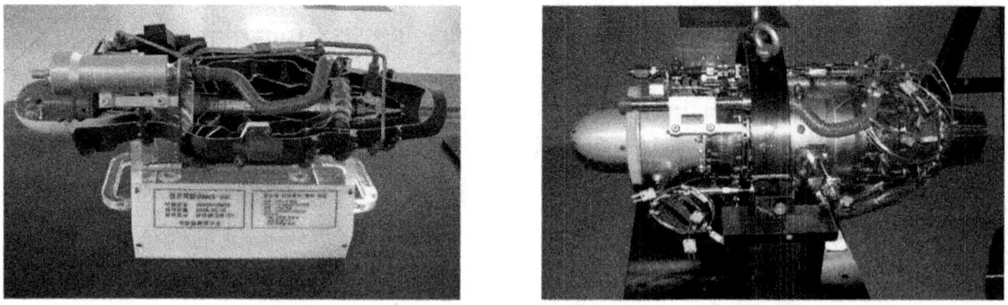

Fig. 10. 667N thrust micro turbojet engine for small scale UAV

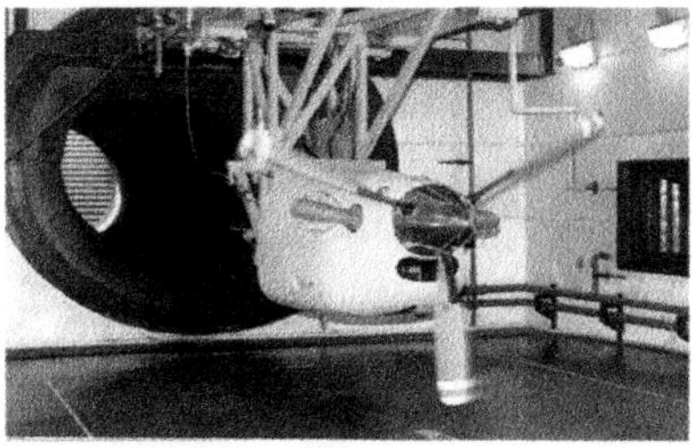

Fig. 11. Korea Basic Trainer K-1 with turboprop engine PWC PT6A-62 and propulsion system integration test

Fig. 12. Conceptual vies of next generation KFX with low bypass turbofan engine

2.5 Korea Aerospace Research Institute

KARI (Korea Aerospace Research Institute) founded in 1989 by Korean government to cater for the research and development of aerospace engineering on aircraft, air-breathing engine, UAV, rocket, launcher vehicle, satellite, etc. It is composed of the aero propulsion department whose main activities are to perform research and development of gas turbine engines, turbo ramjet, scramjet engine, APU, turbo compressor, turbo charger, turbo pump for a liquid rocket engine and propulsion system integration for the KHP, the smart UAV, the small aircraft Bandi, airship, commuter, etc. This department also has some test facilities that supports the research works; such as the gas turbine component test facilities for turbo pump, compressor and combustor, the altitude engine test facility, the shock tube, the supersonic wind tunnel and the ramjet/scramjet test facility.

The propulsion department of KARI classifies R&D area as the gas turbine technology, the air vehicle propulsion system integration and the future propulsion system technology. The gas turbine technology is divided into design, analysis and test of compressor, combustor and turbine, the system integration and performance test, and the development for aircraft and industrial gas turbines. And the air vehicle propulsion system integration is divided into the fuselage/wing/engine

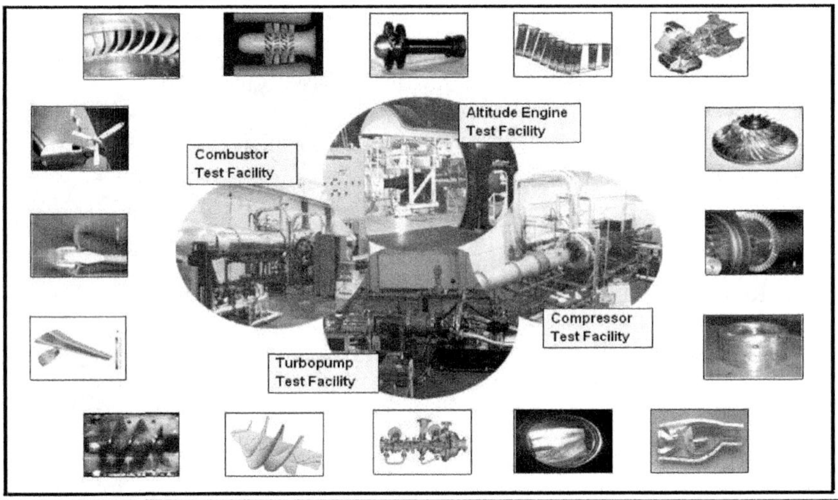

Fig. 13. R&D activities and test facilities of Aero-Propulsion Department of KARI

matching technology, the nacelle system (inlet exhaust nozzle and reverse thruster) and the aircraft subsystems (APU ECS etc). In the future propulsion system technology, there are air turbo-ramjet, ramjet and scramjets. [4]

Figure 13 shows pictures and graphical views on various research and development activities and test facilities of Aero-Propulsion Department of KARI.

Recently, KARI started to develop the next generation small/high loading compressor using MDO(Multi Disciplinary Optimization) technology for

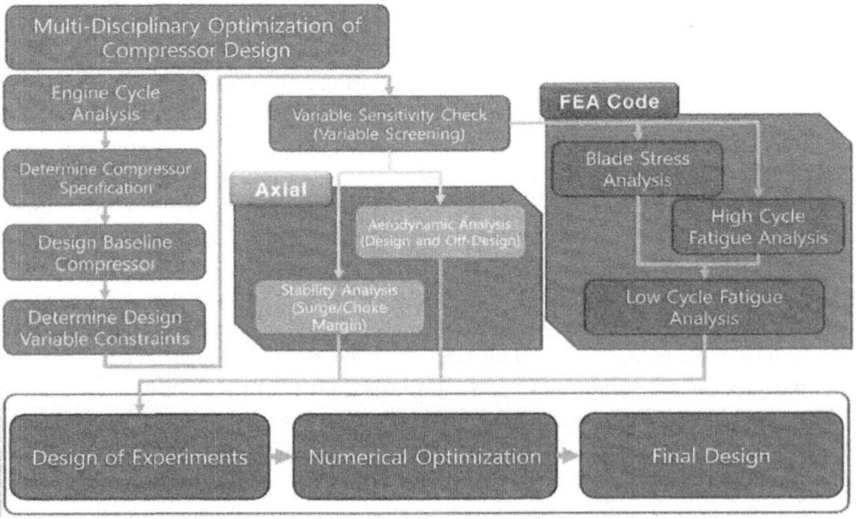

Fig. 14. Development flow of next generation small/high loading compressor using MDO

3000~6000lbf class turbofan engine which has 4 years development plan. Figure 14 shows the flow of development procedure using MDO. This compressor may be used for turbofan engines of VLJ (Very Light Jet) or regional jet class aircrafts.

Figure 15 shows that propulsion system including PW206C turbo shaft engine, tilt rotor, fuel supply system, inlet system, exhaust systems and other engine related system for the Smart UAV being developed by KARI started in 2005 aimed to finish in 2012 prototype UAVs are now being manufactured for ground technology demonstration.

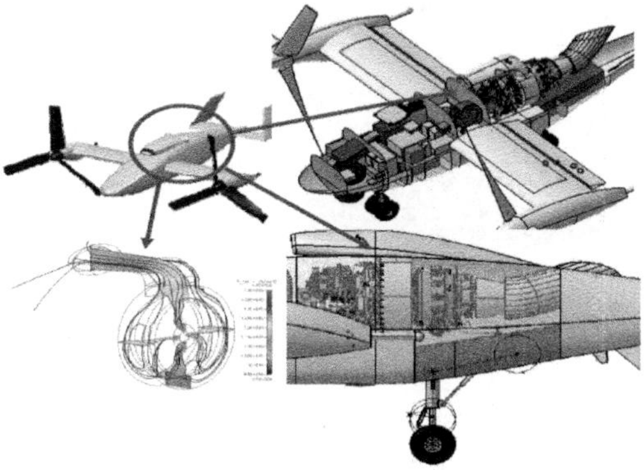

a) Propulsion system of Smart UAV (engine, tilt rotor, fuel, inlet, exhaust systems)

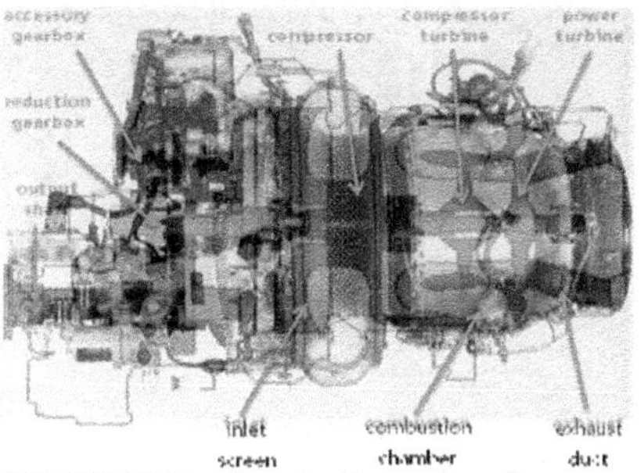

b) PW206C Turbo shaft engine for SMART UAV

c) 3-D view of Smart UAV

Fig. 15. Smart UAV with tilt-rotor and turbo shaft engine (PW206C) for vertical take-off and landing

Figure 16 shows the altitude test facility that can test the gas turbine engine up to maximum altitude of 30,000 ft with maximum inlet air velocity of Mach number 1, maximum air flow of 23kg/s and maximum thrust of 3,000lbf. In addition, KARI has built a new altitude test facility for the T700 class turboshaft engine.

Figure 17 shows the Ramjet and Scramjet test facility that can test the gas turbine engine up to a maximum altitude of 25 km with maximum inlet air velocity of Mach number 5, maximum air flow of 47kg/s and minimum test time of 78sec.

AETF (Altitude Engine Test Facility) KARI

❍ **Overall engine performance test under the simulated environments**

✓ Designed and Built by Sverdrup Technology Inc. from Sep. 1996 to Sep. 1999

Specification	
Altitude	0 ~ 30,000 ft
Mach number	0 ~ 1.0
Inlet temp.	-75 ~ 110 °C
Inlet Pressure	31 ~ 350 kPa
Max. flowrate	23 kg/sec
Max. thrust	3,000 lbf @192 kPa, 286 K
Test cell Dimensions	3.5 m(D) x 6.0 m(L)
DAS channels	Pressure : 160 Temperature : 192 Vibration : 121 Miscellaneous : 288

Fig. 16. Altitude Engine Test Facility of KARI

Ram/SCRamjet Test Facility KARI

❍ **Scope**
- Intake pressure recovery rate
- Supersonic combustion
- Mixing enhancement
- Flame stabilization
- Supersonic Nozzle

❍ **Test Rig**
- Closed flow loop
- Freejet type
- Capacity
  - Operating range : M 2-5, h 0-25km
  - Area of nozzle exit : 200x200mm^2
  - Max. air volume : 40 m^3
  - Max. flow rate : 47 kg/s
  - Min. test time : 78 sec

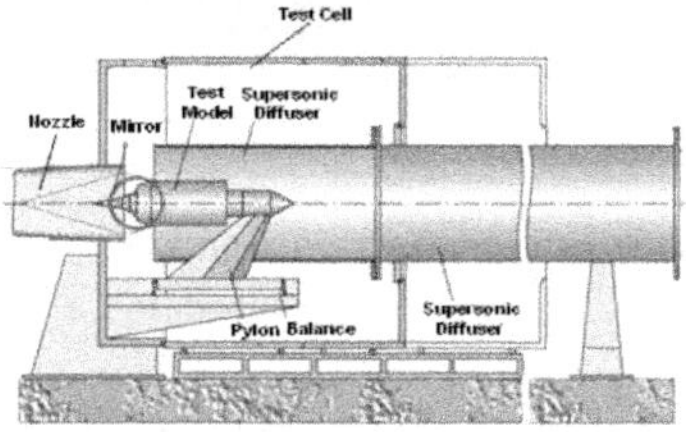

Fig. 17. Ram/Scramjet Test Facility of KARI

2.6 *Universities and Societies Related to Research on Aero-propulsion*

Several universities and societies related to research on aero-propulsion engineering are briefly introduced in the sections below [9].

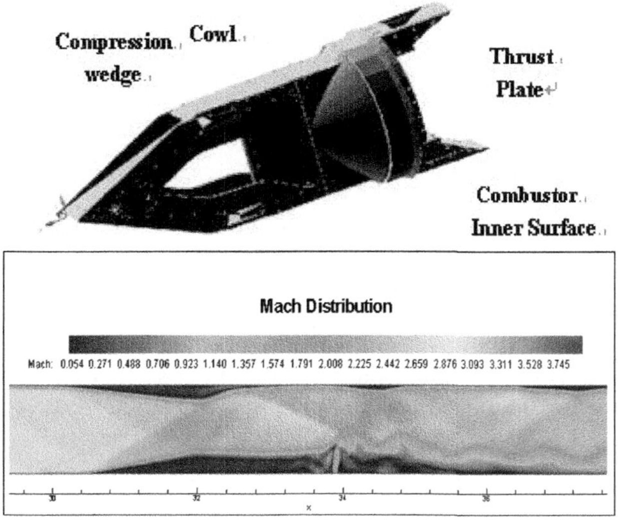

Fig. 18. HyShot Scramjet Engine and CFD analysis studied by SNU

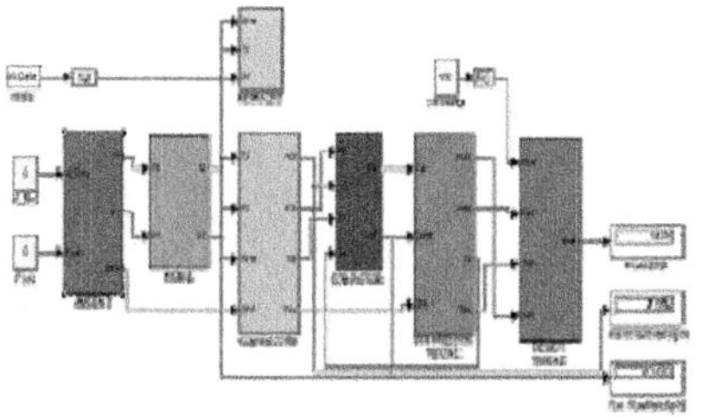

a) Performance model of a turbo-shaft engine

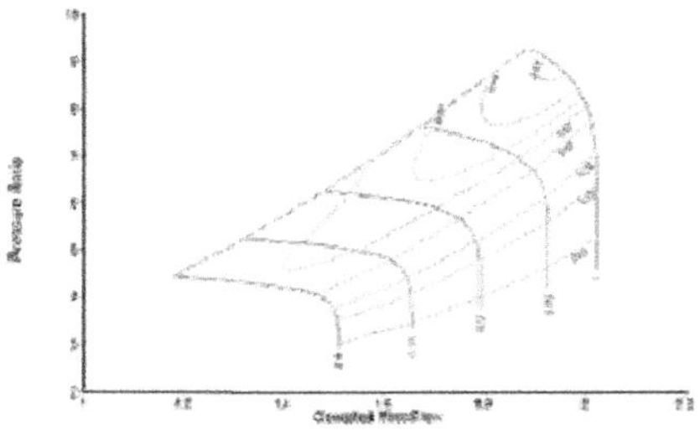

b) Generated compressor map using

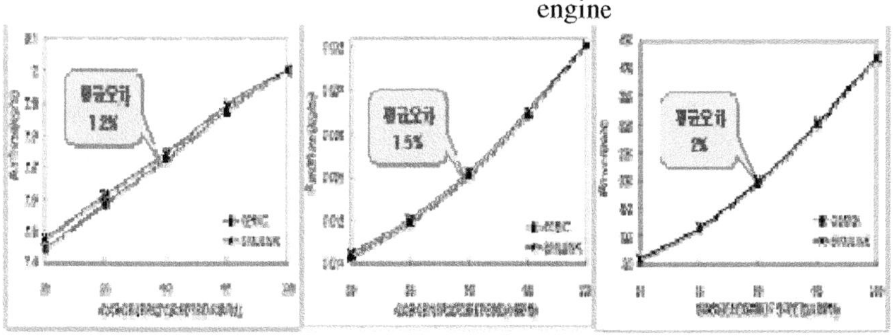

C) Comparison results between SIMULINK model and OEM data

Fig. 19. GUI type engine performance simulation program using SIMULINK and compressor map generated by the hybrid method and performance simulation results using the developed performance model

The high speed aero engine research center of Seoul National University (SNU) started the study on scramjet (see Fig. 18) and high speed combustor in 1980 and ever since they have acquired several test facilities like shock tube, high enthalpy supersonic wind tunnel, etc. And turbo machinery R&D center has developed gas turbine components ranging from compressors, combustor, turbines, etc with its own component test facilities.

The turbo machinery group and RR UTC of Pusan University started the study on gas turbine component design and numerical flow simulation using CFD techniques and test using its sub-scale test facilities in early 2000.

The gas turbine research group of Chosun University started the study on gas turbine performance simulation (see Fig. 19), control and health monitoring system (see Fig. 20) in 1995, and this group developed a micro gas turbine engine test system to demonstrate how on-condition monitoring system works. Related work has recently been proposed by author and Easy Gas Turbine Co., (see Fig. 21).

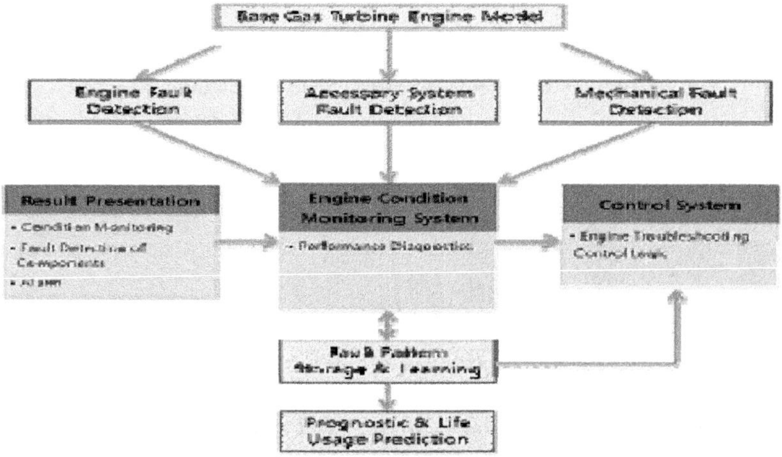

a) Flow of Integrated on-condition EMS

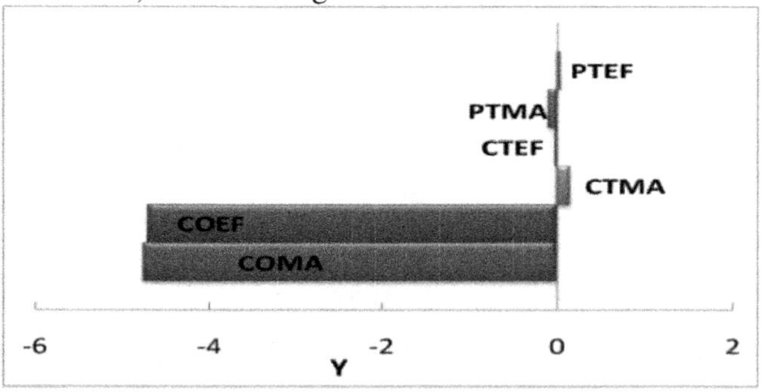

B) Detected contaminant fault of compressor

Fig. 20. Integrated on-board/on-condition engine health monitoring system using Neural Network to be developed by CU

Fig. 21. Micro gas turbine engine test system which can demonstrate the on-condition monitoring system

There are several societies related to aero propulsion engineering in Korea. For instances, The Korean Society of Propulsion Engineers(KSPE), The Korean Society for Aeronautical and Space Sciences(KSAS), The Society for Aerospace System Engineering(SASE) and Korean Aeronautical Engineers' Association(KAEA). Among them, the KSPE mainly focuses on propulsion system including aero and space propulsion. This society publishes a journal for technical papers with 6 issues annually, and it holds 2 domestic conferences annually and an international conference Asian Joint Conference for Power and Propulsion (AJCPP) every two years.

3 Conclusion

This paper aims to introduce the current trend and status in research and development of air-breathing propulsion systems as performed by several major aero propulsion related research institutes, companies, universities and societies in Korea. Even though Korea has a short history in air breathing propulsion system, its technology and engineering has rapidly developed through lots of activities listing from research and development, manufacturing, maintenance and overhaul by research institutes, companies and universities.

According to Korean government's technology road map on the air breathing engine, it is my anticipation that Korea will and has the capability to develop turboshaft engines for civil use on helicopters, a small scale turbofan engines for civil use and special purpose ramjet engine for UAVs already in our next ten years development budget.

Acknowledgments

The author would like to express his thanks for some valuable information provided by KARI and some other companies and institutes.

References

[1] Kong, C., et al.: Aero Gas Turbine Engine, Donmyungsa (2004)
[2] Kong, C.: Propulsion system integration turboprop aircraft for basic trainer. Aircraft Engineering and Aerospace Technology 72-6, 524–536 (2000)
[3] KOTEF, 2007 Technology Road Map for Development of Aerospace Parts, Ministry of Commerce, Industry and Energy, pp. 168-176 (2007)
[4] Yang, S.: An Overview of Activities on Gas Turbine Engines. In: Proceedings of 2006 Seminar of Propulsion Division of KSAS. The Korean Society for Aeronautical and Space Sciences, pp. 4–21 (2006)
[5] Techwin, S.: Technology Winner in Digital World. Aeronautical Engineering & Information, 5–7 (2007)
[6] ADDAMDC, An idea on core technology development related to military aircraft business. In: Proceedings of Aerospace Weapon System Development Seminar 2006, pp. 55–83. Agency for Defense Development (2006)
[7] KAI, Next Generation Trainer for Next Generation Fighters. Brochure for T-50 and KT-1, Korea Aerospace Industries, Ltd. (2007)
[8] KAL, Proposal on Small Aircraft Development for BASA Certification. Presentation for Workshop on Planning of Aircraft Safety Technology Development (2007)
[9] Kong, C.: Introduction to R&D Activities on Gas Turbine Technology of Korean University. In: Proceedings of 2007 Summer Workshop on Rocket Propulsion and Aero Gas Turbine. The Korean Society for Aeronautical and Space Sciences and Korean Society of Propulsion Engineers, pp. 4–21 (2007)

Application of Bioelectrochemical Process (BES) for Electricity Generation and Sustainable Wastewater Treatment

Jung Rae Kim[1]

Abstract. Bioelectrochemical system such as microbial fuel cells (MFCs) and microbial electrolysis cell are an emerging technology which converts biodegradable organic matter to electrical energy or hydrogen using a biofilm on the electrode as the biocatalyst. It has recently been shown that waste-to-energy technology based on MFC can treat organic contaminant in domestic or industrial wastewater and simultaneously produce electricity. The maximum power density increased up to $1kW/m^3$ based on reactor volume. Bioelectrochemical systems may reduce the energy consumption for wastewater treatment by replacing energy intensive aeration of present treatment systems, while generate electrical energy from waste. In addition, the biomass production in MFCs has been reported to be 10-50% of conventional wastewater treatment, leading to reduce environmental impact and disposal costs. Various electrochemically active bacteria metabolize biodegradable organic compounds then discharge electrons to an extracellular electron acceptor for bacterial respiration. These bacteria also transfer electrons to electrodes by direct electron transfer, electron mediators or shuttles, and electrically conductive nanowires. Investigation of bacterial electron transport mechanisms may improve understanding of the biomaterial involved and metabolic pathways as well as improving power from MFCs. Biofuel cell systems require interdisciplinary research ranging from electrochemistry, microbiology, material science and surface chemistry to engineering such as reactor design, operation and modelling. Collaboration within each study and integration of systems might increase the performance and feasibility of BES process for sustainable energy.

1 Overview

Environmental technology has generally focused on waste disposal and the treatment of organic and inorganic contaminants. However, recently this paradigm has shifted toward considering waste as a potential resource for production of energy and useful products with the development of environmental technologies (Iranpour et al. 1999; Angenent et al. 2004; Logan 2004). Conventional biogas

---

[1] Sustainable Environment Research Centre, Faculty of Advanced Technology, University of Glamorgan, Pntryridd, Mid-Glamorgan, CF37 1DL, United Kingdom
jkim@glam.ac.uk

J.H. Lee et al. (Eds.): EU-Korea Conference on Science and Technology, SPPHY 135, pp. 17–23.
springerlink.com

production processes have developed for waste-to-energy technology. Methane production by anaerobic process is a well-known biological process using ubiquitous methanogens for methane production. Among the high-rate anaerobic reactors developed and successfully applied for methane generation processes in recent years, the upflow anaerobic sludge blanket (UASB) reactor has become one of the most popular designs for the biological treatment of effluents and methane production. Anaerobic digestion (AD) technology is on the process of commercialization for the field scale.

Hydrogen is believed to be a sustainable energy source which could replace fossil fuels in the near future (Logan 2004). Biological hydrogen production has been in the spotlight as an environmentally friendly process for the provision of hydrogen age. The technology to recover hydrogen without methanogenesis have been investigated because hydrogen is more valuable than methane in terms of energy content and economic aspects (Oh et al. 2003; Logan 2004; Park et al. 2005).

Recently, bioelectrochemical systems such as microbial fuel cell (MFC) and microbial electrolysis cell (MEC) have been investigated for targeting energy recovery from wastewater treatment. Microbial fuel cells (MFCs) convert various organic substrates into electricity using microorganisms as a biocatalyst without using combustion process. They have advantages in regard to how they directly convert a wide range of organic matter such as acetate, glucose, and domestic and industrial wastewater into electricity without the combined need for gas separation and reforming processes. Microbial electrolysis cell (MEC) use the potential generated from the organic matter by bacterial metabolic degradation, and produce hydrogen by providing additional voltage from power supply.

BES systems are novel bioprocesses which based on combination of conventional biochemical process and electrochemical concept. BES systems have been in the spotlight as an environmentally friendly process for energy recovery and sustainable waste treatment system. In this report, the recent progress of BES system and applications for sustainable environmental technology will be discussed.

2 BES Processes: Microbial Fuel Cells (MFCs) and Microbial Electrolysis Cells (MECs)

2.1 Microbial Fuel Cells (MFCs)

A microbial fuel cell is new type of bioreactor for electricity generation which employs the capability of bacteria to transfer electrons from bacterial respiratory chain to solid electron acceptor (electrode). The schematic diagram of MFC is shown in Fig 1a. MFC consists of anode and cathode chamber which is normally separated by ion exchange membrane. Bacteria on the anode electrode metabolize organic compound and discharge electron and proton. The electron is transported through external circuit as electricity generation while the proton is transferred into cathode electrode through ion exchange membrane. At the cathode electrode, the proton reacts with oxygen and electron into water to complete electrochemical reaction. The MFC is a kind of galvanic cell which uses spontaneous biological reaction. The performance of MFC has been rapidly improved in power production and treatment efficiency (Logan 2008).

2.2 *Microbial Electrolysis Cells (MECs)*

The hydrogen production in a bioelectrochemical system can be possible by electron from exoelectrogenic bacteria in MFC (Fig 1b). The electrons discharged by bacteria in anode electrode are combined with the protons in cathode electrode to produce hydrogen gas by additional power source. Theoretically, the calculated cell voltage for hydrogen production at the cathode is estimated to be 0.114V but it is required about 0.25V in practice due to overpotential. However, supply of voltage is still lower than the voltage required for water electrolysis which is practically over 1.8V, therefore MEC save energy consumption for hydrogen production (Logan 2008). It was reported that hydrogen production rate by MEC has been reported to be 1.1 m^3 H_2 $/m^3$/day with overall energy efficiency of 82% (Cheng and Logan 2007).

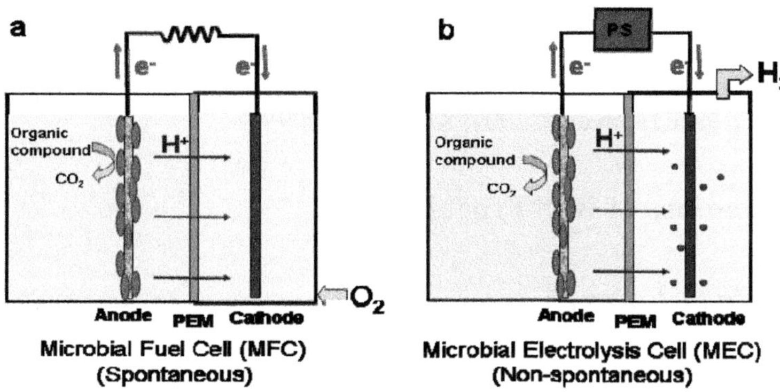

Fig. 1. Schematic diagram of microbial fuel cell (a) and microbial electrolysis cell (b). The bacteria is attached on the anode electrode and the cathode is sparged by oxygen for reduction reaction of proton and electron to make water (MFC) or kept in anoxic condition (MEC). Additional power supply (PS) provide potential for cathode reduction of proton

2.3 *Improvement of Design of BES Systems*

Early MFCs were typically designed and operated as a two-chamber system, where the chambers were separated by a cation exchange membrane (CEM). However, various reactor configurations have been examined to increase MFC performance. Liu and Logan (Liu and Logan 2004) tested a single chamber microbial fuel cell (SCMFC) lacking a proton exchange membrane and compared it to the same system containing a CEM. They reported that removing the membrane and using an air-cathode increased the maximum power density to 494 ± 21 mW/m^2 (12.5 ± 0.5 mW/L; liquid volume) as compared to 262 ± 10 mW/m^2

(6.6 ± 0.3 mW/L) with the membrane. In addition, reducing the electrode spacing and optimizing ion conductivity can increase power due to the decrease in the internal resistance (Liu et al. 2005; Cheng et al. 2006). Rabaey et al. (Rabaey et al. 2003) obtained a maximum power output of 4.31 W/m^2 (664 mV, 30.9 mA) using a flat two-chambered MFC and ferricyanide as a cathode electrolyte, with 81% of electron transfer efficiency.

An early application of a continuous MFC reactor was a flat plate system which had a serpentine channel achieving a plug flow like design (Min and Logan 2004). MFCs have also been designed using a tubular single chamber design (Rabaey et al. 2005), and a porous anode that allows advective flow between the anode and cathode chamber (Cheng et al. 2006). He et al. (He et al. 2005) proposed an upflow microbial fuel cell (UMFC) that was a modified upflow anaerobic sludge blanket (UASB) reactor for MFC operation with a maximum power density of 170 mW/m^2. Recently, stacked MFCs, have been introduced which connect each flat reactor in parallel and series to increase the cell potential. The power density of this stack was 258 W/m^3 using a ferricyanide cathode that is significantly improved from single unit of fuel cell (Aelterman et al. 2006). It was reported cassette type MFC produce 129 W/m^3 of power density (Shimoyama et al. 2008). Recently the maximum power density increased up to 1kW/m^3 based on reactor volume (Fan et al. 2007). Modular tubular microbial fuel cell with membrane electrode assembly cathode has been suggested as post treatment polishing process for high organic loading biogas production system (Kim et al. 2009; Kim et al. 2009).

3 Applications of BES Processes

3.1 Waste-to-Energy

In order to remove organic contaminant, present wastewater treatment systems have employed activated sludge system which consume great amount of electrical energy (14.7 TWh/yr in Europe) for aeration and pumping while produce biomass production which cause secondary disposal problem. It is expected that bioelectrochemical systems reduce the energy consumption for wastewater treatment by replacing energy intensive aeration of present treatment systems, while generate electrical energy from waste. MFC has been known to treat various organic contaminants in domestic or industrial wastewater such as swine and recalcitrant wastewater (Kim et al. 2008). Energy recovery from wastewater has been estimated to be 1kW/1m3 wastewater (Rabaey and Verstraete 2005). In addition, the biomass production in MFCs has been reported to be 10-50% of conventional wastewater treatment, leading to reduce environmental impact and disposal costs.

MFC has also been suggested as a complementary process with conventional AD and BioH_2 systems such as polishing stage (Pham et al. 2006; Rozendal et al. 2008). Combination of BES process with the energy recovery system could allow for improved sustainability of wastewater treatment systems. Control strategy using in-situ operational parameter monitoring of combined system would be required to maximize energy recovery and treatment efficiency.

3.2 *Bioensor for In-situ Monitoring of Organic Contaminant*

BOD (Biological oxygen demand) has been used as standard for contaminant as it represents a biodegradable organic compound in wastewater. However it requires time and separated system for analysis. MFC type biosensor using bacterial electron transfer mechanism is another advantage of MFC. BOD sensor using electrical response of bacteria may replace conventional BOD5 system as its electrical response to biodegradable organic contaminant shows linear in the range 5 to several hundreds mg/l (Kim et al. 2006). Therefore it makes feasible for in situ monitoring and controlling of biological system including wastewater treatment.

3.3 *Bacterial Electron Transfer Mechanism: Nanowire and Electron Shuttle*

It has been known that electrochemically active bacteria transfer electron to electrode by direct electron transfer, electron mediator or shuttle, and electrically conductive nanowire. Diverse bacterial species in the MFC may use these mechanisms in combination according to their location and environment in the biofilm. Electrically conductive nanowire is particularly interesting as it transfer electron to remote electron acceptor efficiently. Bacterial cells have been known to extrude polysaccharide strands and to localize iron precipitation in proximity to the cell membrane. It has been suggested that the conductive biological nanowires can serve as an electron transfer conductors from the cell surface to the Fe(III) oxides. In support of this mechanism, a mutant strain with a depleted pilin-encoded domain showed poor Fe(III) reduction and growth. These functional pili were shown to be highly conductive, indicating that bacteria could accomplish the electron transfer to outside of the cell (Reguera et al. 2005). Although nanowires may explain the efficient electron transfer of *Geobacter* sp. to solid electron acceptors that are difficult to access, its structure and function that make it conductive and whether it is inductive or constitutive has not been fully studied. Investigation of electron transport mechanisms may improve understanding of the involved biomaterial and metabolic pathway as well as producing power of MFC.

4 Conclusions

Bioelectrochemical systems are emerging technologies which are closely related to environmental aspects such as sustainable bioenergy production and bioremediation. While their efficiency and power are relatively low compared to chemical fuel cells at present, further studies to improve fuel cell design, identify better electrode materials and optimize process parameters will enhance the current density and allow practical application of BESs. In order to increase the economic feasibility of this technology, new designs and constructions approaches should examine decreasing the use of expensive materials in BESs, such as platinum on the cathode and costly ion-exchange membranes (Logan 2004; Pham et al. 2006;

Kim et al. 2007). It is poorly understood how electrochemically active bacteria can attach on an electrode and how they can form an electron transfer system to an electrode surface. Therefore, this must be an additional focus of current research as well as improving reactor design.

BES is interdisciplinary research ranging range from electrochemistry, microbiology, material science and surface chemistry to engineering such as reactor design and operation, and system control and modelling. Collaboration of each study and integration of system might increase the performance and feasibility of BES process for sustainable energy production.

References

[1] Aelterman, P., Rabaey, K., Pham, H.T., Boon, N., Verstraete, W.: Continuous Electricity Generation at High Voltages and Currents Using Stacked Microbial Fuel Cells. Environmental Science & Technology 40(10), 3388–3394 (2006)

[2] Angenent, L.T., Karim, K., Al-Dahhan, M.H., Wrenn, B.A., Domiguez-Espinosa, R.: Production of bioenergy and biochemicals from industrial and agricultural wastewater. Trends in Biotechnology 22(9), 477–485 (2004)

[3] Cheng, S., Liu, H., Logan, B.E.: Increased power generation in a continuous flow MFC with advective flow through the porous anode and reduced electrode spacing. Environmental science & technology 40(7), 2426–2432 (2006)

[4] Cheng, S., Logan, B.E.: Sustainable and efficient biohydrogen production via electrohydrogenesis. Proceedings of the National Academy of Sciences of the United States of America 104(47), 18871–18873 (2007)

[5] Fan, Y., Hu, H., Liu, H.: Enhanced Coulombic efficiency and power density of air-cathode microbial fuel cells with an improved cell configuration. Journal of Power Sources 171(2), 348–354 (2007)

[6] He, Z., Minteer, S.D., Angenent, L.T.: Electricity generation from artificial wastewater using an upflow microbial fuel cell. Environmental Science and Technology 39(14), 5262–5267 (2005)

[7] Iranpour, R., Stenstrom, M., Tchobanoglous, G., Miller, D., Wright, J., Vossoughi, M.: Environmental engineering: energy value of replacing waste disposal with resource recovery. Science 285(5428), 706–711 (1999)

[8] Kim, B.H., Chang, I.S., Moon, H.: Microbial fuel cell-type biochemical oxygen demand sensor. In: Grimes, C.A., Dickey, E.C., Pishko, M.V. (eds.) Encyclopedia of Sensors x, pp. 1–12. American Scientific Publishers (2006)

[9] Kim, J.R., Cheng, S., Oh, S.E., Logan, B.E.: Power generation using different cation, anion and ultrafiltration membranes in microbial fuel cells. Environmental Science and Technology 41(3), 1004–1009 (2007)

[10] Kim, J.R., Dec, J., Bruns, M.A., Logan, B.E.: Removal of Odors from Swine Wastewater by Using Microbial Fuel Cells. Applied and Environmental Microbiology 74(8), 2540–2543 (2008)

[11] Kim, J.R., Premier, G.C., Hawkes, F.R., Dinsdale, R.M., Guwy, A.J.: Development of a tubular microbial fuel cell (MFC) employing a membrane electrode assembly cathode. Journal of Power Sources 187(2), 393–399 (2009)

[12] Kim, J.R., Premier, G.C., Hawkes, F.R., Rodríguez, J., Dinsdale, R.M., Guwy, A.J.: Modular tubular microbial fuel cells for treatment and energy recovery at low organic loading (2009) (submitted for publication)

[13] Liu, H., Cheng, S., Logan, B.E.: Power generation in fed-batch microbial fuel cells as a function of ionic strength, temperature, and reactor configuration. Environmental Science and Technology 39(14), 5488–5493 (2005)
[14] Liu, H., Logan, B.E.: Electricity generation using an air-cathode single chamber microbial fuel cell in the presence and absence of a proton exchange membrane. Environmental Science and Technology 38(14), 4040–4046 (2004)
[15] Logan, B.E.: Extracting hydrogen and electricity from renewable resources. Environmental Science and Technology 38(9), 160A–167A (2004)
[16] Logan, B.E.: Microbial fuel cells. Wiley-Interscience, Hoboken (2008)
[17] Min, B., Logan, B.E.: Continuous electricity generation from domestic wastewater and organic substrates in a flat plate microbial fuel cell. Environmental Science and Technology 38(21), 5809–5814 (2004)
[18] Oh, S.E., Van Ginkel, S., Logan, B.E.: The relative effectiveness of pH control and heat treatment for enhancing biohydrogen gas production. Environmental science & technology 37(22), 5186–5190 (2003)
[19] Park, W., Hyun, S.H., Oh, S.E., Logan, B.E., Kim, I.S.: Removal of headspace CO2 increases biological hydrogen production. Environmental science and technology 39(12), 4416–4420 (2005)
[20] Pham, T.H., Rabaey, P., Aelterman, P., Clauwaert, L., Schamphelaire, D., Boon, N., Verstraete, W.: Microbial fuel cells in relation to conventional anaerobic digestion technology. Engineering in Life Sciences 6(3), 285–292 (2006)
[21] Rabaey, K., Clauwaert, P., Aelterman, P., Verstraete, W.: Tubular microbial fuel cells for efficient electricity generation. Environmental Science and Technology 39(20), 8077–8082 (2005)
[22] Rabaey, K., Lissens, G., Siciliano, S.D., Verstraete, W.: A microbial fuel cell capable of converting glucose to electricity at high rate and efficiency. Biotechnology letters 25(18), 1531–1535 (2003)
[23] Rabaey, K., Verstraete, W.: Microbial fuel cells: novel biotechnology for energy generation. Trends in Biotechnology 23(6), 291–298 (2005)
[24] Reguera, G., McCarthy, K.D., Mehta, T., Nicoll, J.S., Tuominen, M.T., Lovley, D.R.: Extracellular electron transfer via microbial nanowires. Nature 435(7045), 1098–1101 (2005)
[25] Rozendal, R.A., Hamelers, H.V.M., Rabaey, K., Keller, J., Buisman, C.J.N.: Towards practical implementation of bioelectrochemical wastewater treatment. Trends in Biotechnology 26(8), 450–459 (2008)
[26] Shimoyama, T., Komukai, S., Yamazawa, A., Ueno, Y., Logan, B.E., Watanabe, K.: Electricity generation from model organic wastewater in a cassette-electrode microbial fuel cell. Applied Microbiology and Biotechnology 80(2), 325–330 (2008)

Trend of Mathematical Models in Microbial Fuel Cell for Environmental Energy Refinery from Waste/Water

Sung Taek Oh[1]

Abstract. A microbial fuel cell (MFC) is a device to use for bio electrochemical energy production. Electrophilic bacteria produce electrons in their metabolic pathway and the electrons can be extracted and concentrated on electrode by the electric potential difference (i.e. Galvanic cell). The bio-electrode may provide new opportunities for the renewable energy in waste water/swage treatment plants.

For the MFC technology to be adopted by industry requires a strategy for scaling up, which, in turn, requires a better understanding of the entire fuel cell system and the potential bottlenecks in the generation of electricity. Mathematical modelling has a role to play in synthesising our knowledge of fuel cells, identifying the limiting factors in electricity generation and informing scale-up strategies. The range of interaction that can occur in a microbial fuel cell makes it difficult to foresee all of the feedbacks between the biology, chemistry and physics in a fuel cell in any new design. Thus, theoretical modeling can highlight potentially important and previously overlooked mechanisms that warrant further investigation through experimentation. The most efforts in the past have almost exclusively been concentrated on electron transport mechanism from catalytic microorganism to electrode in anode.

Here, this paper gives a comprehensive review on recent developments, advancement, and challenges for the techniques in supporting process control of microbial fuel cell. This may help many environmental researchers and modelers to get current information of modeling and simulation in biotechnology and environmental technology.

1 Overview of Microbial Fuel Cell

In the past 20 years, chemical fuel cells have been continuously developed for direct electricity generation from fuel. However, the advantage of the fuel cells is partially limited by (a) the limited viability and high cost of the catalysts, (b) the highly corrosive electrolytes and (c) the elevated operating temperatures. These problems can be solved by the microbial fuel cell (MFC), which is replaced to microorganisms the active catalytic component in the specified electrode compartment. The MFC typically consists of two chambers, an anaerobic anode chamber

[1] Department of Civil Engineering, University of Glassgow, United Kingdom
sungtaek25@hotmail.com

J.H. Lee et al. (Eds.): EU-Korea Conference on Science and Technology, SPPHY 135, pp. 25–30.
springerlink.com

and an aerobic cathode chamber separated by an ion conducting membrane. Anaerobic microorganisms at the anode oxidise organic matter (i.e. fuel) and transfer electrons to the anode that pass through an external circuit. Protons migrate through the solution across the membrane to the cathode where they combine with oxygen and electrons to form water. This device may provide new opportunities for the renewable energy production in waste/water treatment plants.

Recently, the MFC device is becoming an attractive prospect for simultaneously treating wastewater and generating electricity. Liu *et. al.* (2004) were the first demonstrated that domestic wastewater could be used as the substrate in a fuel cell without artificial electron mediator. They developed a single chamber MFC in which the cathode does not need to be immersed in water, rather it was bonded directly to a proton exchange membrane so that oxygen in the air can directly react at the electrode. This significantly altered the economics of MFCs, which previously had been laboratory curiosities, fed with single carbon sources as a substrate, requiring high mass transfer of oxygen to the cathode, and expensive chemical mediators to shuttle electrons to the electrodes. Wastewater is free and needs to be treated and the greatest expense in conventional wastewater treatments, such as the activated sludge process, is aeration. Having air-fed cathodes passively supplying oxygen to wastewater could obviate this requirement. Thus, the combination of reduced aeration costs, free fuel and electricity production make wastewater fed MFCs a very attractive proposition.

2 Mathematical Model in Microbial Fuel Cells

For the MFC technology to be adopted by industry requires a strategy for scaling up, which, in turn, requires a better understanding of the entire fuel cell system and the potential bottlenecks in the generation of electricity. Mathematical modelling has a role to play in synthesising our knowledge of fuel cells, identifying the limiting factors in electricity generation and informing scale-up strategies. The range of interaction that can occur in a microbial fuel cell makes it difficult to foresee all of the feedbacks between the biology, chemistry and physics in a fuel cell in any new design. Even in well studied systems we tend to use spatially averaged macroscopic variables such as mean concentrations or potential gradients across the external circuit to characterise their behaviour. Thus, theoretical modeling can highlight potentially important and previously overlooked mechanisms that warrant further investigation through experimentation.

As a consequence, several mechanistic models of MFCs (Zhang and Halme 1995; Kato-Marcus, Torres et al. 2007; Picioreanu, Head et al. 2007) have recently been produced. They were concentrated on bio-catalytic activities, which are substrate (i.e. fuel) oxidation and its product (i.e. electron) transport from catalytic microorganism to electrode. Currently biochemical shuttle and electric conductive network mechanisms are addressed for the electron transport.

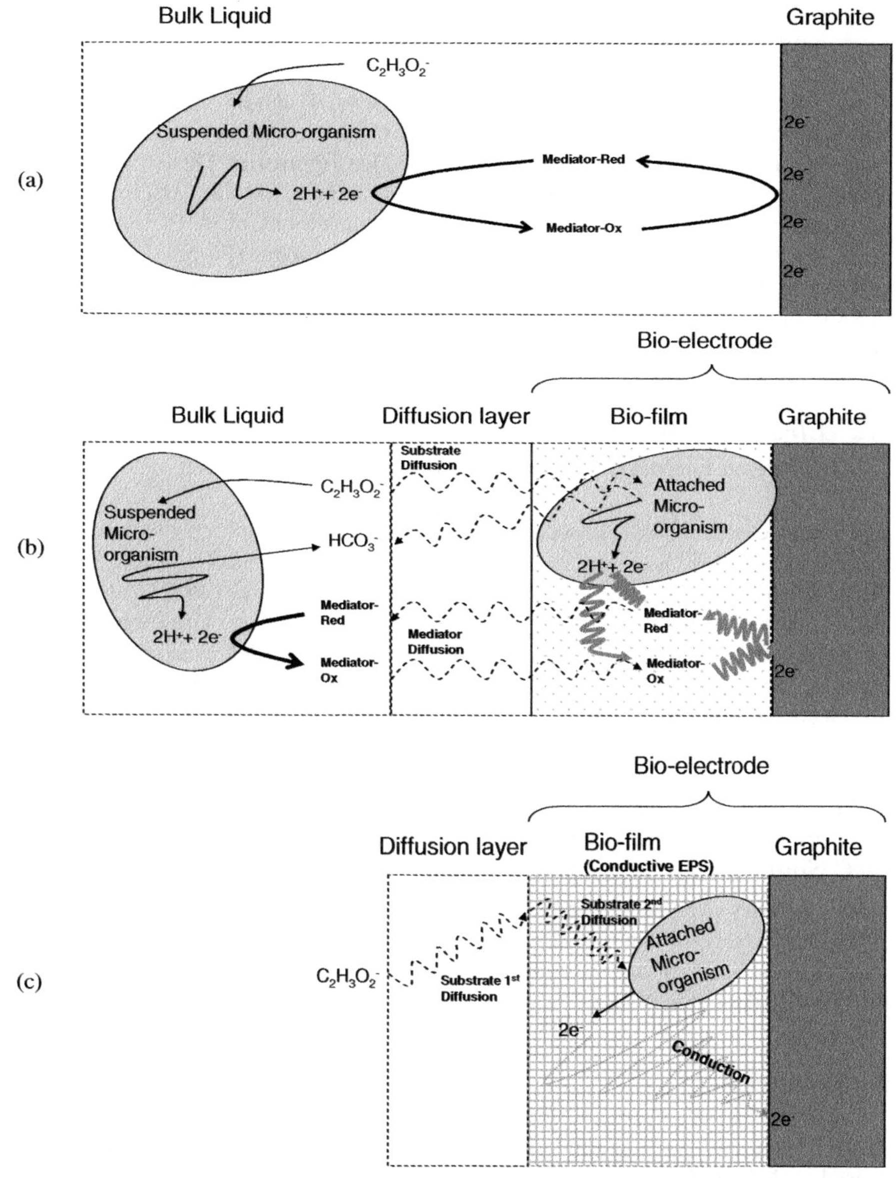

Fig. 1. Schematic diagrams of previously proposed electron transfer mechanisms in microbial fuel cells. (a) electron shuttling (Zhang and Halme 1995) (b) shuttling and diffusion limitation through an attached biofilm (Picioreanu, Head et al. 2007) (c) electrons conducted through extracellular polymeric substances (Kato-Marcus, Torres et al. 2007)

2.1 Electron Shuttle Transport between Catalytic Microorganism and Electrode

Zhang and Halme (1995) assumed that suspended anodic bacteria produce a redox biochemical mediator (electron shuttle), which would be enriched around electrode, that transfers electrons to the electrode (Figure 1a). Their model demonstrated the potential importance of electron shuttles as the electron transfer mechanism. Recently, the idea has been further development by Picioreanu, Head *et. al.*'s (2007), by superimposing a spatial biofilm model (Figure 1b) (Kreft, Picioreanu et al. 2001; Alpkvist, Picioreanu et al. 2006). Their model demonstrated that the biological production of the redox electron shuttle and its diffusion to the anode may be the major limiting factors in electricity production.

2.2 Electric Network Transport of Conductive Substances between Catalytic Microorganism and Electrode

The redox electron shuttle mechanism is not the only proposed electron transfer mechanisms in MFCs. Many researchers (Lee, Phung et al. 2003; Kim, Park et al. 2004; Gorby, Yanina et al. 2006; Kato-Marcus, Torres et al. 2007) observed that electric current is produced in electron shuttle-less condition. These empirical observations have been interpreted as conduction driven by electrical potential gradients (Reguera, Nevin et al. 2006). The path that the electrons trace out on route to the electrode remains a matter of debate and ongoing research. The most prominent current theory is that they are conducted through a network of bacterial *pili*, extracellular cytochromes and/or conductive extracellular polymeric substances (EPS). Kato-Marcus *et al* (2007) used a mathematical model to demonstrate a trade-off between electric conductivity and biomass production in a conducting MFC biofilm (Figure 1c). High conductivity ($1 < \kappa_{bio} < 100\ \mathrm{S\,m^{-1}}$) promotes the metabolic activity of the electrophilic bacteria, but the increase of biomass can create strong heterogeneities in the substrate and local potentials that serve to reduce the current density. They suggested that this limitation can be alleviated by increasing the specific detachment rate.

3 Prospective

A third alterative mechanism for electron transfer has remained relatively unexplored on MFCs. The conductive electrolyte between catalytic microorganism and electrode may be the aqueous mediator (electron shuttle) solution and/or the network of conductive substances in biofilm, but the electrical potential gradient (i.e. Electric field) induces electron transfer through the conductive interface (i.e. between catalytic microorganism and electrode). Thus what appears to be conduction of electrons through the biofilm matrix could equally be explained by the electric field. The two previous models of conduction can both be described using electric field and electric conductivity that is established in the vicinity of the anode. The prospective model can serve to highlight the importance of the electrochemical thermodynamics and its interaction with biological reaction kinetics and may offer a tool for the quantitative manipulation of microbial fuel cell.

References

[1] Bergel, A., Feron, D., et al.: Catalysis of oxygen reduction in PEM fuel cell by seawater biofilm. Electrochemistry Communications 7(9), 900–904 (2005)

[2] Biffinger, J.C., Pietron, J., et al.: A biofilm enhanced miniature microbial fuel cell using Shewanella oneidensis DSP10 and oxygen reduction cathodes. Biosensors and Bioelectronics 22(8), 1672–1679 (2007)

[3] Bullen, R.A., Arnot, T.C., et al.: Biofuel cells and their development. Biosensors and Bioelectronics 21(11), 2015–2045 (2006)

[4] Cheng, S.-A., Liu, H., et al.: Optimization of air cathode used in single-chamber microbial fuel cells. Preprints of Extended Abstracts presented at the ACS National Meeting, American Chemical Society, Division of Environmental Chemistry 44(2), 1514–1516 (2004)

[5] Gorby, Y.A., Yanina, S., et al.: Electrically conductive bacterial nanowires produced by Shewanella oneidensis strain MR-1 and other microorganisms. Proceedings of the National Academy of Sciences of the United States of America 103(30), 11358–11363 (2006)

[6] Kato-Marcus, A., Torres, C.I., et al.: Conduction-based modeling of the biofilm anode of a microbial fuel cell. Biotechnology and Bioengineering (2007)

[7] Kim, B.H., Park, H.S., et al.: Enrichment of microbial community generating electricity using a fuel cell type electrochemical cell. Appl. Microbiol. biotechnology 63, 672–681 (2004)

[8] Kim, J.R., Jung, S.H., et al.: Electricity generation and microbial community analysis of alcohol powered microbial fuel cells. Bioresource Technology 98(13), 2568–2577 (2007)

[9] Lee, J., Phung, N.T., et al.: Use of acetate for enrichment of electrochemically active microorganisms and their 16S rDNA analyses. FEMS Microbiology Letters 223(2), 185–191 (2003)

[10] Liu, H., Ramnarayanan, R., et al.: Production of Electricity during Wastewater Treatment Using a Single Chamber Microbial Fuel Cell. Environmental Science and Technology 38(7), 2281–2285 (2004)

[11] Min, B., Kim, J., et al.: Electricity generation from swine wastewater using microbial fuel cells. Water Research 39(20), 4961–4968 (2005)

[12] Oh, S.-E., Logan, B.E.: Proton exchange membrane and electrode surface areas as factors that affect power generation in microbial fuel cells. Applied Microbiology and Biotechnology 70(2), 162–169 (2006)

[13] Oh, S.-E., Min, B., et al.: Characterization of design factors affecting power output in a microbial fuel cell. In: Abstracts of Papers, 228th ACS National Meeting, Philadelphia, PA, United States, August 22-26, ENVR-195 (2004)

[14] Oh, S., Min, B., et al.: Cathode Performance as a Factor in Electricity Generation in Microbial Fuel Cells. Environmental Science and Technology 38(18), 4900–4904 (2004)

[15] Prasad, D., Sivaram, T.K., et al.: Microbial fuel cell constructed with a micro-organism isolated from sugar industry effluent. Journal of Power Sources 160(2), 991–996 (2006)
[16] Rabaey, K., Lissens, G., et al.: A microbial fuel cell capable of converting glucose to electricity at high rate and efficiency. Biotechnology Letters 25(18), 1531–1535 (2003)
[17] Rabaey, K., Verstraete, W.: Microbial fuel cells: novel biotechnology for energy generation. Trends in Biotechnology 23(6) (2005)
[18] Reguera, G., McCarthy, K.D., et al.: Extracellular electron transfer via microbial nanowires. Nature 435, 1098–1101 (2005)
[19] Venkata Mohan, S., Saravanan, R., et al.: Bioelectricity production from wastewater treatment in dual chambered microbial fuel cell (MFC) using selectively enriched mixed microflora: Effect of catholyte. Bioresource Technology, Corrected Proof: 496 (in Press)
[20] Walker, A.L., Walker, J.C.W.: Biological fuel cell and an application as a reserve power source. Journal of Power Sources 160(1), 123–129 (2006)
[21] Wingard, L.B., Shaw, C.H., et al.: Bioelectrochemical fuel cells. Enzyme and Microbial Technology 4(3), 137–142 (1982)
[22] You, S., Zhao, Q., et al.: A microbial fuel cell using permanganate as the cathodic electron acceptor. Journal of Power Sources 162(2), 1409–1415 (2006)
[23] Zhang, X.-C., Halme, A.: Modelling of a microbial fuel cell process. Biotechnology Letters 17, 809–814 (1995)
[24] Zhang, X.-C., Ranta, A., et al.: Direct methanol biocatalytic fuel cell–Considerations of restraints on electron transfer. Biosensors and Bioelectronics 21(11), 2052–2057 (2006)
[25] Zhao, F., Harnisch, F., et al.: Application of pyrolysed iron(II) phthalocyanine and CoTMPP based oxygen reduction catalysts as cathode materials in microbial fuel cells. Electrochemistry Communications 7(12), 1405–1410 (2005)
[26] Zhao, F., Harnisch, F., et al.: Challenges and Constraints of Using Oxygen Cathodes in Microbial Fuel Cells. Environmental Science & Technology 40(17), 5193–5199 (2006)

Optimization Methodology of Low Carbon Mixed Energy Systems Using the Bees Algorithm

Ji Young Lee[1,*] and Jae Min Kim[2]

Abstract. This paper proposes to decide the optimal capacity of the mixed system for thermal and electric power to minimize total capital cost and total CO_2 emission while they are satisfying with total thermal demand and total load demand. The Bees Algorithm is used as a multi-objective solver for this work. Optimal Pareto solutions obtained from approximate 0% to 90% CO_2 reduction rates are compared with traditional systems like boiler and grid power and those solutions also suggest optimal capacities of the mixed system in each CO_2 reduction rate.

1 Introduction

Scientists say global warming is already happening and predict that there will be more extreme weather phenomena and rising sea levels if the planet's temperature rises more than 2C above the pre-industrial level. They blame deforestation and the burning of fossil fuels for the warming effect because these two factors increase carbon dioxide (CO_2), which is the main greenhouse gas, in the atmosphere. In addition, soaring oil and gas prices have made many countries' reliance on imported fossil fuels much more costly. Furthermore the EU's package on climate change has kept 20-20-20 targets early this year and it focuses on three areas which are a 20% cut in emissions of greenhouse gases by 2020 compared with 1990 levels, a 20% increase in the share of renewables in the energy mix and a 20% cut in energy consumption. Therefore CO_2 reduction and development of renewables are not optional any more and they are globally main key issue now, especially heavy industrial plants such as power plants, oil refineries and steel mills which together account for almost half the EU's CO_2 emissions. Therefore this study targets on power plants to tackle global warming using mixed energy systems and proposes the new optimization methodology of low carbon energy systems. The target is not only electric power plants but also thermal plants. Moreover total capital cost of the system is also optimized simultaneously. Therefore this study is handled as multi-objective optimization problem and the Bees Algorithm is used as its solver.

---

[1] Cardiff University, Manufacturing Engineering Centre, Cardiff, United Kingdom
LeeJ5@cardiff.ac.uk
[2] Strathclyde University, Departmet of Mechanical Engineering,
Glasgow, United Kingdom,
min@esru.strath.ac.uk
[*] Corresponding author.

J.H. Lee et al. (Eds.): EU-Korea Conference on Science and Technology, SPPHY 135, pp. 31–39.
springerlink.com

2 Background

Traditional fossil-fuelled electric power plants have been centralized and the aim of the classical environmental/economic dispatch problem (EEDP) [1] has been to minimize the total fuel cost and total emission gases such as NO_x and SO_2. However, this system could be no longer considered singly due to developing renewable systems. Hernandez-Aramburo et al [2] and Mohamed and Koivo [3, 4] introduced MicroGrid (MG) system which combined with a variety of power sources. The MG system in [2] consists of two reciprocating gas engines, a combined heat and power (CHP) plant, a photovoltaic (PV) array and a wind generator. The optimization is aimed at reducing the fuel consumption rate of the system while constraining it to fulfill the local energy demand (both electrical and thermal) and provide a certain minimum reserve power. A penalty is applied for any heat produced in excess of demand. While the system in [2] was only considering as an economic dispatch problem because it was not considering environmental factors at all, the MG system in [3, 4] was applied as the environmental/economic problem because the model took into consideration the operation and maintenance costs as well as the emissions NO_x, SO_2 and CO_2 reduction. The MG system in [3, 4] consists of a wind turbine, a micro turbine, a diesel generator, a PV array and fuel cell. However, this model was only considered total load demand as a constraint. Many researchers as above have been interested in operating cost rather than capital cost. However, initial capital cost of renewable systems is considerably expensive compared to traditional power systems therefore a decision maker (DM) or an engineer cannot easily decide their installation and capacity. In addition, the capital cost can be no longer considered singly due to raising environmental restriction. Therefore this paper is to determine each capacity of heat and power system which minimize total capital cost and total CO_2 emission while constraining it to meet the thermal and electricity demand simultaneously using the Bees Algorithm.

3 The Bees Algorithm

Many artificial intelligent algorithms have been developed for many decades. However, recently population-based algorithms, especially swarm-based optimization algorithms, are more popular than direct search algorithms in many cases. Swarm-based optimization algorithms (SOAs) mimic nature's methods to drive a search towards the optimal solution. A key difference between SOAs and direct search algorithms is that SOAs use a population of solutions in every iteration instead of a single solution. As a population of solutions is processed in each iteration, the outcome is also a population of solutions. SOAs include the Ant Colony Optimization (ACO) algorithm [5], the Genetic Algorithm (GA) [6] and the Particle Swarm Optimization [7] (PSO) algorithm. More recently, honey bee algorithms have been developed and they mimic various behavior of honey bee. Especially, foraging behavior is more popular among researchers for optimization problems and several bee algorithms inspired by this behavior have been developed. Artificial Bee Colony (ABC) [8] algorithm and A Bee Colony Optimization

(BCO) algorithm [9] are similar and they have been applied to many combinatorial problems such as container loading problem, Travel Salesman Problem (TSP), etc. BeeAdHoc [10] and BeeHive algorithm [11] were especially developed for networking routing problem. In this work, the Bees Algorithm was used because it is easy to implement and it has been proved its superiority and robustness in various optimization areas. Especially Pham et al [12] applied this algorithm for multi-objective Environmental/Economical dispatching problem (EEDP) and its results was superior to Non-dominated sorting Genetic Algorithm-Ⅱ (NSGA-Ⅱ) [13] and Strength Pareto Evolutionary Algorithm (SPEA) [1]. Therefore this work modified [12] and 'random selection neighborhood search' in [12] is also adopted for neighborhood search. The Bees Algorithm was developed by Pham et al [14] in 2006 and it was inspired by the natural foraging behavior of honey bees to find the optimal solution. The foraging process begins in a colony by scout bees being randomly sent to search for promising flower patches. When they return to the hive, those scout bees that found a patch which is rated above a certain quality threshold perform a dance known as the 'waggle dance' to advertise their found patches and recruit unemployed bees. More follower bees are sent to more promising patches. As a result, the colony gathers food much quickly and efficiently. The Bees Algorithm mimicked this mechanism and random search as well.

4 The Proposed System Model

The aim of this study is to design a low carbon town which is demanded electricity and thermal like Table 1. Electricity and thermal demand are pick demand among hour-based annual demand and the system takes them into consideration as constraints. These demands also affect the maximum capacity of systems except photovoltaic (PV), solar thermal and boiler. Total building areas were calculated by summation of whole building area and they affect the maximum capacity of PV and solar thermal system. Fig. 1 depicts configuration of the proposed system.

Table 1. Demand and total building area

Electricity demand (kW/hr)	Thermal demand (kW/hr)	Total building area (m^2)
51,000	129,000	652,848

This approach is to find optimal solutions sequentially. At first optimal solutions for thermal system will be found with thermal demand, then optimal solutions for whole system will be found with load demand. If output power from heat and power system among optimal solutions of thermal system exceeds the total load demand, this solution will be discarded for electric power system, which means that there is no need to install electric power system separately. With only thermal system, it is satisfying with thermal demand and load demand simultaneously. However, if output power from this heat and power system among optimal solutions of thermal system is less than total load demand, this solution will be

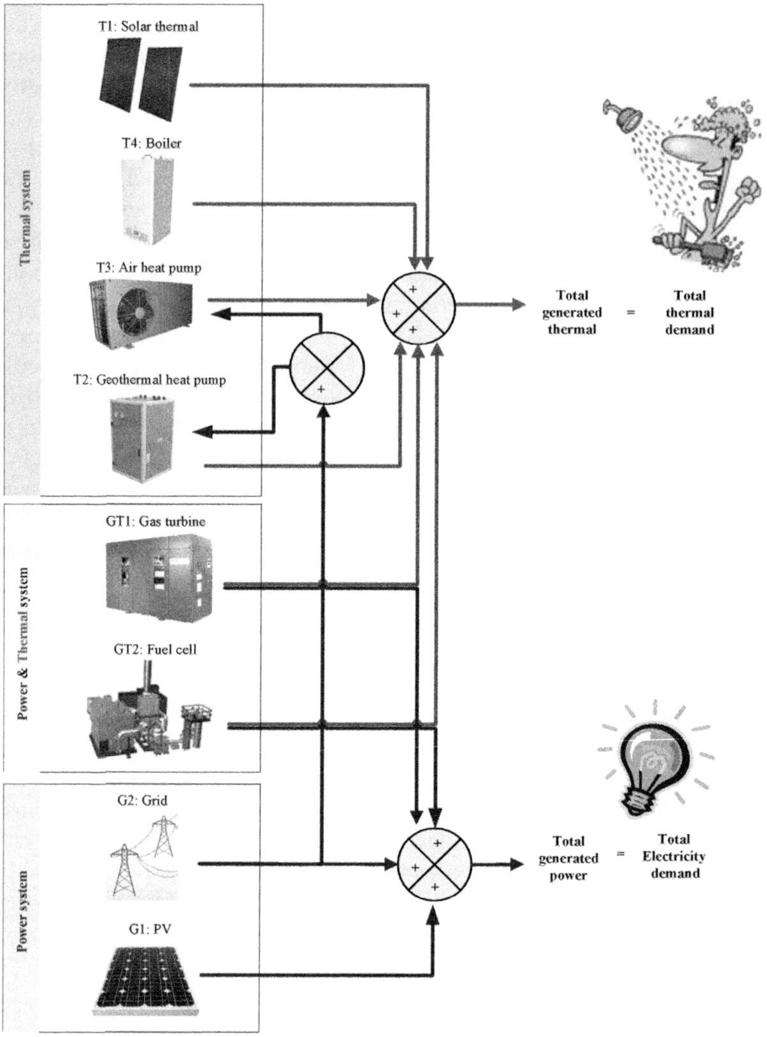

Fig. 1. System configuration

conducted to find optimal solutions of electric power system, while the capacity of whole thermal system (T1, T2, T3, T4, GT1, and GT2) is already fixed from optimal thermal Pareto solutions. Fig. 2 shows its flowchart.

Table 2 shows all system parameters used in this study. Most of unit costs are the average value from market price surveyed in Korea. For example, unit costs of T1 and G1 come from Korea Energy Management Corporation (KEMCO) and Unit cost of G2 is from Korea Electric Power Corporation (KEPCO) [15]. However, unit cost of GT1 [16] and GT2 [17] are from United States Environmental Protection Agency (EPA) respectively and the average unit cost is taken for this work. CO_2 factors of GT1 and T4 which use natural gas are from Carbon Trust

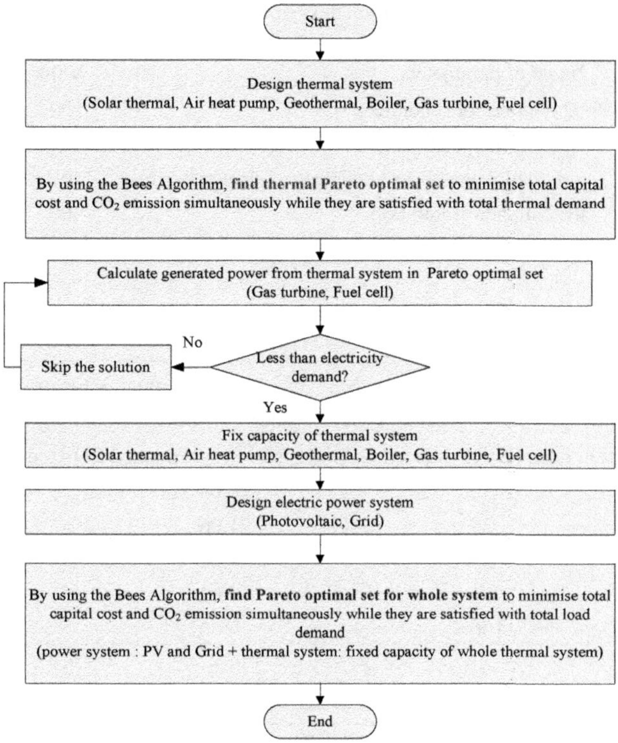

Fig. 2. Flowchart of the proposed optimization methodology

Table 2. System parameters

	G1	G2	T1	T2	T3	T4	GT1	GT2
Thermal efficiency (%)	-	-	12	90	65	85	85	98
Power efficiency (%)	15	100	-	-	-	-	100	100
Unit Cost (kWon/kW)	10,578	72.E-3	3,144	916	1,270	25	1,643	6,852
CO_2 Factor (kg/kW)	0	0.448	0	0.448	0.448	0.185	0.185	0
Min. Capacity (MW)	0	0	0	0	0	0	0	0
Max. Capacity (MW)	272	101	821	21	29	426	76	82
COP	-	-	-	3.4	3.4	-	-	-
Power/Heat ratio	-	-	-	-	-	-	0.5	1.25

[18]. However, CO_2 factors of T2 and T3 which use grid power are from KEPCO [15]. Table 3 is parameters of the Bees Algorithm and Table 4 shows produced CO_2 quantity regarding CO_2 reduction rates when electric power and thermal are supplied by traditional way such as natural gas boiler and grid power while they are satisfying with electricity demand and thermal demand respectively. These values are like absolute CO_2 quantity to compare with CO_2 produced by proposed mixed energy system.

Table 3. Parameters of the Bees Algorithm

Name of parameters	Value
n : Number of scout bees	100
ngh : Initial patch size	10
nm : Number of bees allocated to selected patches	10
itr: Number of iteration	400
Accuracy (MW)	0.1

Table 4. CO_2 reduction rates and produced CO_2

	Thermal system: Boiler	Thermal & Electricity system: Boiler + Grid power
CO_2 reduction rates (%)	Produced CO_2 (ton/hr)	Produced CO_2 (ton/hr)
0	28.08	50.92
25	21.06	38.19
50	14.04	25.46
75	7.02	12.73
100	0	0

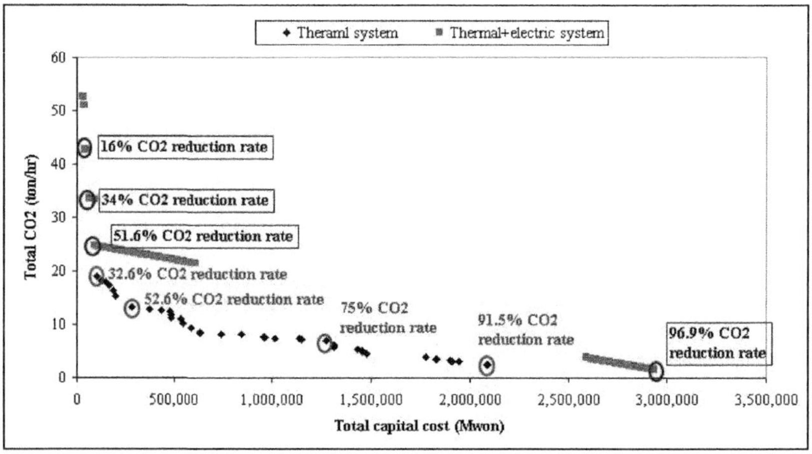

Fig. 3. Pareto front

5 Results

Optimal capacities of power and thermal systems were determined by using the Bees Algorithm. To obtain the average performance of the algorithm, 10 runs were carried out. Fig. 3 shows Pareto-optimal fronts among 10 runs. Pareto front

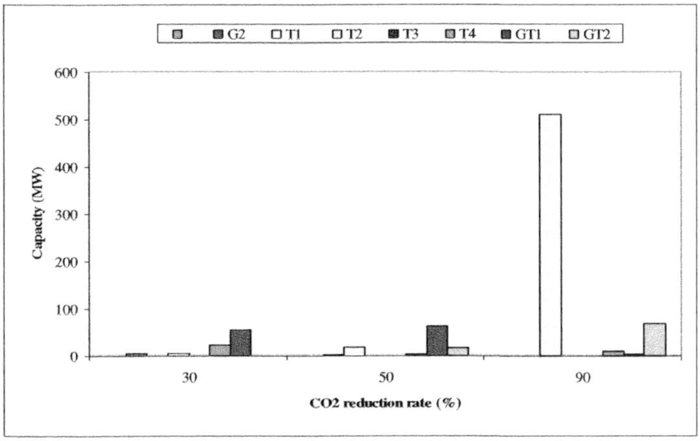

Fig. 4. Graphs of thermal system regarding CO_2 reduction rates

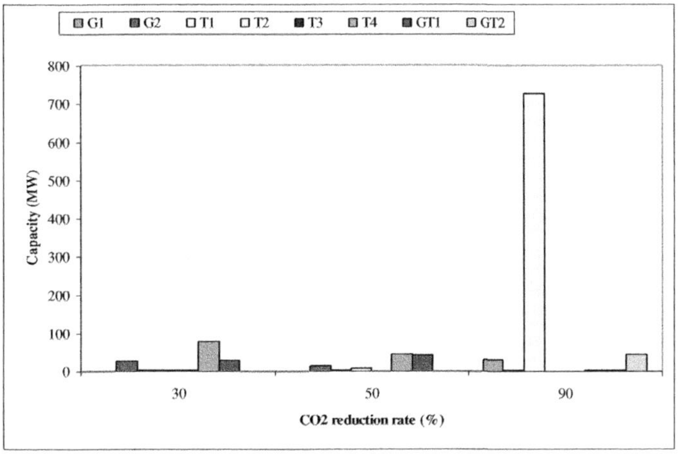

Fig. 5. Graphs of 'Thermal + electric' system regarding CO_2 reduction rates

of diamond markers represent thermal system which is satisfying with thermal demand and is more than electricity demand, while Pareto front of square markers represents 'thermal + electric' system. Both Pareto fronts are satisfying with thermal demand and electricity demand, although Pareto front of thermal system might overproduce electricity.

While Fig. 4 shows combination of thermal system, Fig. 5 shows combination of 'thermal + electric' system. When CO_2 reduction rate is less than 50% in Fig. 4 GT1 dominates the most of system. However, when CO_2 reduction is around 90% in both Fig. 4 and 5 the capacity of T1 is sharply increased and it dominates the most of system. However, T4 takes a large position in Fig. 5 when CO2 reduction rate is less than 50%. Fig. 6 shows comparison of total capital cost regarding CO_2 reductions. When CO_2 reduction rate is less than 50%, the installation of

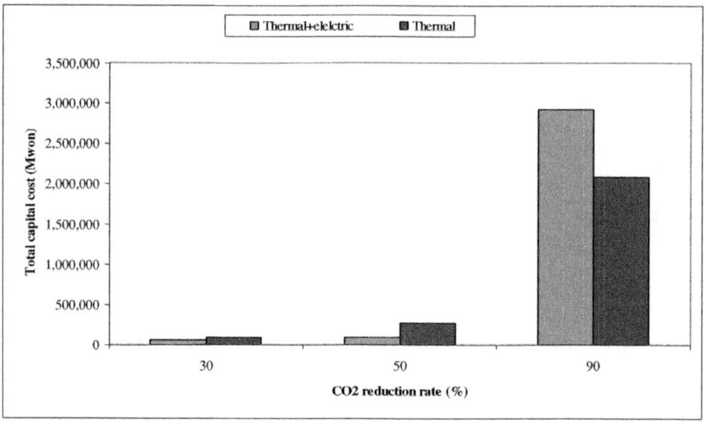

Fig. 6. Graphs of total capital cost regarding CO_2 reduction rates

'thermal + electric' system is slightly cheaper than the installation of thermal system alone. However, when CO_2 reduction rate is around 90%, the installation of thermal system alone is much cheaper than the installation of whole systems.

6 Conclusions

This paper proposed to sequentially decide the optimal capacity of the mixed system for thermal and electric power to minimize total capital cost and total CO_2 emission simultaneously while they are satisfying with total thermal demand and total load demand. The Bees Algorithm was used as a multi-objective solver and simulation results obtain a good diversity of Pareto optimal sets. Some optimal solutions from approximate 0% to 90% CO_2 reduction rates were compared with traditional systems like boiler and grid power and those solutions also suggest optimal capacities of the mixed system in each CO_2 reduction rate. Therefore, this work has confirmed the suitability of the Bees Algorithm for solving the multi-objective problem, simultaneously achieving financial savings and reducing the emission of greenhouse gases into the atmosphere.

In the further, after the mixed system is installed with each optimal capacity, Environmental/Economic dispatching is required again to minimize total running cost and greenhouse gases. However, output of G1, T1, T2 and T3 is almost fixed by whether condition. Therefore in order to minimize total running cost and total CO_2 emission during operation, output of T4, GT1, GT2 and G2 must be optimized by an optimization algorithm again.

References

[1] Abido, M.A.: Environmental/Economic Power Dispatch Using Multiobjective Evolutionary Algorithms. IEEE Transactions on power system 18 (2003)

[2] Hernandez-Aramburo, C.A., Green, T.C., Mugniot, N.: Fuel Consumption Minimization of a Microgrid. IEEE transactions on industry applications 41, 673–681 (2005)

[3] Mohamed, F.A., Koivo, H.N.: System Modelling and Online Optimal Management of MicroGrid Using Multiobjective Optimization, pp. 148–153. IEEE, Los Alamitos (2007)
[4] Mohamed, F.A., Koivo, H.N.: MicroGrid Online Management and Balancing Using Multiobjective optimization, pp. 639–644. IEEE, Los Alamitos (2007)
[5] Dorigo, M., Stutzle, T.: Ant Colony Optimization. MIT Press, Cambridge (2004)
[6] Goldberg, D.E.: Genetic Algorithms in Search, Optimization and Machine Learning. Addison-Wesley Longman, Reading (1989)
[7] Eberhart, R., Kennedy, J.: Swarm Intelligence. Morgan Kaufmann, San Francisco (2001)
[8] Karaboga, D., Basturk, B.: On the performance of artificial bee colony (ABC) algorithm. Applied soft computing 8, 687–697 (2008)
[9] Chong, C.S., Low, M.Y.H., Sivakumar, A.I., Gay, K.L.: A Bee Colony Optimization Algorithm to Job Shop Scheduling. In: The 2006 Winter Simulation Conference (2006)
[10] Wedde, H.F., Farooq, M., Pannenbaecker, T., Vogel, B., Mueller, C., Meth, J., Jeruschkat, R.: BeeAdHoc: An Energy Efficient Routing Algorithm for Mobile Ad Hoc Networks Inspired by Bee Behavior. In: GECCO 2005, Washington DC (2005)
[11] Wang, X., Liang, G., Huang, M.: A Beehive Algorithm Based QoS Unicast Routing Scheme with ABC Supported. In: Xu, M., Zhan, Y.-W., Cao, J., Liu, Y. (eds.) APPT 2007. LNCS, vol. 4847, pp. 450–459. Springer, Heidelberg (2007)
[12] Pham, D.T., Lee, J.Y., Darwish, H., Soroka, A.: Multi-objective Environmental/Economic Power Dispatch using the Bees Algorithm with Pareto optimality. In: Innovative Production Machines and Systems (2008)
[13] King, R.T.F.A., Rughooputh, H., Deb, K.: Evolutionary Multi-objective Environmental/Economic Dispatch: Stochastic Versus Deterministic Approaches. In: Evolutionary Multi-Criterion Optimization (2005)
[14] Pham, D.T., Ghanbarzadeh, A., Koc, E., Otri, S., Rahim, S., Zaidi, M.: The Bees Algorithm-A Novel Tool for Complex Optimisation Problems. In: Innovative Production Machines and Systems (2006)
[15] Electric Rates Table (2009), `http://www.kepco.co.kr/eng/` (cited July 1, 2009)
[16] Technology Characterization: Gas Turbines (2008), `http://www.epa.gov/CHP/documents/catalog_chptech_gas_turbines.pdf` (cited June 22, 2009)
[17] Technology Characterization: Fuel Cells (2008), `http://www.epa.gov/CHP/documents/catalog_chptech_fuel_cells.pdf` (cited June 22, 2009)
[18] Greenhouse Gas Conversion Factors (2008), `http://www.carbontrust.co.uk/resource/conversion_factors/default.htm` (cited July 1, 2009)

Technology Co-evolution Analysis in the Energy Sector

Sungjoo Lee[1,*] and Byungun Yoon[2]

Abstract. This paper suggests the method that can describe the co-evolutionary patterns in the energy sectors. Technologies that have facilitated the growth of other technologies should get the priority in the R&D investment, if other conditions are almost the same. In the suggested method, LVC equations were applied to the patents relating to energy technologies. Then a network showing the interactions between technologies in their evolution process is visualised. Research findings will provide numerous implications for policy-making and strategic planning for energy technology development.

1 Introduction

Renewable and sustainable energy technology is indispensable for the future [1]. To meet the goal of sustainable energy development, there has been a growing body of R&D efforts and thus effective investment in promising and emerging technologies especially is required when total budget is constrained [2]. Especially, in today's knowledge-based economy, the growth of one technology relies greatly on the application of new science and technology in the other sectors and the relations between technologies are regarded as an important factor that drives innovation [3]. Nevertheless, the existing studies on energy policy hardly deal with the technological interactions in development process, which can be useful sources to support R&D strategy and policy-making in the energy sector.

This paper investigates the interactive nature of relationships between energy technologies in terms of their innovations and diffusion. For the purpose, the USPTO database was used and the Lotka-Volterra Competition (LVC) equations [4], were applied to the patent data, resulting in a patent interaction network. The number of patent applications is a good indicator for innovation activity [5] and

[1] Ajou University, Department of Industrial & Information Systems Engineering, Suwon, Kyungki-do, South Korea
sungjoo@ajou.ac.kr

[2] Dongguk University, Department of Industrial & Systems Engineering, Seoul, South Korea
postman@dongguk.edu

* Corresponding author.

J.H. Lee et al. (Eds.): EU-Korea Conference on Science and Technology, SPPHY 135, pp. 41–48.
springerlink.com

the LVC equations are one of the most well-known models for technology competition analysis. The network derived from the analysis was then employed to investigate the interaction patterns among energy technologies, allowing co-evolution process across them to be analysed [6]. Research findings will provide numerous implications for policy-making and strategic planning for energy technology development. The remaining of this article is organised as follows. In Section 2, the overall research framework is introduced and in Section 3, the patent data is analysed and related implications are provided. Finally, Section 4 contains concluding remarks and points up future research issues.

2 Research Methodology

The overall research methodology is described in Figure 1. First, the technology sectors relating to energy should be defined and USPC (United States Patent Classification) codes are assigned to each of them. Then the data on annual patent applications in each code are collected from the USPTO database. The next step is to assign a type of interaction (pure competition, prey-predator, mutualism, commnsalism, amensalism and neutralism) to each pair of technologies by applying LVC equations to the growth of patents in numbers. Based on the interaction patterns between technologies over time, the technology co-evolution in the energy sector can be analysed, providing some meaningful implications for the management of energy technology.

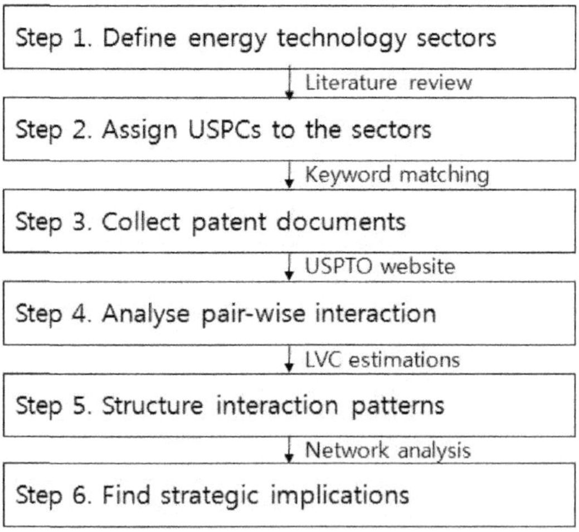

Fig. 1. Overall research process

3 Analysis and Results

3.1 Energy Technology and the Relevant Patents

Kajikawa et al. [3] identified the top ten clusters in energy research, which was used in this research to define 10 energy sectors. By matching keywords used to describe the characteristics of clusters and USPC titles, we could assign relevant patents to the sectors. The matching results are shown in Table 1.

Table 1. Concordance table between energy sectors and USPCs:

No	Cluster	Keywords	USPCs
1	Combustion(Com)	Flame, turbulent, reaction mechanism, soot, kinetics	431
2	Coal(Col)	Liquefaction, gasification, coal char, combustion	260
3	Battery(B)	Elecochemistry, lithium-ion batteries, electrode, capacitor	136
4	Petroleum(P)	Asphaltene, resin, combustion, pylolysis, tar sand	196, 208, 320
5	Fuel Cell(F)	Proton exchange, membrane, methanol, crossover	44
6	Wastewater(W)	Pollution, waste disposal, biomass, textile dye	210
7	Heat pump(H)	Heat pumping, heat transfer, absorption, hysteresis	48, 165, 237, 432
8	Engine(E)	Carnot engine, heat engine, thermoeconomics	123
9	Solar cell(S)	Photovoltaic, silicon, organic, thin film, solar energy	-
10	Power system(PS)	Synchronous machine, circuit, motor, lord	60
	Others	204, 250, 376, 903, 976	

As a result, 14 USPCs were assigned to nine sectors except one that has no relevant USPC. Also, five USPCs that do not directly related to the 10 sectors but have something to do with energy technology were identified and grouped in "Others". The titles of USPC are summarised in Appendix 1. Then the data on the annual number of USPTO patent applications between 1976 and 2009 were gathered for each class. Figure 2 shows the cumulative number of patent applications in the energy sector.

3.2 LVC Estimations and Pair-Wise Interactions

On the assumption that all technologies in the energy sector are closely related to each other, competing for the limited R&D resources, LVC equations can be applied in this research. Initially, LVC equations were suggested to describe the interaction of biological species competing for the same resources [4] and then applied to display parasitic and symbiotic relations or emerging and declining

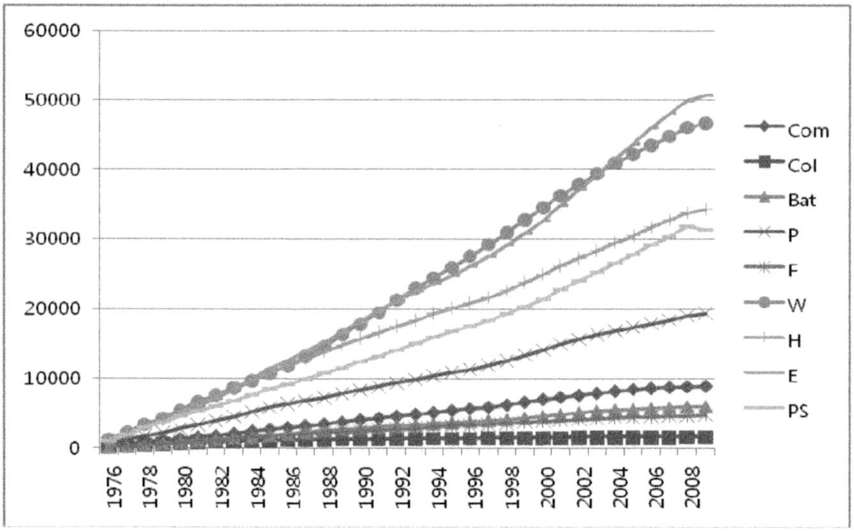

Fig. 2. The number of patents in the energy sector

competitors, allowing intuitive understanding of the factors that drive co-evolution [7]. Subsequently, interesting analogies were drawn between biological ecology and technological development, and it was Lee et al. [6] who used the equations to patents analysis for the first time. In this vein, the LVC equations enable one to examine the technological diffusion process through interaction, whether it is competition or complementation.

To understand the basic Lotka-Volterra model, let's assume that two species S1 and S2 were each living alone in a limited environment. The model suggests that the two species would increase in numbers according to the following equations when both are together in the same environment.

$$\frac{dN_1(t)}{dt} = (r_1 + a_1 N_1 + b_1 N_2) N_1$$
$$\frac{dN_2(t)}{dt} = (r_2 + a_2 N_2 + b_2 N_1) N_2 \tag{1}$$

Where, N_i : the population of S_i (i=1,2)
r_i : the rate constant of S_i (i=1,2)
a_i: the coefficient of its own growth of S_i (i=1,2)
b_i: the coefficient of competition of S_i (i=1,2)

As When this model is used in patent analysis, S_i should be the technology T_i, and N_i be the number of patents relevant to T_i as technology development is measured by the degree of patenting activities in that technology area. For all pairs of technologies, parameter b_i was estimated. Here, the equations (1) needed to be transformed to be a discrete and linear model for empirical analysis according to

Table 2. LVC estimation results:

	Com	Col	B	P	F	W	H	E	PS
Com	-	8.38E-05	0	0.001598	0	-0.00108	0.000969	-0.00118	0
Col	-0.00087	-	-0.00099	-0.00125	-0.0018	-0.00088	-0.00117	-0.00088	-0.00077
B	0	0.000142	-	0.0014	0.002236	-0.00093	0.003356	-0.0008	0
P	-0.00064	4.23E-05	-0.00026	-	0	-0.00042	0	-0.0005	0
F	-0.00081	0.000232	-0.0019	0	-	-0.00074	0	-0.0009	-0.00045
W	0.000197	1.36E-05	0.000122	0.0002	8.67E-05	-	0.00014	0	0.000216
H	-0.00025	2.82E-05	-0.00039	0	0	-0.00019	-	-0.00021	0
E	0.00019	1.31E-05	9.63E-05	0.00024	0.000104	0	0.000161	-	0.000264
PS	0	1.38E-05	0	0	0	-0.00042	0	-0.00045	-

the study by Kim et al. [8]. If $b_1(b_2)$ has a positive value, it means that the growth of $N_1(N_2)$ has been facilitated by the growth of $N_2(N_1)$. Table 2 summarise the results of b_i estimation.

For example, the signs of the values in the cell indicate the types of interaction between two technologies. For examples, the growth of coal industry (Col) may have affected negative effects on the growth of combustion technology (Com), since the estimation result shows a minus value (-0.00087).

3.3 Co-evolutionary Patterns

Using the estimation results as an input for network analysis, co-evolutionary patterns in the energy sector can be visualised. Ucinet 5.0 was used for the analysis and the cut-off value was set to 0.0009, which seems to give the best results in terms of visualisation and interpretation. It means that a link between two technologies will be generated if the degree of interaction between them is over 0.0009. Figure 2 shows the combination of two networks, one showing positive interactions and the other showing negative interactions.

Overall, technologies in the energy sector have advanced based on the complex interactions with each other. The negative effects dominate the positive effects, as we can easily imagine, most energy technologies are competing technologies, not complimentary ones, doing the same functions, Therefore they tend to compete with each other and the R&D funding moves from one area to the other areas. In detail, the figure indicates that Power system (PS) has developed independently, while Battery (B), Coal (Col) and Combustion (Com) have advanced with intimate interactions with other areas. In particular, Coal is characterised as "negatively affecting" and the other two have mixed effects on the other areas. Therefore, the government should be careful when investing on Coal technology, as there might be some unexpected side-effect of the investment. Another R&D funding may be required to the areas that will be affected by the growth of Coal technology.

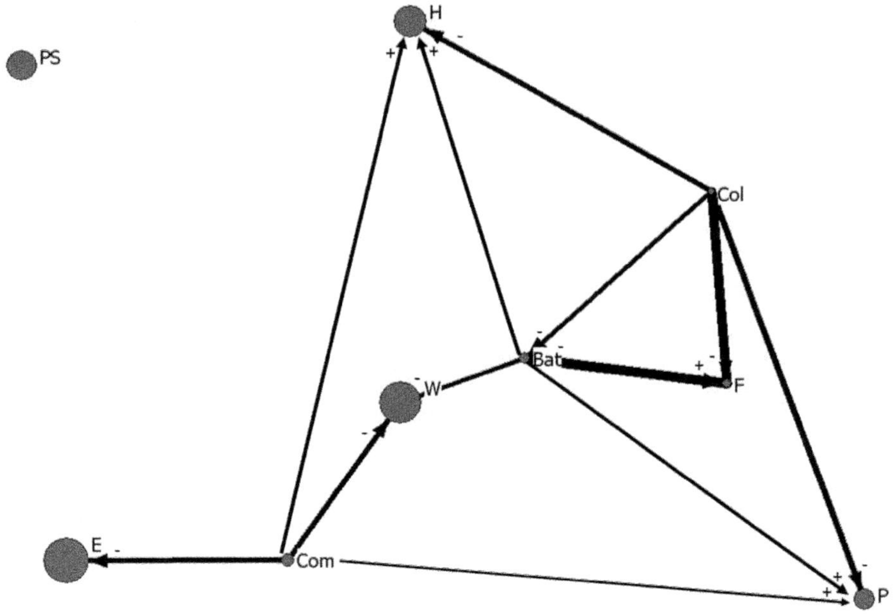

Fig. 3. Co-evolutionary patterns in the energy sector

Combustion (Com) has facilitated the growth of H (Heat pump) and Petroleum (P), working as base technology, but has prohibited the growth of Engine (E) and Wastewater (W). Table 3 summarises the type of interaction between two technologies in their evolution process, which can support strategic decision-making on R&D investments.

Table 3. Interaction types between technologies:

Interaction types	Technology pairs
prey-predator	Bat-F
commnsalism	Com →H, Bat → H, Bat → P, Com → P
amensalism	Com → E, Com → W, Bat → W, Col → Bat, Col → H, Col → F, Col → P

4 Conclusions

For the effective development of energy technology within the limited resources, it is required to make strategic decisions on R&D investment in promising and emerging technologies. As a supporting tool to identify such technologies, this paper suggested the method that can describe the co-evolutionary patterns in the energy sectors. Technologies that have facilitated the growth of other technologies should get the priority in the R&D investment, if other conditions are almost the same. In the suggested method, LVC equations were applied to the patents relating to energy technologies. Then a network showing the interactions between technologies in their evolution process is visualised, which can support the strategic decision-makings and policy-makings.

In spite of its meaningful implications, this research is subject to several limitations. Firstly, the concordance of USPC and industry is to some extent arbitrary, and more sophisticated matching is needed for further analysis. Secondly, to apply the LVC equations, it is assumed that only two technologies should be in the competition system. The growth or decline of certain technology may be affected by competitor in the system and also b other technologies external to the system. Nevertheless, this type of analysis facilitates the understanding of energy technology and can give information about how technologies have co-evolved in terms of technological growth or technological diffusion. It also gives second-hand information on the possibilities of types of interactions between technologies. Finally, this study is merely descriptive in nature and needs additional work before attempting to derive policy implications. Such tasks require more work and thus are reserved for future research.

References

[1] Jefferson, M.: Sustainable energy development: performance and prospects. Renewable Energy 31, 571–582 (2006)

[2] Nemet, G., Kammen, D.: U.S. Energy research and development: declining investment, increasing need, and the feasibility of expansion. Energy Policy 35, 746–755 (2007)

[3] Kajikawa, Y., Yoshikawa, J., Takeda, Y., et al.: Tracking emerging technologies in energy research: toward a roadmap for sustainable energy 75, 771–782 (2008)

[4] Lotka, A.: Elements of physical biology. Williams and Wilkins, Baltimore (1925)

[5] Ernst, H.: Patent information for strategic technology management. World Patent Information 25(3), 233–242 (2003)

[6] Lee, S., Kim, M., Park, Y.: ICT co-evolution and Korean ICT strategy: an analysis based on patent data, Telecommunications Policy (2009) doi: 10.1016/j.telpol.2009.02.004

[7] Bazykin, A.: Nonlinear dynamics of interacting populations. In: Khibnik, A., Krauskopf, B. (eds.) World Scientific Series on Nonlinear Science, A11. World Scientific, Singapore (1998)

[8] Kim, J., Lee, D., Ahn, J.: A dynamic competition analysis on the Korean mobile phone market using competitive diffusion model. Computers & Industrial Engineering 51(1), 174–182 (2006)

Appendix

Appendix 1. Class number and titles:

Number	Title
44	Fuel and related compositions
48	Gas: heating and illuminating
60	Power plants
123	Internal-combustion engines
136	Batteries: thermoelectric and photoelectric
165	Heat exchange
196	Mineral oils: apparatus
204	Chemistry: electrical and wave energy
208	Mineral oils: processes and products
210	Liquid purification or separation
237	Heating systems
250	Radiant energy
260	Chemistry of carbon compounds
320	Electricity: battery or capacitor charging or discharging
376	Induced nuclear reactions: processes, systems, and elements
431	Combustion
432	Heating
903	Hybrid electric vehicles (hevs)

Optimal Operation System of the Integrated District Heating System with Multiple Regional Branches

Ui Sik Kim[1], Tae Chang Park[2], Lae-Hyun Kim[3], and Yeong Koo Yeo[4,*]

Abstract. This paper presents an optimal production and distribution management for structural and operational optimization of the integrated district heating system (DHS) with multiple regional branches. A DHS consists of energy suppliers and consumers, district heating pipelines network and heat storage facilities in the covered region. In the optimal management system, production of heat and electric power, regional heat demand, electric power bidding and sales, transport and storage of heat at each regional DHS are taken into account. The optimal management system is formulated as a mixed integer linear programming (MILP) where the objectives is to minimize the overall cost of the integrated DHS while satisfying the operation constraints of heat units and networks as well as fulfilling heating demands from consumers. Piecewise linear formulation of the production cost function and stairwise formulation of the start-up cost function are used to compute nonlinear cost function approximately. Evaluation of the total overall cost is based on weekly operations at each district heat branches. Numerical simulations show the increase of energy efficiency due to the introduction of the present optimal management system.

1 Introduction

A district heating system (DHS) is a complex system consisting of a large number of energy suppliers and consumers, district heating pipelines and heat storage facilities in a region. A DHS plays an important part in covering the heating demands in downtown and suburban area. DHSs can be characterized by reduction

[1] Hanyang University, Department of Chemical Engineering, Seoul, South Korea
uisik119@gmail.com

[2] Hanyang University, Department of Chemical Engineering, Seoul, South Korea

[3] Seoul National University of Technology, Department of Chemical Engineering, Seoul South Korea

[4] Hanyang University, Department of Chemical Engineering, Seoul, South Korea
ykyeo@hanyang.ac.kr

* Corresponding author.

J.H. Lee et al. (Eds.): EU-Korea Conference on Science and Technology, SPPHY 135, pp. 49–60.
springerlink.com

of energy consumption, increase of energy efficiency and decrease of generation of pollutants. Hence the subject of optimal operation of DHSs has a significant economical potential. DHSs fulfill a significant part of energy demand especially in Nordic countries such as Iceland, Poland, Finland, Denmark, Norway, etc. Korea began to employ DHSs in 1987. In contrast to other countries the heat source used in DHSs mainly consists of fossil fuels in Korea. For this reason the energy supply by DHSs still suffers from economics and environmental contamination problems. To overcome these problems it is recommended to use waste materials as a heat source and to increase energy efficiency by the optimal operation of heat generation systems and heat distribution networks. In this work, a model for the optimal operation of a DHS to increase energy efficiency is developed. Since a heat generation and a network distribution system are main constituents of a DHS, it is obvious that the optimization of a DHS should be based upon the optimization of the heat generation system and of the network distribution system to give the integrated optimal operating system. More specifically, it is suggested to partition the optimization of the entire system into a scheduling among different heat producing units followed by a control problem for the distribution network in order to make the solution of the optimization problem feasible.

The main role of an optimal operation system for a DHS is to minimize the operation cost of the DHS or to maximize the profits of the DHS by generating electricity while satisfying the constraints of the system as well as fulfilling heating demands from consumers. Nordic countries have been showing active research activities in this area for last decade. Due to the existence of time delays in the heat supply, the operation of a DHS is highly dependent upon time-varying demands of customers and the energy distribution network. Moreover, the heat storage as well as the heart loss to the environment should be considered in the operation. To take into account all of these effects, a nodal method was presented to resolve the problem of determination of the supply temperature and the amount of the heat supply. This method is based on the modeling of the heat generation system and the energy distribution network and was applied to simulate the heat flow and the temperature distribution in a DHS [1]. As a heat source of DHS, a cogeneration heat plant (CHP), peak load boiler (PLB), an incinerator and geothermal heat generation can be used.

Typical optimization problem for a DHS consists of modeling of the heat generation system and the energy distribution network and employment of MILP algorithm [3]. Results of evaluation of manufacturing costs in a DHS were reported to show the optimal heat generation and optimal operation of a heat distribution network [4]. As the energy efficiency is increased due to the utilization of energy distribution networks in DHSs, the optimization of energy distribution networks has been paid much attention from many researchers. In the formulation of optimization problems, models for generation and consumption of electricity and heat, a model for fuel transmission to generation plants, a flow transportation model within district heating pipeline networks and a heat storage model are incorporated to give a mixed integer linear programming (MILP) problem so that consumers, suppliers, designers and operators can make decisions in different situations [5]. Recently many researchers tried to apply optimized models in the planning and scheduling of new DHSs as well as in the operation of existing DHSs. They used linear programming (LP) model in the planning and scheduling of DHSs including CHP to determine optimal operating costs while satisfying demands of heat and electricity of regional customers [7].

In the construction of an optimal operation system for a DHS, consideration of heat generation and storage facilities, constraints on the operation of distribution networks and satisfaction of consumers is equivalent to the constraints in the optimization of large scale electricity generation systems which were employed to supply energy prior to DHSs. Many researchers investigated the optimal operation of electricity generation systems where the primary purpose is to minimize costs of electricity generation while fulfilling demands from consumers. A piecewise linear function was used to compute operation costs of electricity generation plant incorporating start-up and shutdown costs. The nonlinear start-up cost function is modelled by using the stairwise approximation formulation [2, 6, 8-10].

In this paper an integrated optimal operation system is developed for the heat generation systems and heat distributed networks of the DHSs located in Seoul and Gyeonggi-do, Korea. Most of the results on the optimization study for DHSs reported so far are concerned about single DHS while an integrated DHS consisting of multiple regional DHSs (branches) is considered in this work. The overall cost of actual operation of DHSs is compared with that of optimal operation of DHSs by employing present integrated optimal operation system to analyze energy efficiency. In the optimization, much stress was laid on how to take into account heat demands from various regional branches and how to incorporate constraints on heat generation and storage facilities. Hourly numerical simulations are performed for a selected week based on the present MILP model.

2 Optimization

2.1 Description of KDHC

A DHS consists of energy suppliers and consumers, district heating pipelines network and heat storage facilities in the covered region. The integrated DHS considered in this work consists of 24 regional DHSs among Each regional DHS is interconnected via 32 distribution networks. Fig. 2 shows the structure of the interconnected DHS networks.

2.2 Formulation of KDHC Optimization Problem

The primary objective of the optimal operation system for the integrated DHS is to minimize operation cost or to maximize profit obtained especially from the sales

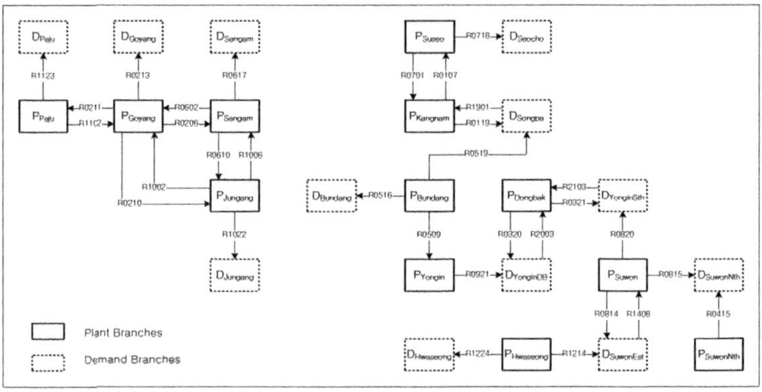

Fig. 1. A schematic diagram of heat distribution network of KDHC.

of electricity while fulfilling heat demand from consumers and satisfying various constraints consisting of operating conditions of heat generation and storage facilities. Most of the results on the optimization study on district energy systems reported so far are concerned about single DHS while an integrated DHS consisting of multiple regional DHSs (branches) is considered in this work. From the optimization we get the optimal heat load in heat generation and storage facilities at each regional DHS, the optimal amount of transmission of heat in distribution networks as well as overall operating cost. The optimal operation system for the integrated DHS proposed in this work is constructed in terms of MILP and the objective function can be represented as following:

$$\text{Minimize} \quad \sum_{t \in T} \left(\sum_{p \in P} OC(p,t) - ES(p,t) \right) \tag{1}$$

subject to (Constraints of the operation system of KDHC)

where T is set of the intervals of the optimization period. p denotes each plant branch and $OC(p,t)$ and $ES(p,t)$ represent operating cost of each plant branch and sales of electricity respectively.

2.2.1 Calculation of the Operation Cost and the Electricity Sales

The overall cost of the integrated DHS is defined as the difference between the sum of operating cost of total regional DHSs and the amount of total electricity sales at those regional branches. The operating cost of a regional DHS is defined as the sum of the operating costs of CHP and heat generation facilities. The transportation cost of heat among regional DHSs is not included in the computation of the total operating cost. The operating cost of a regional DHS is given by

$$OC(p,t) = \sum_{j\in J} Cost_j(t) + \sum_{u\in U} Q_u(t)\cdot C_u(t) + \sum_{h\in H} Q_h(t)\cdot C_h(t) + \sum_{i\in I} Q_i(t)\cdot C_i(t)$$
$$where,\quad Cost_j(t) = Cost_j^{Oper}(t) + Cost_j^{Start}(t) \qquad \forall j\in J,\ p\in P,\ t\in T \tag{2}$$

where $Cost_j(t)$ represents the operating cost of CHP ($Cost_j^{Oper}(t)$) with the start-up cost($Cost_j^{Start}(t)$) and $Q_u(t), Q_h(t), Q_i(t)$ and $C_u(t), C_h(t), C_i(t)$ denote the amount of heat generation and the unit operating price of each unit and each hanjun supply and each incinerator respectively. The amount of electricity sales at each regional DHS can be represented as

$$ES(p,t) = \sum_{j\in J} P_j(t)\cdot SMP(t) \qquad \forall p\in P,\ t\in T \tag{3}$$

where $P_j(t)$ denotes the amount of electricity generated at CHPs and $SMP(t)$ is the marginal unit price. Values of terms in Eqs. (2) and (3) are dependent upon the CHPs and heat generation facilities.

2.2.2 Piecewise of the Opertion Cost Function

The operating cost of a CHP without start-up cost can be represented in the form of a nonlinear function depending upon the amount of heat generation. Fig. 3 shows the profile of the operating cost of a CHP. In order to avoid nonlinearity exhibited by the operating cost of a CHP, the operating interval can be divided into subintervals having equal marginal cost to get piecewise linear forms as shown in Fig. 2.

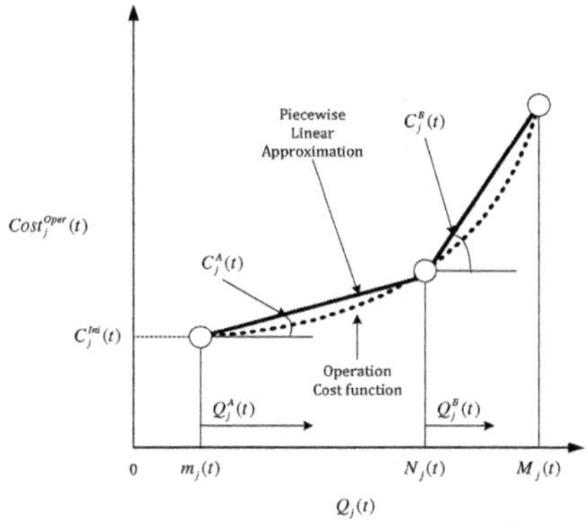

Fig. 2. Operating cost function of a CHP.

Therefore we can see that the cost function of a CHP shown in Fig. 3 is represented by Eq. (4)

$$Cost_j^{Oper}(t) = C_j^{Ini}(t) \cdot Y_j(t) + C_j^A(t) \cdot Q_j^A(t) + C_j^B(t) \cdot Q_j^B(t) \qquad \forall j \in J, t \in T \qquad (4)$$

where $C_j^A(t)$ and $C_j^B(t)$ represent limit costs at each heat generation subinterval, and $C_j^{Ini}(t)$ denotes the minimum operating cost in the operation of CHP. In Fig. 3, $m_j(t)$ denotes the minimum heat generation during CHP operation and $M_j(t)$ is the maximum heat generation. $N_j(t)$ is an intermediate break point between linear approximations with different limit cost depending upon the heat generation. Therefore we can see that $Q_j(t)$, the amount of heat generated from a CHP, is given by Eqs. (5) to (7).

$$Q_j(t) = Q_j^A(t) + Q_j^B(t) + m_j(t) \cdot Y_j(t) \qquad \forall j \in J, t \in T \qquad (5)$$

$$Q_j^A(t) \le [N_j(t) - m_j(t)] \times Y_j(t) \qquad \forall j \in J, t \in T \qquad (6)$$

$$Q_j^B(t) \le [M_j(t) - N_j(t)] \times Y_j(t) \qquad \forall j \in J, t \in T \qquad (7)$$

where j denotes each CHP at the corresponding DHS, and t is time. $Y_j(t)$, representing the operational status of a CHP, is a binary variable taking 0 or 1. In Eqs. (5) to (7), $Q_j^A(t)$ is 0 at $m_j(t)$ and then increases up to $N_j(t)$. Similarly, $Q_j^B(t)$ is 0 at $N_j(t)$ and then increases up to $M_j(t)$. Both $Q_j^A(t)$ and $Q_j^B(t)$ are subject to the following constraints.

2.2.3 Stairwise of the Start-Up Cost Fnction

The start-up cost of a CHP cost can be represented in the form of a nonlinear(exponential) function depending upon the number of hours a CHP has been offline. Fig. 3 show the profile of the start-up cost of a CHP. In order to avoid nonlinearity exhibited by the start-up cost of a CHP, the stairwise linear cost function of a CHP is presented by Eq. (8).

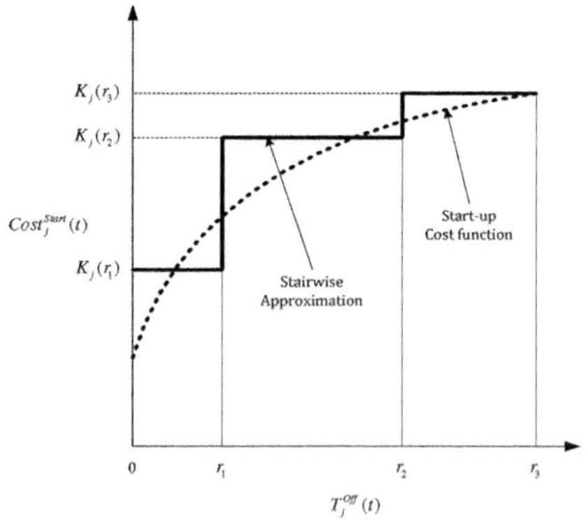

Fig. 3. Start-up cost function of a CHP

$$Cost_j^{Start}(t) = \sum_{r=1}^{ND} K_j(r) \cdot w_j(r,t) \qquad \forall j \in J, t \in T \tag{8}$$

$$\sum_{r=1}^{ND} w_j(r,t) = X_j(t) \qquad \forall j \in J, t \in T \tag{9}$$

$$T_j^{Off}(t-1) = \sum_{r=1}^{ND-1} r \cdot w_j(r,t) + \alpha(t) \qquad \forall j \in J, t \in T \tag{10}$$

$$ND \cdot w_j(ND,t) \le \alpha(t) \le M_j^{Off} \times [w_j(ND,t) - X_j(t) + 1] \qquad \forall j \in J, t \in T \tag{11}$$

$$w_j(r,t) \in \{0,1\} \qquad \forall r \in R,\ j \in J,\ t \in T \tag{12}$$

where ND and M_j^{Off} are the number of discrete intervals of the start-up cost function and the maximum number of hours CHP can be off. R is a set of the discrete intervals of the start-up cost function for CHP. $X_j(t)$ denotes binary variable that is equal to if a CHP is start-up at the beginning of hour t. If a CHP has been off for r hours, $Cost_j^{Start}(t)$ is a constant of value $K_j(r)$.

If a CHP is started-up when it has been off for hours ND or longer, $Cost_j^{Start}(t)$ is equal to $K_j(ND)$.This is modelled through a new binary variable $w_j(r,t)$, which is equal to 1 if a CHP is started-up at hour t and has been offline for r hours. If a CHP is started-up, Eq. (9) forces only one of these binary variables to be equal to 1. Every Eq. (10) relates variables $w_j(r,t)$ with the time counter $T_j^{off}(t)$ through a dummy variable, $\alpha(t)$, whose limits are imposed Eq.(10) and (11). Variable $\alpha(t)$ is used either when CHP is off at t hour or when CHP is started-up at t hour and has been off for ND hours or longer. Lastly, Eq. (12) states that variables $w_j(r,t)$ are binary.

2.2.4 Time Couner

The off-time counter is an important variable in expressing the start-up costs. Mathematically, its constraints are given by Eqs. (13) to (16).

$$T_j^{Off}(t) \le T_j^{Off}(t-1) + 1 \qquad \forall j \in J, t \in T \tag{13}$$

$$T_j^{Off}(t) + \left(M_j^{Off} + 1\right) \times Y_j(t) \ge T_j^{Off}(t-1) + 1 \qquad \forall j \in J, t \in T \tag{14}$$

$$T_j^{Off}(t) - M_j^{Off} \times [1 - Y_j(t)] \le 0 \qquad \forall j \in J, t \in T \tag{15}$$

$$T_j^{Off}(t) \ge 0 \qquad \forall j \in J, t \in T \tag{16}$$

where $T_j^{Off}(t)$ denotes time periods that a CHP has shut-down at hour t.

2.2.5 Logical Status of Commitment

Eqs. (17) and (18) express the logic of start-ups and shutdowns of a CHP.

$$X_j(t) - Z_j(t) = Y_j(t) - Y_j(t-1) \qquad \forall j \in J, t \in T \tag{17}$$

$$X_j(t) + Z_j(t) \le 1 \qquad \forall j \in J, t \in T \tag{18}$$

where $Z_j(t)$ denotes a binary variable which is equal to 1 if a CHP is shut-down at the beginning of hour t.

2.3 Constraints of KDHC Optimization Problem

2.3.1 Heat Demands of Consumer

The fulfillment of heat demand from consumers by adjusting the amount of heat generation and heat supply through distributed networks as well as heat storage and transmission is the basic operating constraint in the optimizations of the integrated DHS. This condition can be summarized as Eqs. (19) and (20). If a DHS is a plant branch, there have to satisfy Eq. (19). Otherwise, have to satisfy Eq. (20).

Plant branches:

$$\sum_{j\in J} Q_j(t) + \sum_{u\in U} Q_u(t) + \sum_{h\in H} Q_h(t) + \sum_{i\in I} Q_i(t) + \sum_{a\in A} \Delta Q_a(t) + \sum_{n\in N} Q_n^{In}(t) - \sum_{n\in N} Q_n^{Out}(t) = 0 \tag{19}$$

$$where, \quad \Delta Q_a(t) = Q_a(t) - Q_a(t-1)$$

$$\forall j \in J, u \in U, h \in H, i \in I, a \in A, n \in N, p \in P, t \in T$$

Demands branches:

$$D^{Heat}(d,t) = \sum_{n\in N} Q_n^{In}(t) - \sum_{n\in N} Q_n^{Out}(t) \qquad \forall d \in D, t \in T \tag{20}$$

where J,U,H,I,A,N are sets of each CHP, heat unit(PLB_{SO}, PLB_{WG}, GRB), hanjun supply, incinerator, accumulator, network. $D^{Heat}(d,t)$ is the heat demand of each demand branch d at hour t, and $Q_u(t)$ denotes the amount of heat generation from all heat generation facilities in each DHS, and $Q_n^{In}(t)$ and $Q_n^{out}(t)$ denote the amount of heat transmission through interconnected distribution networks respectively. $\Delta Q_a(t)$ represents changes in the heat accumulator. Application of Eqs. (19) and (20) depend upon the characteristics of each DHS. For example, P_{Suwon} in Suwon area does not accept electricity outside the area and the corresponding term ($Q_h(t)$) is excluded from Eq. (19).

2.3.2 Heat Generation Limit and Hat Transfer Limit

The heat facilities used to supply heat to fulfill demands of customers include heat units and hanjun supplies and incinerators. To fulfill the supply shortage of each branch, the heat distribution networks in DHSs also transfer heat to each other. In each heat facility and network, there exist the minimum and maximum amount of hourly heat generation(transmission) and which consist of an operating range to be satisfied. The operational status of a heat facility (network) is represented by an integer 0 (not being operated) or 1 (being operated). The overall constraints for heat facilities(networks) can be represented as Eq. (21).

$$m_k(t) \cdot Y_k(t) \le Q_k(t) \le M_k(t) \cdot Y_k(t) \qquad \forall k \in J \,\&\, U \,\&\, H \,\&\, N,\, t \in T \tag{21}$$

where $Y_u(t)$ denotes the operating status (i.e., on/off status), and $m_u(t)$ and $M_u(t)$ represent the minimum and maximum hourly heat generation(transmission) for the heat unit(network) respectively. In general, the value of $m_u(t)$ is greater than 0.

2.3.3 Het Ramp Limit

The amount of heat generation in heat units should not show abrupt variations for safe operations. The variations in the amount of heat generation in heat units should be confined within certain range which is usually given by the maximum hourly generation multiplied by safety factor as shown in Eq. (22)

$$-S_k(t) \cdot M_k(t) \le Q_k(t) - Q_k(t-1) \le S_k(t) \cdot M_k(t) \qquad \forall k \in J \,\&\, U,\, t \in T \tag{22}$$

where $S_k(t)$ denotes the safety factor for the corresponding CHP and heat unit.

2.3.4 Power Generation Limit

The amount of the electricity generated from a CHP which produces heat and electricity simultaneously is a function of heat generated as given by Eq. (23).

$$P_j(t) = G_R(t) \cdot Q_j(t) + G_C(t) \cdot Y_j(t) \qquad \forall j \in J,\, t \in T \tag{23}$$

where $G_R(t)$ is the gradient of the linear function representing the relationship between the amount of heat production and electricity generation, and $G_C(t)$ is the interception of y-axis.

2.3.5 Heat Storage Limit and Ramp Limit and Safety Standards of Acc

There exist the minimum and maximum amount of heat that can be stored in a heat storage unit. Thus the constraint on a heat storage unit can be written as Eq. (24).

$$m_a(t) \le Q_a(t) \le M_a(t) \qquad \forall a \in A,\, t \in T \tag{24}$$

The constraints on the variations of the amount of heat stored in a heat storage unit are given by Eq. (25).

$$-S_a^Q(t) \le Q_a(t) - Q_a(t-1) \le S_a^Q(t) \qquad \forall a \in A,\, t \in T \tag{25}$$

where $S_a^Q(t)$ represents a permissible amount of possible variations. A heat storage unit has to store a certain amount of heat at a specific time for a day to prepare for abnormal or urgent situations. This condition can be written as Eq. (26).

$$S_a^{Lower} \le Q_a(d) \le S_a^{Upper} \qquad \forall a \in A,\, t \in T \tag{26}$$

where d represents a specific time for a day.

2.3.6 Minimum up Time and Minimum down Time

Once the operation of a heat facility begins, the operational status should be maintained at least for a certain time period defined as the minimum up time (Mut). Similarly, if the operation of a heat unit is terminated, the un-operational status should be maintained at least for a certain time period defined as the maximum down time (Mdt). Each heat facility has its own Mut and Mdt to be satisfied as represented by Eqs. (27) to (32) which behave as constraints on heat facilities.

If a heat facility k is on at time 0:

$$Y_k(t) = 1, \qquad \forall k \in J \,\&\, U \,\&\, H,\, t \in [1, 2, ..., Mut_k - T_k^{On}(0)] \tag{27}$$

$$\begin{aligned} &[Y_K(t) - Y_k(t-1)] + [Y_K(t+m-1) - Y_k(t+m)] \le 1 \\ &\quad \forall k \in J \,\&\, U \,\&\, H \\ &\quad \forall t \in [Mut_k - T_k^{On}(0) + 1, Mut_k - T_k^{On}(0) + 2, ... T_{last}],\ m \in [1, 2, ..., Mut_k - 1] \end{aligned} \tag{28}$$

$$\begin{aligned} &[Y_K(t-1) - Y_k(t)] + [Y_K(t+m) - Y_k(t+m-1)] \le 1 \\ &\quad \forall k \in J \,\&\, U \,\&\, H,\, t \in T,\ m \in [1, 2, ..., Mdt_k - 1] \end{aligned} \tag{29}$$

If a heat facility k is off at time 0:

$$Y_k(t) = 0, \qquad \forall k \in J \,\&\, U \,\&\, H,\, t \in [1, 2, ..., Mdt_k - T_k^{Off}(0)] \tag{30}$$

$$\begin{aligned} &[Y_K(t-1) - Y_k(t)] + [Y_K(t+m) - Y_k(t+m-1)] \le 1 \\ &\quad \forall k \in J \,\&\, U \,\&\, H \\ &\quad \forall t \in [Mdt_k - T_k^{Off}(0) + 1, Mdt_k - T_k^{Off}(0) + 2, ... T_{last}],\ m \in [1, 2, ..., Mdt_k - 1] \end{aligned} \tag{31}$$

$$\begin{aligned} &[Y_K(t) - Y_k(t-1)] + [Y_K(t+m-1) - Y_k(t+m)] \le 1 \\ &\quad \forall k \in J \,\&\, U \,\&\, H,\, t \in T,\ m \in [1, 2, ..., Mut_k - 1] \end{aligned} \tag{32}$$

where $T_k^{On}(t)$ and $T_k^{Off}(t)$ represent the duration times for heat generation facilities to be on/off status at hour 0. Eqs. (27) and (30) ensure that the heat facility will satisfy the minimum up time and the minimum down constraints if it has been on/off at hour 0 for fewer hours than the minimum up time and the minimum down time.

2.3.7 Interconnected Network Lmit

In the integrated DHS considering in this work, there are 32 interconnected networks. The specific network lines of all networks are defined as the couple network between two regional DHSs. For example, the interconnected networks R0614 and R1406 in Fig. 2 between P_{Suwon} and D_{SuwonEst} are network lines between these two areas. In this case, the constraint on couple network is represented as Eq. (33).

$$Y_c^{In}(t) + Y_c^{Out}(t) \le 1 \qquad \forall c \in C,\, t \in T \tag{33}$$

where $Y_c^{In}(t)$ and $Y_c^{Out}(t)$ denote the status of heat transmission through interconnected distribution couple networks respectively.

2.4 Simulation and Results

Except for P_{Paju} and $D_{Hwaseong}$ branches, almost DHSs of KDHC were selected to apply the optimization system presented in this work. The effectiveness of the optimization system for the integrated DHS developed in this work could be evaluated by estimating total overall cost. The overall cost between the operation of DHS based on the proposed optimization system and actual operation of DHS were compared for a week. The optimization problem consists of 88,892 constraints and 36,504 variables. CPLEX 11.0 of ILOG, Inc. was used to solve the optimization problem. Results of numerical solution of the optimization problem were compared with actual operation data. As shown in Fig. 4, we can see much reduction in the overall cost for actual operation period. For example, the weekly overall cost obtained from the optimization system is 11,536,760,000 won which is well compared with 18,150,088,000 won of actual operation.

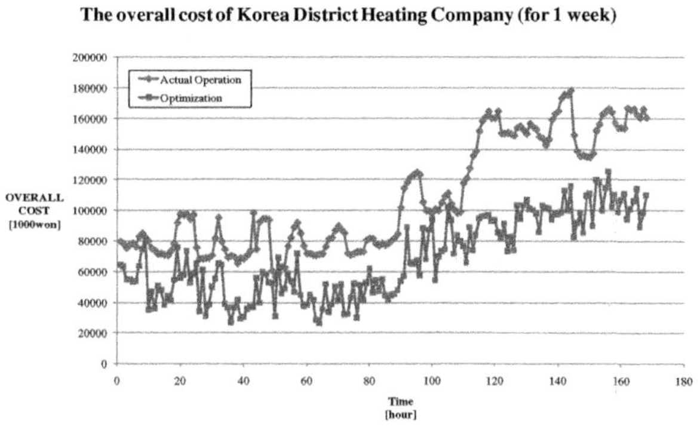

Fig. 4. The overall cost of optimization and actual operation for a week.

3 Conclusion

The operation of regional district heating systems (DHSs) needs to be optimized in order to improve the economics and efficiency. In contrast to the optimization of each regional DHS separately, the optimization problem of the integrated DHSs in relatively large area is tackled in this work. The optimal operation system for the integrated DHSs was developed in the form of an MILP configuration and was investigated to validate the effectiveness of the optimal operation system. In the optimization, the total overall cost of the integrated DHSs is minimized while heat demands from customers and operation constraints of heat facilities and interconnected networks are fulfilled. From the results of numerical simulations, it was found that the optimal operating system for the integrated DHSs shows reduction of energy cost compared to conventional operation.

Acknowledgements. This work has been supported by the Korea Energy Management Corporation and Ministry of Knowledge Economy of Korea as a parts of the Project of "Construction of Energy Network using Wide Area Energy" (2007-E-ID25-P-02-0-000) in "Energy conservation Technology R&D" project. The authors wish to thank them for their support.

References

[1] Benonysson, A., et al.: Operation optimization in a district heating system. Energy conversion and management 36(5), 297–314 (1995)

[2] Arroyo, J., Conejo, A.: Optimal response of a thermal unit to and electricity spot market. IEEE Trans. Power Syst. 15(3), 1098–1104 (2000)

[3] Aringhieri, R., Malucelli, F.: Optimal operations management and network planning of a district heating system with a combined heat and power plant. Annals of Operations Research 120, 173–199 (2003)

[4] Sjödin, J., Henning, D.: Calculating the marginal costs of a district heating utility. Appl. Energy 78, 1–18 (2004)

[5] Söderman, J., Pettersson, F.: Structural and operational optimization of distributed energy systems. Applied Thermal Engineering 26, 1400–1408 (2006)

[6] Carrion, M., Arroyo, J.: A computationally efficient mixed-integer linear formulation for the thermal unit commitment problem. IEEE Transactions on power systems 21(3), 1371–1378 (2006)

[7] Naser, A.T., Husam, A.: A linear programming model to optimize the decision-making to managing cogeneration system. Clean Techn. Environ. Policy 9, 235–240 (2007)

[8] Delarue, E., D'haeseleer, W.: Adaptive mixed-integer programming unit commitment strategy for determining the value of forecasting. Applied Energy 85, 171–181 (2008)

[9] Afshar, K., et al.: Cost-benefit analysis and MILP for optimal reserve capacity determination in power system. Applied Mathematics and Computation 196, 752–761 (2008)

[10] Smajo, B., Mesur, H., Muris, D.: Hydrothermal self-scheduling problem in a day-ahead electricity market. Electric Power Systems Research 78, 1579–1596

A Research on the Application Method for Renewable Energy Complex System for School Buildings

Ji-Yeon Kim[1,*], Sung-Hee Hong[2], Hyo-Soon Park[2], Sung-Sil Kim[3], and Jae-Min Kim[4]

Abstract. This research aims to study a new optimum renewable energy application method that can cover the minimum energy and operating costs within a range of school budgets. Deriving the optimum application method is expected to maximize the cooling/heating and water heating energy savings of educational facilities. Therefore, this research examined the renewable energy utilization technique diffusion expansion method and the optimum method. The first optimum plan was introduced with a 174kW multi-type geothermal heat pump + a 94m² solar collectors + a 249.4kW highly efficient electronic cooling/heating device (EHP). On the other hand, the second optimum plan was introduced as a 255.2kW multi-type geothermal heat pump + a highly efficient 168.2kW electronic cooling/heating device (EHP)

1 Introduction

The government of Korea has established a law that requires the utilization of renewable energy by public organizations, based on the principle of energy rationalization promotion in public organizations in Prime Minister Order No. 2008-3 (June 12, 2008). The mandatory renewable energy system installation is for newly-built buildings with gross dimensions of over 3,000 m^2. It requires that public organizations invest 5 percent of their total building construction costs in the renewable energy installation. After the revision of the Renewable Energy Development, Utilization, and Diffusion Promotion Act on March 13, 2008, it has been

[1] Department of Architectural Engineering, Inha University/KIER, South Korea
jykim@kier.re.kr
[2] High Efficiency Energy Research Department, Korea Institute of Energy Research, Daejeon, South Korea
shhong@kier.re.kr, hspark@kier.re.kr
[3] LG Electronics, Seoul 150-875, South Korea
rosini@lge.com
[4] Department Mechanical Engineering University of Strathclyde, 75 Montrose St. Glasgow, G1 1XJ, United Kingdom
min@esru.strath.ac.uk
[*] Corresponding author.

J.H. Lee et al. (Eds.): EU-Korea Conference on Science and Technology, SPPHY 135, pp. 61–72.
springerlink.com

carried out since 2009, while expanding the target buildings to include school buildings and 'expanded/remodeled' buildings.

As newly-built schools have gross dimensions of almost over 3,000 ㎡, they need to install a renewable energy system. The school building research conducted showed that majority of the schools prefer the EHP or GHP system as their cooling/heating system. Of these two systems in which independent control is possible, the schools with lower electronic power unit costs chose EHP as their cooling/heating method for cost reduction.

Connecting the EHP system, as the main cooling/heating method, to a multi-type water/air geothermal heat pump will further reduce cooling/heating costs while a solar collector or Photovoltaic will reduce the hot water and electricity loads.

Even if a high-tech renewable energy equipment is introduced and even if the policy is advanced, however, energy reduction could not be expected and it would be difficult to build the system if there are no applicable capacity standards for the renewable energy including with respect to the size of the education facility, and if the optimum system for the building and its economic feasibility would not be analyzed and evaluated.

In this regard, this research aims to come up with an optimum heating/cooling and renewable energy application method that can cover the minimum energy and operating costs within a given range of school budgets.

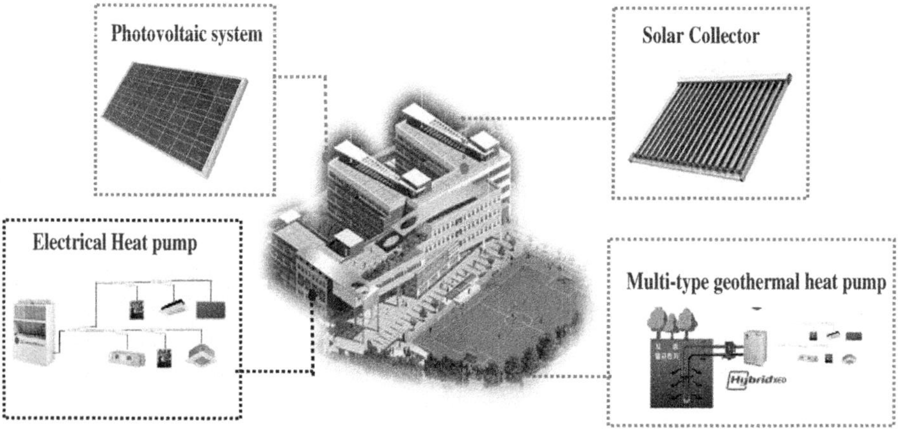

Fig. 1. Cooling/heating and Renewable energy systems

2 Research Method

The research flow chart is shown in Figure 2, and is composed of the majority of the inputs and schedules in the actual school research.

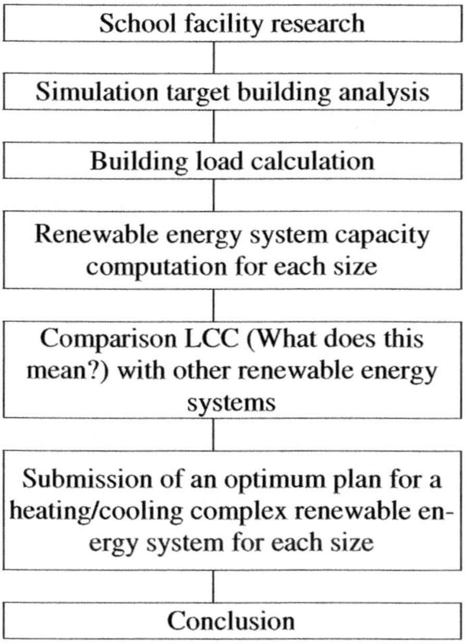

Fig. 2. The research flowchart

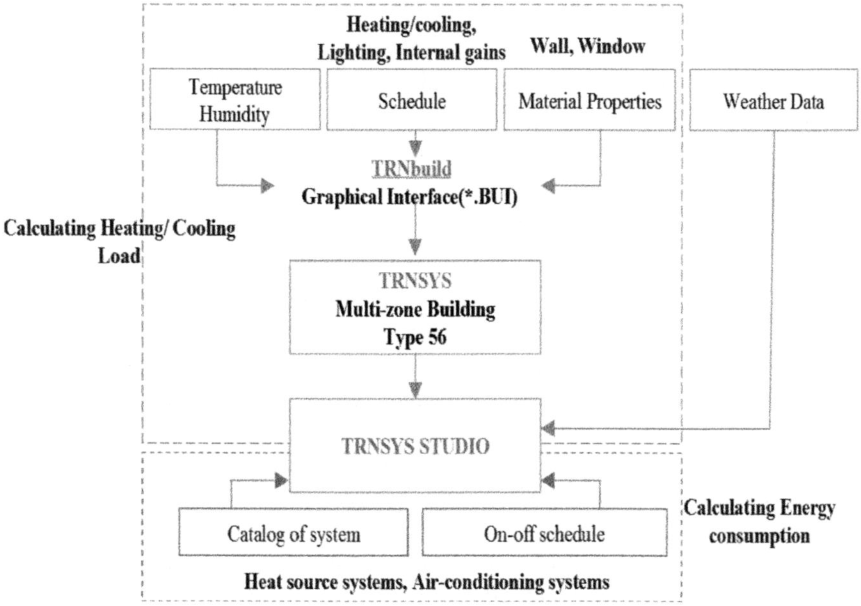

Fig. 3. TRNSYS simulation concept chart

Unlike most of the simulation programs in the market today, the users of the proposed program can perfectly simulate and monitor all the relationships among the system components through 'TRNSYS'. Besides, the time gap for the simulation would be freely set up from specific hours to certain seconds. As each system component is composed of a module structure, the user can compose the system freely, such as the pump, chiller, cooling top, and solar collector.

The program offers approximately 60 standard components such as HVAC equipment, a multi-zone model, several controllers, and a solar collector, and provides a climate file processor, a schedule manager, and results output equipment.

In particular, TRNSYS is one of the building capacity simulation programs that can interpret the temperature level control and simulate the actual dynamic response of a building. In this regard, this project assumed that TRNSYS would be appropriated as the building heat capacity simulation program for what, and interpreted the data generated from this study using the said program.

3 Computation of Cooling/Heating Capacity in School

3.1 Classroom Size

The standard school size is defined as 1-15 classes and below a gross dimension of 7,000 m^2. It has been seen, however, that the actual dimension module of a class is 67.5 m^2, for which an indoor facility with the capacity of 11.6 kW is required. The computation of the cooling/heating capacity of a 12-classroom school based on the abovementioned standard led to the analysis that a small school must have a cooling/heating facility with a capacity of approximately 423.4 kW.

3.2 Cooling/Heating System

Since a geothermal heat pump system could not be installed in all classrooms with building construction costs of 5 percent of the standard, the existing cooling/heating system should instead be installed. A research on school facilities showed that the majority of schools have their adopted the EHP or GHP system as cooling/heating system, and that the schools that were using central cooling/heating methods such as absorptive cooling/heating equipment or 'district heating' were hoping or planning to replace them.

Besides, for a small school, it is expected that a system that can be independently controlled such as EHP and GHP would be appropriated, and that l arge-capacity absorptive cooling/heating would be inappropriate. Therefore, the classrooms that have not adopted a geothermal heat pump system were

Table 1. Actual dimension and capacity computation table for a school with 12 classrooms

Section	Room Name	Dimensions (m^2)	Number	Total Dimensions (m^2)	Needed Capacity (kW)
General classrooms	Classroom	67.5	12	810	139.2
	Small lecture room	33.75	2	67.5	11.6
	Medium-sized lecture room	67.5	2	135	23.2
Management rooms	Teachers' room	67.5	1	67.5	46.4
	Principal's room	67.5	1	67.5	
	Administration room	67.5	1	67.5	
	Infirmary	56	1	56	
Special rooms	Music room	101.25	1	101.25	23.2
	Preparation room	33.75	1	33.75	
	Art room	101.25	1	101.25	23.2
	Preparation room	33.75	1	33.75	
	Science room	101.25	1	101.25	23.2
	Preparation room	33.75	1	33.75	
	Domestic room	101.25	1	101.25	23.2
	Preparation room	33.75	1	33.75	
Support rooms	Computer room	67.5	1	67.5	11.6
	Library	168.75	1	168.75	29
	Language laboratory	67.5	1	67.5	11.6
	Dining room	350	1	350	58
Total				2,464.8	423.4

*1HP=2500kcal/hr=2.9kW.

encouraged to calculate their cooling/heating needs by applying EHP and GHP. The features of each cooling/heating equipment are as follows.

Table 2. Comparison of the characteristics of each cooling and heating system

	Multi-type Geothermal Heat Pump	Water-Water Geothermal Heat Pump	EHP	GHP	Absorption Chiller
Cooling COP	5.00	4.11[1)]	3.81	1.00	1.00
Heating COP	5.42	4.24	3.23	1.20	0.80

1) Cooling capacity: Standards of heat source water - inlet temperature 25°C, cooled water outlet temperature 7°C. Heating capacity: Standards of heat source water - inlet temperature 12°C, warm water outlet temperature 50°C.

2) Cooling/heating COP: Capacity data of a renewable energy company and cooling/heating machinery companies.

Table 3. Photovoltaic module and vacuum pipe-type collector specifications

Section	Photovoltaic Module	Section	Vacuum pipe-type Collector
Rated max. output (Wp)	175	Heat collecting panel number (pcs)	18
Open voltage(V)	44.50	Gross area(m^2)	3.41
Short-circuit electricity(A)	5.36	Valid heat collection area(m^2)	3.00
Max. operating voltage(V)	35.65	Heat medium volume(L)	2.4
Max. operating electricity(A)	4.91	Heat collection amount (Kcal/day . m^2)	2,280
Module efficiency(%)	14	Max. temperature(℃)	295
Size(mm) L×W×T	1570×78×40	Vacuum pipe length(mm)	1,500

3.3 Renewable Energy System

Since the general activities within a school are carried out during the day, the electricity and hot water loads also occur during the day. As the Photovoltaic system can lower the electricity peak loads, and the solar heating system can immediately cope with the hot water loads, they are referred to as the renewable energy systems that are suitable for installation in school facilities.

Table 4. Comparison of cooling/heating energy and water heating energy costs

EHP	Multi-type Geothermal Heat pump	Solar collector (m²)									
(kW)	(kW)	0	28	56	94	124	154	184	214	244	281
8.2	255.2	13,621									
		0.0									
191.4	232	13,985	13,422								
		3.0	0.0								
220.4	203	14,149	13,595	13,138							
		6.6	3.4	0.0							
249.4	174	14,435	13,825	13,355	12,845						
		10.1	7.0	3.9	0.0						
278.4	145	15,182	14,628	14,106	13,610	13,355					
		13.7	10.5	7.4	3.2	0.0					
307.4	116	15,424	14,813	14,343	13,852	13,601	13,493				
		17.2	14.1	11.0	6.7	3.4	0.0				
336.4	87	15,650	15,092	14,626	14,078	13,884	13,771	13,768			
		20.8	17.7	14.5	10.3	6.9	3.6	0.0			
365.4	58	15,916	15,358	14,888	14,344	14,094	13,985	13,969	14,102		
		24.3	21.2	18.1	13.8	10.5	7.1	3.8	0.0		
394.4	29	16,313	15,755	15,289	14,741	14,495	14,382	14,370	14,429	14,620	
		27.9	24.8	21.6	17.4	14.0	10.7	7.3	4.0	0.0	
3.4	0	16,456	15,827	15,357	14,865	14,615	14,506	14,489	14,553	14,660	14,869
		31.5	28.3	25.2	20.9	17.6	14.2	10.9	7.5	4.2	0.0

Energy cost (thousand won)

Photovoltaic system (kW)

3.4 Optimum Inducement Simulation

To select the suitable renewable energy application method, the energy costs were compared based on 5 percent of the building construction costs with the combination of a solar collector, a multi-type geothermal heat pump, and a Photovoltaic system. Within 5 percent of the building construction costs, the solar collector and the multi-type geothermal heat pump were applied and the Photovoltaic system was applied with the remaining costs.

The energy cost was reduced as the dimension of the collector increased. Moreover, when a collector of the same size as what was installed, the energy cost dropped to cover the other system with the multi-type heat pump rather than the combination of the multi-type geothermal heat pump and the Photovoltaic system. Since more heat energy is producible through a solar collector and its heat efficiency is higher than that of the energy reduced through the application of a geothermal heat pump or the electronic energy generated through a Photovoltaic system, the gas unit cost was deemed to be higher than the electricity unit cost.

Therefore, considering the installation of solar heat, the geothermal and photovoltaic system is the method that reduced energy most. The heat produced from a solar heat system was emphasized at specific times during the day, however, as the amount of the collected solar heat was larger than the low or no load capacity in the day. Thus, the system operation would be stopped and the operating temperature in the heat-collecting part would increase due to the compressed heat temperature increase, after which the heat collection efficiency would drop. In this regard, it would be appropriate to install a multi-type heat pump with the remaining installation costs after selecting an appropriate capacity heat collector.

Moreover, when applying a multi-type heat pump, it is deemed that the energy reduction effect would be greater if focus is given to applying it to a room with a larger cooling/heating capacity. Therefore, it would be effective to apply a geothermal heat pump to classrooms, infirmaries, principals' rooms, administration rooms, teachers' rooms, and dining rooms, where the frequency of occupation or the number of occupants is larger than in other rooms.

4 Comparative Evaluation of the Alternatives

The standard gross dimension of a small school is 6,000 m², and it has an approximate construction cost of 5.9 billion Korean won. When investing 5 percent of the construction cost, which is 290 million won, the submitted alternative to the renewable energy equipment is as follows. The optimum size of the solar hot water system is 94 m², and the capacity of a multi-geothermal heat pump system is 174 kW, which can be applied when installing a multi-type geothermal heat pump with the remaining costs.

Moreover, this research proposes that a highly efficient electronic EHP and GHP would cover the cooling/heating load capacities, while excluding 'large-capacity' absorptive cooling/heating facilities, as the cooling/heating capacity of a small school is also small. This research further proposes the application of the full 5 percent of the construction costs to a multi-type geothermal heat pump (Alternative 4) and a Photovoltaic system (Alternative 5), and compares all other proposals.

Table 5. System application capacity for each alternative

Section	Alt. 1	Alt. 2	Alt. 3	Alt. 4	Alt. 5
Cooling/heating method 1	Multi-type geothermal heat pump system 174kW	Multi-type geothermal heat pump system 174kW	Water-water geothermal heat pump 174kW	Multi-type geothermal heat pump system 255.2kW	EHP 423.4kW
Cooling/heating method 2	EHP 249.4kW	GHP 249.4kW	GHP 249.4kW	EHP 168.2kW	Photovoltaic system 31kWp
Hot Water method	Solar collector 94m² +boiler 100,000 kcal/h	Solar collector 94m² + boiler 100,000 kcal/h	Solar collector 94m² +boiler 100,000 kcal/h	Boiler 100,000 kcal/h	Boiler 100,000 kcal/h

4.1 Comparison of Energy Cost

The energy cost difference between Alternatives 1 and 2 is the energy cost difference between the highly efficient EHP and GHP systems. As the EHP system has a higher COP than the GHP system, it is deemed that the GHP system, due to its comparatively higher gas unit cost, has higher energy costs, since gas is its energy source. As the energy cost difference between Alternatives 2 and 3 emanates from the energy cost difference between the multi-type geothermal heat pump and the water-water geothermal heat pump, it is deemed that the energy cost will increase due to the additional indoor-side circulating pump's power demand and comparatively lower COP.

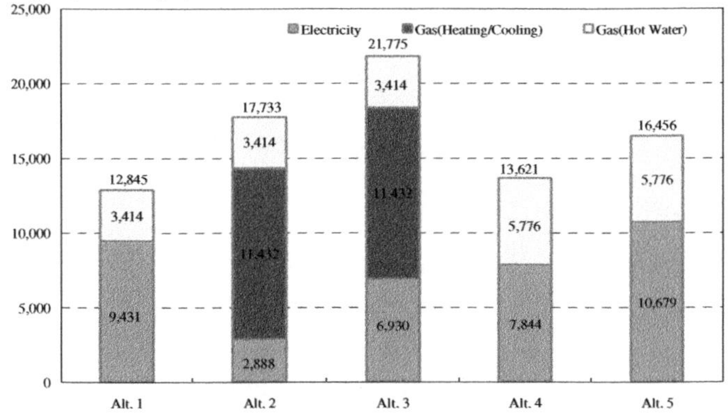

Fig. 4. Comparison of the energy costs of the alternatives

When installing a multi-type geothermal heat pump (174 kW) with the remaining costs based on a 94m² solar -collecting panel, the annual energy cost is 12,747,000 won, which is the lowest cost; and even when applying multi-type geothermal heat pump as the mandatory installing cast, it is deemed that the energy efficiency will be high enough.

4.2 Comparison of Initial Investments

Among the early investments in each alternative, Alternative 4 had the lowest required cost, in which a multi-type geothermal heat pump and a highly efficient

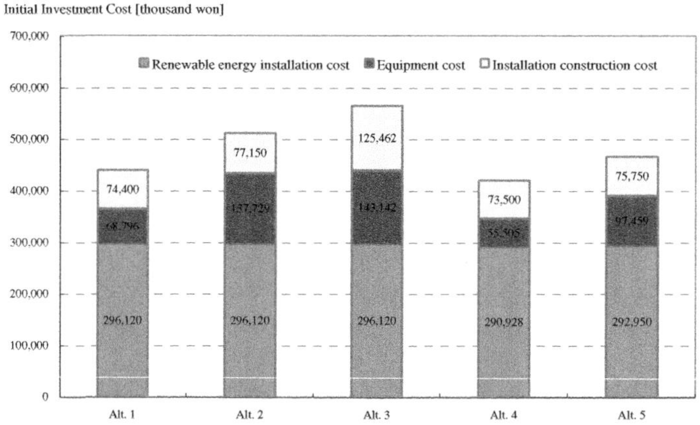

Fig. 5. Comparison of the initial investment costs of the alternatives

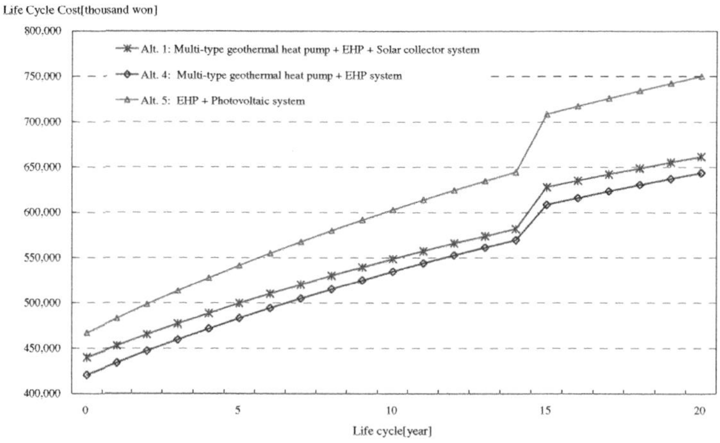

Fig. 6. Comparison.of LCC

EHP were applied. The total initial investment cost was 439,316,000 won, 19,383,000 won less than that of Alternative 1, since the installation capacity of the multi-type geothermal heat pump was reduced due to the installation cost of the solar collector, and the EHP installation capacity was larger. As for Alternative 3, the GHP installation cost or equipment cost was higher than the cost of the EHP, so the initial investment in it was the highest.

4.3 Comparison of LCC

This research used the initial investment cost calculated above for the LCC evaluation. The maintenance cost of the renewable energy system was not considered, and the maintenance cost of the EHP system was pegged at 0.75 percent of the equipment cost. The computation of the total life cycle cost of Alternatives 1, 4, and 5 showed that the life cycle cost of Alternative 4 was the lowest and that Alternative 1 had the lowest energy cost.

Thus, the collection of the investment costs would be impossible compared to Alternative 4. In With respect to the medium-sized and large schools, the gap in their life cycle costs was almost the same as what, but the investment cost collection would be impossible due to the equipment replacement time. Since the energy cost and the initial investment cost were higher in Alternative 5, its total life cycle cost was the highest.

5 Conclusion

The comparison and evaluation of the complex systems, after categorizing the education facilities as small, medium-sized, and large, were linked to a multi-geothermal heat pump system and EHP, and other renewable energy and cooling/heating systems within the 5 percent cost range of the building construction cost, which is the mandatory investment ratio for renewable energy. The energy costs increased in the following order: with the multi-type geothermal heat pump + the vacuum-type solar collector system < the multi-type geothermal heat pump + the highly efficient EHP < the multi-type geothermal heat pump + the GHP + the vacuum-type solar collector system or the highly efficient EHP + the Photovoltaic system < the water-water geothermal heat pump + the GHP.

The comparison of the initial investment costs is as follows: the cost of the multi-type geothermal heat pump + the highly efficient EHP < the cost of the multi-type geothermal heat pump + the highly efficient EHP + the vacuum-type solar collector system < the cost of the highly efficient EHP + the Photovoltaic system < the cost of the multi-type geothermal heat pump + the GHP + the vacuum-type solar collector system < the cost of the water-water geothermal heat pump + the GHP. The LCC evaluation showed that the multi-type geothermal heat pump + the highly efficient EHP are the appropriate systems for school buildings.

The best system for a small 6,000-m^2 school with 12 classrooms is the 174-kW multi-type geothermal heat pump + the 94-m^2 solar collector + the highly efficient

EHP, and the second best system was the 255.2-kW multi-type geothermal heat pump + the highly efficient 168.2-kW EHP.

With respect to the energy and facility investment aspects, the multi-type geothermal heat pump should increase the energy efficiency of education facilities. Moreover, when a cooling/heating complex system that is linked with a highly efficient EHP is installed with the mandated renewable equipment, savings are expected on the annual energy and maintenance costs due to the resulting low electricity unit costs and the reduced peak load capacity.

References

[1] Cane, D., Garnet, J., Caneta Research Inc.: Commercial/Institutional Heat Pump Systems in Cold Climates (2000)

[2] Forbes, R.: Code for Sustainable Homes: An Evaluation of Low Carbon Dwellings, University of Strathclyde Department of Mechanical Engineering Energy Systems Research Unit (2007)

[3] Cane, R.L.D., Clemes, S.B.: A comparison of measured and predicted performance of a ground-source heat pump system in a large building. ASHRAE Transactions (2005)

[4] U.S. Department of Energy. Annual Energy Consumption Analysis Report for Rechland Middle School (2005)

[5] Alabama Universities-TVA Research Consortium, Cost Contaminant for Ground-Source Heat Pumps (1995)

[6] Ozgener, O.: Modeling and performance evaluation of ground source heat pump systems, Building and Environment (2007)

Green Architecture
Analysis of the Tendency of Green Architecture in France

Seung-Ho Lee[1]

Abstract. Today, the environment has become a main subject in lots of science disciplines and the industrial development due to the global warming. This paper presents the analysis of the tendency of Green Architecture in France on the threes axes: Regulations and Approach for the Sustainable Architecture (Certificate and Standard), Renewable Materials (Green Materials) and Strategies (Equipments) of Sustainable Technology. The definition of 'Green Architecture' will be cited in the introduction and the question of the interdisciplinary for the technological development in 'Green Architecture' will be raised up in the conclusion.

1 What Is Green Architecture?

What is the Green Architecture? Today, we use the green colour based on the concept of 'ecology' or 'friendly environment'. It is popular to define green architecture as a sustainable building. The last is a general term that describes environmentally-conscious design techniques in the field of architecture. In the broad context, sustainable architecture seeks to minimize the negative environmental impact of buildings by enhancing efficiency and moderation in the use of materials, energy, and development space. Most simply, the idea of sustainability, or ecological design, is to ensure that our actions and decisions today do not inhibit the opportunities of future generations. [1]

This ecologically conscious approach involves the specific design with vast knowledge on the appropriate materials and equipments, the environment-friendly site management during construction and the optimal energy management for low-emitting.

2 Regulations and Approach for the Sustainable Architecture in France

2.1 HQE (Haute Qualité Environnementale or High Quality Environmental Standard)

HQE is a code for sustainable architecture in France, based on the principles of sustainable development first set out at the 1992 Earth Summit. The standard is

[1] Architect DPLG / Civil Engineer / Master of 'Materials for the Architecture',
Cabinet d'Architecture Seung-Ho LEE, France
archi.shl@free.fr

J.H. Lee et al. (Eds.): EU-Korea Conference on Science and Technology, SPPHY 135, pp. 73–80.
springerlink.com

controlled by the Paris based *Association pour la Haute Qualité Environnementale (ASSOHQE)*. It is a private initiative which makes a commercial offer of engineering to improve the conception or the renovation of buildings and cities by limiting as much as possible their environmental impact.

Since 2004, the HQE association has engaged a dynamics of certification in order to provide the possibility, recognized by an independent third party the environmental quality of their approach and their realization, to the project owners. Having no vocation to be an organisation of the certification, it confided to AFNOR CERTIFICATION, which is auditing methods with an attention of setting up this 'Certifications NF Ouvrage Démarche HQE®'. [2]

Table 1. 14 Targets of HQE® with 4 families divided

Managing the impacts on the external environment	**Creating a pleasant interior environment**
ECO-CONSTRUCTION : · Harmonious relationship between buildings and their immediate environment · Integrated choice of construction methods and materials · The avoidance of nuisance by the construction site	COMFORT : · Hydrothermal control measures · Acoustic control measures · Visual attractiveness · Measures to control smells
ECO-MANAGEMENT : · Minimizing energy use · Minimizing water use · Minimizing construction waste · Minimizing building maintenance and repair	HEALTH : · Hygiene and cleanliness of the interior spaces · Air quality controls · Water quality controls

2.2 RT 2005 (Thermal Regulation 2005)

RT 2005 is the French Thermal Regulation that applies actually to the new buildings (with some exceptions) in France metropolitan. This regulation sets out a maximum energy consumption of the new buildings for the heating, the ventilation, the air conditioning, the production of domestic hot water and the lighting. [3]

Three elements are regulated by the RT 2005 and it is necessary to respect simultaneously the rule 3 C

- La consommation d'énergie (The energy consumption)
- Le confort d'été (The summer comfort)
- Les caractéristiques thermiques de divers éléments du bâtiment (The thermal characteristics of diverse elements of the building)

2.3 *Information and Communications Technology HPE Label (Haute Performance Energétique or High Energy Performance)*

HPE is a complement in the regulations RT2005 for the new buildings. It is a set of French official labels which reports energy, sanitary performances and environmental of a building at the level of its conception and of its maintenance. It is the qualitative initiative which integrates all the activities involving the conception, the construction, the functioning and the maintenance of a building (housing, public building, tertiary or industrial building). The energy performance is defined according to the "global cost" including the energy balance, the cycles of maintenance and renewal. But the label HPE does not take into account the energy costs of manufacturing of materials and equipments, construction, maintenance of the building and the recycling. [4]

The label contains five degrees (with five stars) and gives the right to an improved financing or to bonuses or fiscal advantages. The label HPE 2005 and THPE 2005 were available actually by RT 2005.

- HPE 2005 : Maximal consumption reduced to 10% than RT2005
- HPE EnR 2005 : Maximal consumption reduced to 10% than RT2005, with sustainable energy using
- THPE 2005 : Maximal consumption reduced to 20% than RT2005
- THPE EnR 2005, Maximal consumption reduced to 30% than RT2005, with sustainable energy using
- BBC 2005 : Maximal consumption to 50 kWh/m²/year (about 50% than RT2005).

HPE : Haute Performance Energétique, High Perfomance Energizing
THPE : Très Haute Performance Energétique, Very High Perfomance Energizing
BBC : Low Consumption Building

3 Materials

3.1 *Wood*

The wood is a rapidly renewable material. And also it does not require any high energy consumption for the production. The controlled forestry development may limit the emission of CO2.

The innovations of new by-products of the wood were going to be developed on two main axes:

- Engineered wood with metal connector: Reduce the waste and small section wood became useful
    - o Laminated Veneer Lumber (LVL)
    - o Wood I-joists
    - o Glue-lam
    - o Manufactured Trusses

- Non toxic treatment of wood: Reduce the chemical product use
  - Choice of naturally resistant wood
  - Retified wood (Thermal treatment)
  - Thermal oiled wood

3.2 Earth

In spite of the long history of use of earth, as a material of construction, many architects revisit the earth in the contemporary architecture. The use of earth for construction divides in two categories: non baked earth and the terra-cotta.

- Non baked earth:
  - Adobe (Mud brick)
  - Rammed earth
  - Wattle and daub
  - Bauge (Stacked clay with vegetable fiber)
- Terra-cotta (Baked earth)
  - Alveolar terra-cotta masonry unit
  - Terra-cotta wall cladding panel
  - Big dimension structural terra-cotta panel including insulation (60 X 280cm, Th. 30cm)

3.3 Green Insulations

The insulation of building is a key point to reduce the energy consumption. Mineral wools require much energy consumption for the manufacturing. Polystyrene or polyurethane boards are by-products of petroleum. New green insulations are produced by a simple transformation and they are naturally biodegradable.

- Vegetable insulation:
  - Cannabis fiber
  - Straw
  - Cork
  - Wood fiber
  - Cotton
- Organic insulation
  - Sheep wool
  - Down feather

3.4 Miscellaneous Green Materials

- Mud plaster: This type of plaster replaces cement plaster. Its porosity allows the breath of wall and the regularization of humidity. The natural colour does not require synthetic pigments.

- Lime plaster-cannabis fiber masonry unit: It is a composite masonry block who may replace cement masonry unit.
- Vegetable mortar: This mortar is consisted of calibrated vegetable particles and binders (base of lime and of mineral powder).
- Green Paint and wood stain: These type of paint not continents the formaldehyde. Because it use the organic binders as casein or starch of corn with natural pigments.

4 Equipments

4.1 Solar Panels

4.1.1 Solar Hot Water System with Vacuum Tubes

The vacuum tubes increase the efficiency even in winter. The regulation by micro processor controller optimizes electric consumption.

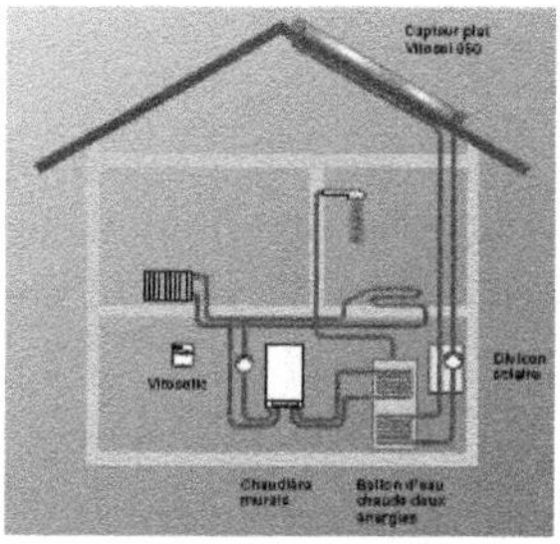

Source of figures [2]

4.1.2 Photovoltaic Solar Cells

High efficiency solar cell is a device that converts light directly into electricity by the photovoltaic effect. It can generate electricity at higher efficiencies than conventional solar cells. While high efficiency solar cells are more efficient in terms of electrical output per incident energy (watt/watt), much of the industry is focused on the most cost efficient technologies, i.e.cost-per-watt. Many businesses and academics are focused on increasing the electrical efficiency of cells, and much development is focused on high efficiency solar cells.

4.2 Reversible Thermodynamic Heat Pump

A heat pump is a machine or device that moves heat from one location (the 'source') to another location (the 'sink' or 'heat sink') using mechanical work. Most heat pump technology moves heat from a low temperature heat source to a higher temperature heat sink. [5]

[2] Catalog of Vitosol, Vissmann.

4.2.1 Different Types of Heat Pump

- Air source heat pump (extracts heat from outside air)
  - o air–air heat pump (transfers heat to inside air)
  - o air–water heat pump (transfers heat to a tank of water)
- Geothermal heat pump (extracts heat from the ground or similar sources)
  - o geothermal–air heat pump (transfers heat to inside air) : ground, rock and body of water
  - o geothermal–water heat pump (transfers heat to a tank of water) : ground, rock and body of water

4.2.2 Different Types of Diffusion Device

- Floor panel
- Fan coil unit
- Air handling unit

4.3 Reversible Thermodynamic Recuperator for CMV (Controlled Mechanical Ventilation)

This device is a ventilation block with heat exchanger. New cold exterior air is pre-heated by warm exhausting interior air in winter and the inverse procedure with air-conditioning in summer. The rate of recuperation may reach to 70%.

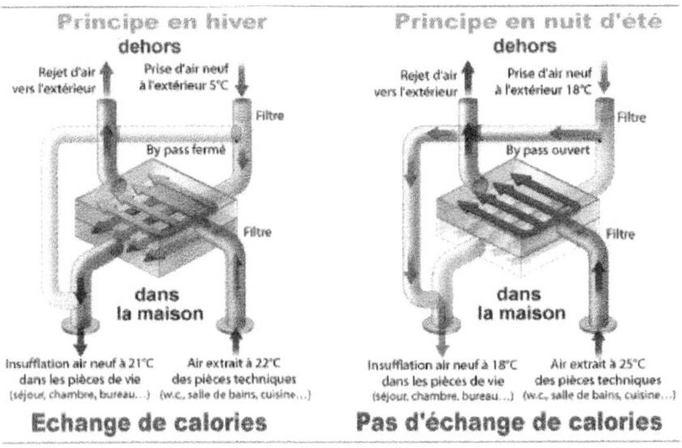

Source of figures [3]

4.4 Ground-Coupled Heat Exchanger

Earth Tubes (also known as ground-coupled heat exchangers, earth cooling tubes or earth warming tubes) use near constant subterranean temperature of the earth to warm or cool air for residential, agricultural or industrial uses. They are often a viable and economical alternative to conventional heating, cooling or heat pump systems since there are no compressors, chemicals or burners and only blowers are required to move the air.

[3] http://www.econology.fr/vmc-ventilation/vmc-double-flux.html

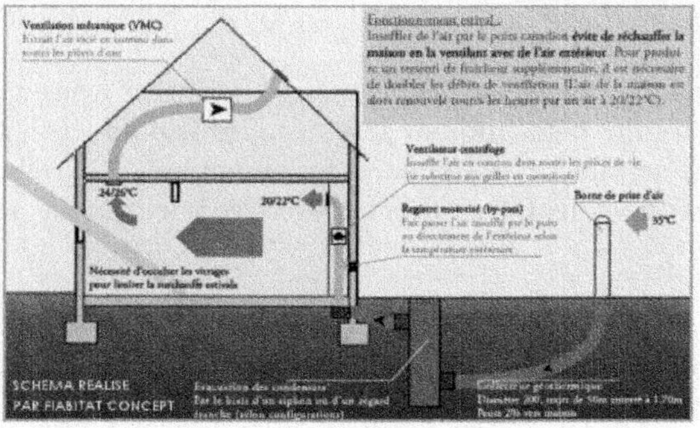

Source of figures [4]

4.5 *Gas Condensing Boiler*

A condensing boiler is a water heating device designed to recover energy normally discharged to the atmosphere through the flue. It can do this through the use of a secondary heat exchanger which most commonly uses residual heat in the flue gas to heat the cooler returning water stream or by having a primary heat exchanger with sufficient surface for condensing to easily take place.

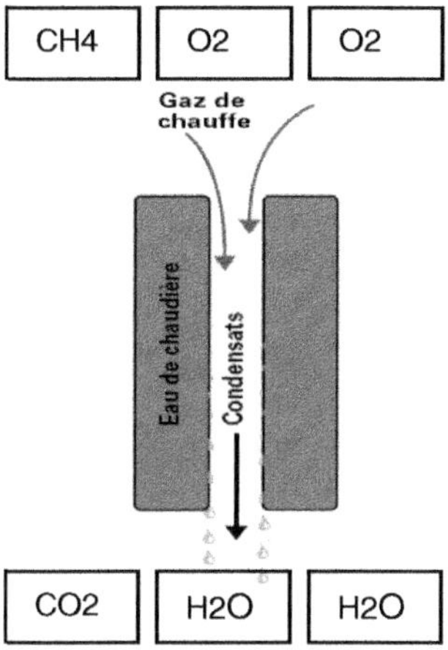

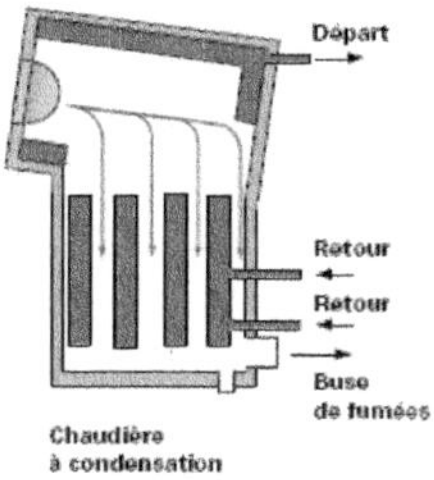

Source of figures [5]

5 Conclusion

From the Design Phase, during the Construction and Until the Utilization

Sustainable architecture requires the integration of vast knowledge of environment-friendly products and the specific knowhow of construction from the phase of conception on a project. This last one can reduce the possible nuisance and the pollution during the construction and optimize the energy consumption in utilization of building, while respecting the comfort and the health of the occupants.

[4] Figure designed by Fiabitat Concept, http://www.fiabitat.com/puits-canadien.php

[5] Vissmann catalog,

Interdisciplinary Development

The narrow collaboration is required between engineers and scientists who are working in different disciplines for the innovation of new sustainable products and systems. The interdisciplinary of architect is a key point for an optimized sustainable architecture design and a harmonious integration of diverse devices in the architecture.

References

[1] Doerr Architecture, Definition of Sustainability and the Impacts of Buildings, http://www.doerr.org/services/sustainability.html
[2] Certifications NF Ouvrage Démarche HQE®, http://www.assohqe.org/documents_certifications_hqe.php
[3] Art. 2 du Décret n° 2000-1153 du 29 novembre (2000) French decree
[4] 28/08/06 : Le label HPE. Un complément à la nouvelle réglementation RT2005 pour les bâtiments neufs, http://www.logement.gouv.fr/article.php3?id_article=5838
[5] The Systems and Equipment volume of the ASHRAE Handbook, ASHRAE, Inc., Atlanta, GA (2004)

An Analysis of the Plans to Reduce Demand for Energy and Introduce Renewable Energy Systems in Innovation Cities

Hyo-Soon Park[1,*], Sung-Hee Hong[1], Ji-Yeon Kim[2], and Jong-Hun Hyun[3]

Abstract. Innovation cities, which meet the best innovation conditions for close cooperation with public institutions and industry-university researchers, were promoted to make them characteristic of local cities and to make local cities independent. The plan to establish an independent base of local cities has to consider the economic conditions and quality of life in the cities, and sustainable development. First, balanced city planning is needed to build environment-friendly and sustainable cities. Energy-efficient buildings should be designed to address energy and environmental problems. In this paper, the energy demand plan and the method of introducing renewable energy systems in innovation cities were analyzed. It was found that the reduction ratios of energy demand are greatly imbalanced between innovation cities, and that only the Gwangju-Jeonnam innovation city is planning to use renewable energy to supply 5% of its total energy demand.

1 Introduction

Problems related to energy and the environments are emerging as the most difficult tasks we human beings have ever faced. An urgent response to the problems that concern energy consumption, which causes the global warming phenomenon, is required.

[1] High Efficiency Energy Research Department, Korea Institute of Energy Research, Daejeon, South Korea
hspark@kier.re.kr, shhong@kier.re.kr
[2] Department of Architectural Engineering, Inha University/KIER, South Korea
jykim@kier.re.kr
[3] Department of Architectural Engineering, Kyungpook University, Daegu, South Korea
mount803@yahoo.co.kr
[*] Corresponding author.

J.H. Lee et al. (Eds.): EU-Korea Conference on Science and Technology, SPPHY 135, pp. 81–89.
springerlink.com

A look at the exploitable resources left for each energy resource throughout the world will show that their supply is dwindling. It is expected that oil and natural gas will be depleted in 40 and 60 years, respectively, while the acquisition of energy resources is becoming more difficult. In pursuit of further economic growth and the further enhancement of quality of life, however, energy consumption is expected to continuously increase.

In April 2006, the oil price hit 64 dollars (based on Dubai oil) due to the tension in the Middle East since 2000, and it will increase to 100 dollars per barrel in the near future, according to economic experts. These oil price increases would increase the consumption of foreign capital. Thus, the excessive use of coal and oil should be avoided. Doing so would also enable nations to cope with global environmental problems such as global warming.

The government of Korea, to cope with this global environmental concern, is planning to reduce the energy being used in buildings in the country, which account for approximately 25 percent of Korea's total energy demand, and to replace the source of the supply of 5 percent of the country's total energy demand with renewable energy.

Innovation cities have been promoted that will transform local cities into independent futuristic cities equipped with optimum innovation conditions, where public organizations and industry, academia, and research centers can cooperate closely and where there is a high-quality government environment such as for dwellings, education, health services, and culture. The method of establishing the independent foundation of innovation cities should be planned not only considering the economic aspect but also the enhancement of quality of life and sustainable development. Therefore, to build environment-friendly and sustainable cities, balanced urban planning is required above all, and the foundation for the cities' energy independence should be established. Moreover, their plan should comply with the national policy that addresses energy and environmental problems with the design of energy-efficient buildings. Therefore, this research will analyze the renewable energy promotion plan and the energy demand reduction plan for each innovation city, and will evaluate the planned energy utilization methods of innovation cities.

2 Research Method

This research will compare and evaluate the energy demand and the renewable energy promotion plan of the Multifunctional Administrative City and each innovation city, analyze the application feasibilities, and introduce supplemental opinions. This research will be carried out as follows.

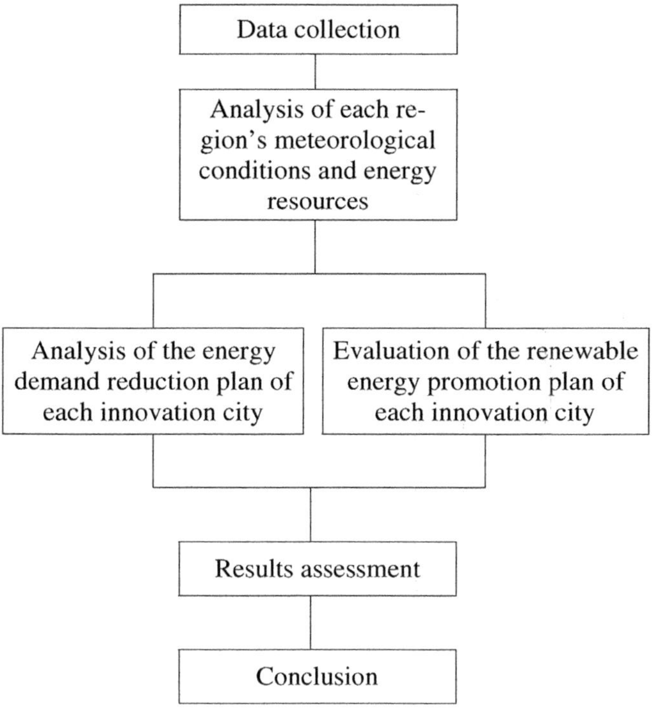

Fig. 1. Research flowchart

3 The Energy Plan of the Innovation Cities

3.1 The Concept of an Innovation City

The government organized the Innovation City Location Selection Committee based on such criteria as the possibility of developing the site into an innovation hub, the feasibility of its urban development, and its region's growth prospects, then selected the locations of the innovation cities and confirmed the final locations in February 2006. The dimensions, project duration, and execution of the plans for the ongoing Multifunctional Administrative City and the 10 innovation cities are shown in Table 1.

3.2 Metrological Data Analysis

In planning for renewable energy utilization and energy reduction, it is critical to scrutinize the climate resources of the target region. Among several resources that are classified as renewable energy resources, the radiation is regarded as the most appropriate characteristic that can be evaluated objectively, and velocity and the annual average temperature are also included. For the regions, when the researcher

Table 1. The innovation city concept

Table .	Dimension[m²]	Project duration	Project in charge
Multifunctional Administrative City	72,908,221	2005~2012	Korea Land Corporation
Busan	959,000	2007~2012	Busan Metropolitan Corporation
Daegu-sinseo	4,216,496	2005~2012	Korea Land Corporation
Gimcheon, Gyeongbuk	3,803,000	2006~2011	Korea Land Corporation Gyeongbuk Development Corporation
Wonju, Gangwon	3,603,048	2007~2012	Korea Land Corporation Wonju city
Jinju, Gyeongnam	4,062,670	2005~2020	Korea National Housing Corporation
Gwangju, Jeonnam	7,295,000	2007~2012	Gwangju Metropolitan Development Corporation Korea Land Corporation Jeonnam Development Corporation
Ulsan - Ujong	2,877,615	2007~2012	Korea Land Corporation
Jincheon, Chungbuk	6,914,000	2007~2012	Korea National Housing Corporation
Junju, Jeonbuk	9,260,000	2007~2012	Jeonbuk Development Corporation Korea Land Corporation
Jeju - Seogwipo	1,150,939	2007~2012	Korea National Housing Corporation

could not obtain climate data (radiation, sunshine, temperature, relative humidity, and velocity), the climate data of the most adjacent areas were collected instead. The climate data for each region are shown in Table 2.

The 'sunshine percentage' refers to the ratio of the time when the solar beam reaches the earth's surface with the hindrance of clouds or fogs to the time in which the solar energy system could operate effectively. The horizontal radiation of the regions of Junju and Jeju showed values that are lower than the nationwide average. While it is expected that the utilization of solar energy in such regions will cause no major problem, their heat collection efficiencies are expected to be lower than those of the other regions.

Velocity is an important factor in the evaluation of the wind power resources at the location of a wind power plant, and such a plant is possible only when the annual average velocity is over 4 m/s. The annual average velocity of Korea is below 4 m/s. Installing windmill turbines in the regions of Jeju and Busan maybe

Table 2. Meteorological data

Region Name	Total Horizontal Radiation[1)]	Sunshine percentage [%]	Temperature [℃]	Relative Humidity [%]	Velocity [m/s]
Daejeon	3.63	50.4	12.7	69.0	1.7
Busan	3.66	53.6	14.7	65.5	3.7
Daegu	3.56	52.4	14.0	63.0	2.8
Wonju	3.48	49.0	11.2	70.3	1.1
Jinju	3.84	50.2	13.1	70.5	1.8
Gwangju	3.67	49.4	13.7	70.1	2.2
Pohang	3.60	50.9	14.2	63.4	2.8
Cheongju	3.57	51.4	12.4	69.1	1.9
Junju	3.41	47.3	13.2	70.2	1.5
Jeju	3.43	41.8	15.7	71.1	3.6
Average	3.58	49.9	13.0	68.3	2.4

1) The average annual horizontal radiation [kWh/m^2·day]

possible, however, because although they lack wind power resources, they show minimum numerical values for height and topography.

3.3 *Energy Demand Reduction Plan of the Innovation City*

The energy demand reduction plan for each of the innovation cities is shown in Table 3. All the innovation cities use highly efficient gas boilers.

3.4 *Renewable Energy Inducement of Innovation City*

The renewable energies used in the innovation cities are solar heat, Photovoltaic power, and ground heat energies. Solar thermal water heating facilities are being planned for use in individual houses, apartments, central administrative organization buildings, public buildings, welfare facilities, and culture facilities, among others.

Photovoltaic systems are being planned for use in individual houses, apartments, central administrative organization buildings, public buildings, education facilities, welfare facilities, and culture facilities, among others.

Ground heat cooling/heating systems are being planned for use in central administrative organization buildings, public buildings, education facilities, welfare facilities, culture facilities, and gymnastic facilities, among others.

Table 3. The energy demand reduction plan for each innovation city

Section			Multifunctional Administrative City	Daegu-sinseo	Gimcheon-Gyeongbuk	Wonju-Gangwon	Jinju-Gyeongnam	Gwangju, Jeonnam	Ulsan - Ujong	Jincheon, Chungbuk	Jeju - Seogwipo
Energy using enhancement facility	Construction	Building energy efficiency authentication	-	1,093	1,409	1,024	1,289	2,387	-	2,182	312
		Strengthening window insulation	48,309	-	-	-	108	-	2,248	194	230
		Strengthening wall insulation	11,292	-	-	-	-	-	-	-	-
		Top-level energy building	440	-	-	-	-	-	-	-	-
	Equipment	Highly efficient gas boiler	6,390	248	1,855	298	339	963	220	3,086	440
		Balancing the valve	2,175	252	-	826	200	638	164	213	178
		GHP	630	-	-	-	30	52	54	-	-
		controllable warm water distributor for each room	17,640	-	2,606	-	2,270	1,935	590	3,183	313
		Compressor automatic cleaning system	1,791	-	-	-	-	383	-	-	-
		VAVsystem	6,576	-	-	-	726	2,296	-	-	-
		Water saving facility	1,511	63	49	232	116	567	349	38	18
		Temperature controller for heating	-	807	2,606	3,377	-	-	-	3,183	-
		Automatic building control	-	11,803	7,461	2,574	-	-	4,904	10,204	-
	Electricity	VVVF	6,809	768	610	477	845	2,526	666	1,833	274
		Highly effici-ent generator	12,200	2,347	1,421	1,340	1,435	3,507	2,068	2,730	413
		Uninterruptible power supply	1,517	207	157	220	55	565	216	206	125
		Highly efficient street lights	2,439	82,198	157	158	161	256	137	233	-
		LED traffic light	559	134	79	5	57	76	56	354	49

Table 3. (*continued*)

	Highly efficient waste heat recol-lection equip-ment	37,183	2,053	2,298	2,599	747	3,811	2,473	2,679	461
Waste -heat	Waste water heat recollection using heat pump	205	33	19	33	53	139	267	19	30
	Sewage water heat pump	-	-	-	-	25	-	-	-	-
Collective Energy		141,334	23,360	-	31,527	10,632	4,144	14,057	231	-
Total [TOE/Year]		299,000	125,366	20,727	44,690	19,088	24,245	28,469	30,568	2,843
Reduction Rates [%]		22.20	19.70	12.00	22.17	14.10	7.9	18.10	12.40	11.30

Table 4. The renewable energy plan for each innovation city

Section		Multifunc-tional Ad-ministrative City	Daegu-sinseo	Gimcheon-Gyeongbuk	Wonju-Gangwon	Jinju-Gyeong-nam	Gwangju-Jeonnam	Ulsan - Ujong	Jincheon-Chungbuk	Jeju - Seog-wipo
	Heat collection area [m²]	9,944[1]	998	416	1,032	3,670	4,400	2,278	1,400	4,772
Solar Heat	Reduction quantity [Gcal/yr.]	5,595	765	263	822	2,610	3,488	2162	560	2,640
	Reduction amount [million won/ yr.]	3,520	359	123	386	199	275	135	263	196
	Generating capacity [kWp]	16,709	1,210	1,170	126	1,850	30	821	1,060	1,615
Photo-voltaic power	Reduction quantity [MWh/yr.]	22,687	2,695	1,537	252	2,592	42	1,561	1,485	2,146
	Reduction amount [million won/yr.]	2,405	302	149	18	298	4	166	149	227
	Induction capacity [RT]	2,454,214[2]	6,800	9,310	-	741,951[2]	1,987,266[2]	1,083,595[2]	8,840	-
Ground Heat	Reduction quantity [thousand Nm³/yr.]	10.618	3,757	3,266[3]	598	1,884	4,703	2,988	5,412[3]	-
	Reduction amount [million won/ yr.]	9.964	1,406	1,562	-	966	4,446	1,036	1,376	-\

1) Installation Number [ea]
2) Mutual Area [m²]
3) Reduction Quantity [TOE/ year]

3.5 Target Energy Demand Reduction Rate for Each Innovation City

The comprehensive reduction rates in the energy demand reduction plans and the renewable energy promotion plans for the innovation cities are shown in Table 30. The total energy demand reduction rate for the Multifunctional Administrative City is the highest at 24.50 percent, and the rate for Gimcheon, Gyeongbuk is the lowest at 13.40 percent. With respect to the renewable energy promotion rate, the Gwangju/Jeonnam innovation city has the highest at 16.00 percent; but the approximately 14 percent ratio of its disposed material energy does not meet the national policy guideline on the replacement of 5 percent of the total energy mix with renewable energy.

Moreover, the energy demand of all the cities, excluding the Multifunctional Administrative City and the Gangwon/Wonju innovation city, are below 20 percent. The Gwangju/Jeonnam innovation city, however, has already established methods for renewable energy promotion and energy demand reduction through an outsourced research, unlike the other cities, and has prepared a plan to supply renewable energy at approximately 5 percent of its energy mix, excluding the energy demand in the city (What does this mean?), based on the national renewable energy supply policy and bylaws. Besides, it has encouraged all the apartments in the city to have top-level energy efficiency, and it is developing a cooling/heating load reduction technology and energy reduction technologies so that the public organization buildings and other non-residential buildings that have been transferred therein could achieve 25 and 20 percent energy reduction, respectively.

Table 5. Energy reduction ratio for each Innovation cities

Section	Energy Demand Reduction Rates [%]	Renewable Energy Reduction Rates [%]	Total Energy Reduction Rates [%]
Multifunctional Administrative City	22.20	2.30	24.50
Daegu-sinseo	19.70	2.80	22.50
Gimcheon, Gyeongbuk	12.00	1.40	13.40
Wonju, Gangwon	22.17	0.43	22.60
Jinju, Gyeongnam	14.10	1.70	15.80
Gwangju, Jeonnam	7.90 (25.00)	16.00 (5.45)	23.90 (30.45)
Ulsan-Ujong	18.10	1.20	19.30
Jincheon, Chungbuk	12.40	1.00	13.40
Jeju-Seogwipo	11.30	2.43	13.73

* The values in brackets ( ) are from the feasibility study on the energy eco-city composition of the Gwangju/Jeonnam innovation city.

4 Conclusion

This research analyzed and evaluated the energy utilization plans of the Multifunctional Administrative City and eight innovation cities. The reduction plans showed a 12.30 percent maximum regional disparity, and no city satisfied the policy on 5 percent renewable energy supply.

The Gwangju/Jeonnam innovation city is planning to meet the 5 percent target, however, through an outsourced research. The energy utilization plan is the most important information in the district unit plan.

Therefore, for the promotion of sustainable development and the development of environment-friendly cities, energy utilization plans should be formulated and carried out with the consideration of regional circumstances and economic funds acquisition plans.

References

[1] Korea Land Corporation, The energy plan for the development project of Kang-won, Won-ju innovation city (2007)
[2] Korea Land Corporation, The energy plan for the development project of Gyeongnam, Jin-ju innovation city (2007)
[3] Korea Land Corporation, The energy plan for the development project of Gwang-ju, Jeon-nam innovation city (2007)
[4] Korea Land Corporation, The energy plan for the development project of Dae-gu. Shin-seo innovation city (2007)
[5] Korea Land Corporation, The energy plan for the development project of Ul-san. Woo-jeong innovation city (2007)
[6] Korea Land Corporation, The energy plan for the development project of Gyeongbuk. Gim-cheon innovation city (2007)
[7] Korea Land Corporation, The energy plan for the development project of Chungbuk.Jin-cheon innovation city (2007)
[8] Korea Land Corporation, The energy plan for the development project of Multifunctional administrative city (2007)
[9] Korea Land Corporation, The energy plan for the development project of JejuSeogwipo city (2007)
[10] Ministry of Commerce, Industry and Energy, Analysis and Evaluation of Solar Radiation Resources and Reliability Enhancement on Isolation Data in Korea (2007)
[11] Jang, Y.S., Kim, J.Y., Hong, S.H., Park, H.S., Suh, S.J.: An Envelope Database of a Non-residential Building to Cope with Framework Convention on Climate Change. In: Proceedings of the KSES 2005 Autumn Annual Conference, pp. 109–114 (2005)

Wireless Monitoring System for Hybrid Power Generation System

Jin-Seok Oh[1,*], Soo-Young Bae[2], Ji-Young Lee[3], Jun-Ho Kwak[4], Jae-Min Kim[5], and Cameron Johnstone[6]

Abstract. The number of Renewable Energy (RE) systems increases due to the low carbon energy policy in most of industrialised countries. In order to ensure the performance of RE sustainable and reduce maintenance costs, remote monitoring systems of RE plant condition and performance are essential. Wireless monitoring systems are easy to install and operate, but requires high cost in general. Radio Frequency technology is relatively cheap, but its transmission range is limited. It is a challenge to establish long range communication between RE plants and the monitoring centre in cost effective way. In this paper, a novel remote condition monitoring system for RE systems is presented, which exploits RF modems in the multi-point link network to make long range communication. The paper illustrates the overview of the proposed system and the results of the field tests implemented with PV+Wind and PV+Wave hybrid power generation systems at land and sea.

1 Introduction

1.1 Research Collaboration Overview

Building Integrated Renewable Technologies (BIRT) are being promoted as making a step in reducing building energy demands. To be realised, the following challenges need addressing 1) substantive improvements to efficiency and operations of BIRT's; 2) mitigate the potential negative impact on intermittency and

[1] Korea Maritime University, Department of Mechatronics Engineering, Yeongdo-Gu, Busan, South Korea
ojs@hhu.ac.kr
[2] Korea Maritime University, Department of Mechatronics Engineering, Yeongdo-Gu, Busan, South Korea
sybae@hhu.ac.kr
[3] Cardiff University, Manufacturing Engineering Centre, United Kingdom
hannar52@hotmail.com
[4] Korea Maritime University, Department of Mechatronics Engineering, Yeongdo-Gu, Busan, South Korea
haedae55@hhu.ac.kr
[5] University of Strathclyde, Department of Mechanical Engineering, United Kingdom
[6] University of Strathclyde, Department of Mechanical Engineering, United Kingdom
* Corresponding author.

J.H. Lee et al. (Eds.): EU-Korea Conference on Science and Technology, SPPHY 135, pp. 91–98.
springerlink.com

power quality within the electrical supply network; and 3) a reduction in both capital and maintenance costs through apposite technology selection and optimisation of installed capacity at both individual building and community scales. In addition, societal and sociological factors which influence stakeholders (end users, energy suppliers, local authorities and RE providers) to take a more proactive stance in enabling the uptake of sustainable energy solutions need to be encouraged. To address these in a cost effective way, existing Information and Communication technologies (e.g. Internet and wireless telecommunication infrastructures) be exploited for real-time monitoring of RE plant condition and performance. The aim of this proposal is to develop and demonstrate remote condition monitoring and management of BIRT's at both single and multiple buildings scales.

This will demonstrate different levels of stakeholder management and response to e-services based on the sharing of energy generation profiles but controlled via a hierarchy access to the source data. The applicability and commercial viability of the e-services will be identified in different societal contexts (i.e. UK and Korea). The outcomes from this project will be disseminated in both UK and Korea.

1.2 Methodology and Research Merit

This project aims to develop an internet-based remote condition monitoring system for building integrated Wind-PV hybrid energy systems. Through the course of this research, an effective and low cost communication infrastructure will be established between monitoring devices and the back-end e-service system.

All information about the performance of the BIRT's and results of the condition checking routine are available on the Internet for registered users (i.e. stakeholders of the systems) via an e-service. Before commercially demonstrating this system, the following have to be addressed:

1. Cost effective communication infrastructure:
2. Algorithms for condition checking:
3. Central data processing, storage and management

The research will have commercial viability associated with new e-service businesses focusing on carbon reduction strategies in both countries. Korean government has set up a strong low carbon energy policy which encourages Renewable Energy (RE) systems to be installed at various scales. Not only increasing the number of RE installation but also reducing capital and maintenance costs is significant to implement the low carbon energy strategy at both individual building and community scales. Real time monitoring of RE plant condition and performance is an essential measure to improve system efficiency and to reduce operation costs. There are a number of RE monitoring systems reported in previous studies. For example, Eftichios Koutroulis et al. introduced a computer-based data-acquisition system for RE monitoring systems which consists of electronic circuits, a data-acquisition card to collect data and then an interface to a personal computer (PC). Since it is a stand-alone system, however, the data monitored can not be accessed remotely. Kostas Kalaizakis et al. proposed a data acquisition system for remote monitoring of renewable energy systems based on the Client/Server network architecture using serial Radio Frequency (RF) transceivers, RS-232 port and Ethernet communication. The remote monitoring system, however, costs a lot as every RE system has to be equipped with all communication devices (i.e. RF modem, RS-232, Ethernet and PC). In addition, the transmission range is only approximately 200m between the RE plant and data monitoring unit. Main barriers of current remote monitoring systems are, consequently, high costs and short range of wireless communication.

This paper presents a novel remote condition monitoring system for RE systems with low cost and virtually unlimited communication distance by using a multi-point link monitoring technique with RF modems. The paper illustrates the overview of the proposed system and communication algorithms. The field tests were implemented with PV+Wind and PV+Wave hybrid power generation systems at land and sea.

2 Proposed Communication Network Technique

2.1 System Overview

There are a number of wireless communication methods available which can be used for the remote monitoring of RE systems. The cost comparison of various communication methods such as Code Division Multiple Access (CDMA), Wireless Local Area Network (WLAN), Very High Frequency (VHF) and Ultra High Frequency (UHF). While CDMA and WLAN are popular technical options in terms of wide transmission range, their costs of installation and operation are not economical. On the contrary, VHF/UHF is a good alternative to wireless remote communication. One drawback of this technology is, however, the limitation of the transmission range. VHF/UHF cannot be applied for long rage communication as it is. The purpose of this study is to establish a low cost system for the long range remote communication between RE plats and monitoring centre. To this end, short range RF modems are linked in a multi-point link network which allows the long-range communication. Fig. 1 shows a schematic diagram of the multi-point link communication consisting of four units: U_0, U_1, U_2 and U_3. A unit represents a hybrid RE system to communicate. A host PC is located at U_0 while communication between two adjacent units is made via RF modems. To make the long range communication in the multi-point link network, the following features are to be established:

- There must be two units at least within wireless transmission range to make the communication link.
- All units in the communication range are connected in the form of multi-point link network structure (see Fig. 2).
- Each unit has a role for relaying data between units not directly connected.
- Each unit shares the information on linking relationship between its adjacent units and check the status of the communication network.

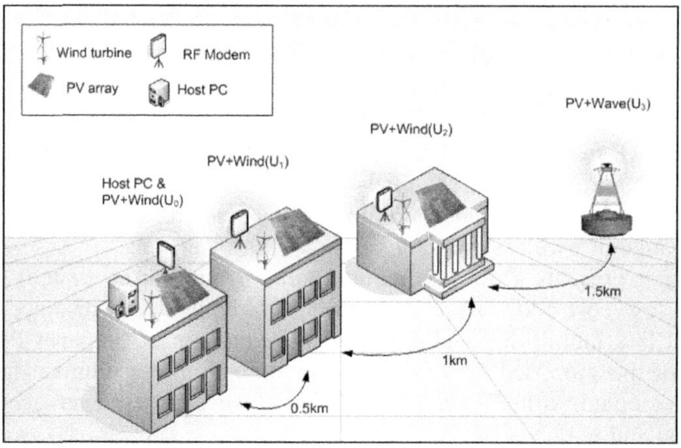

Fig. 1. A schematic diagram of multi-point link communication

2.2 *Network Structure*

Fig. 2 describes topology of the multi-point link network structure to overcome the limitation of communication range in using RF modems. There are five units from U_0 to U_4 which have their own circle representing the communication range of RF modems at each unit. The line between units shows the shortest distance for communication between two adjacent units. In this network, communication is made as follow:

- Each unit periodically sends its unique identification (ID) to its adjacent units to get recognized (By doing this, each unit has information on the IDs of its adjacent units.).
- Each unit sets up database (DB) of its adjacent IDs and transmits it to the Host PC (U_0).
- The Host PC can identify current status of the network by receiving DBs form units and check if there is any faulty unit.
- If a unit wants to send data to another unit in the network, a control program in the Host PC informs the unit about the shorted route via which the unit transmits data (i.e. assignment of intensive adaptation routes).Robotics and Automation

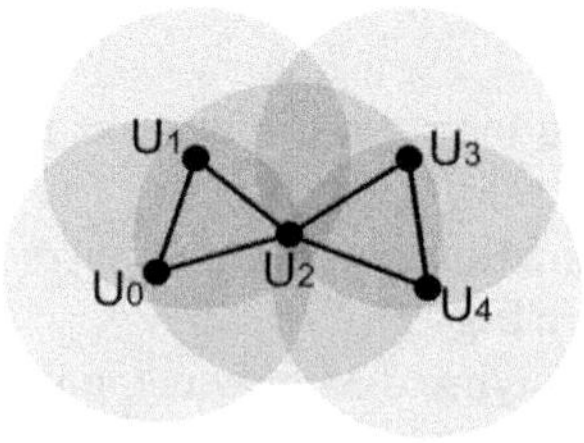

Fig. 2. Topology of multi-point link network structure

While this network structure could give rise to long transmission time, it allows multi-group server/client architecture in the network. For example, when U_0 and U_1 communicate each other, U_3 and U_4 are also able to communicate separately. In order to exploit this feature, polling delay time and data transmission time have to be figured out. In general, the average time of polling time in the Poll & Select technique increases in proportion to the number of units and size of transmission data. The average time of polling can be calculated with the following equation (1).

$$T_{avr} = \frac{T_{poll} \times N}{1 - M_r \times T_D} \quad (1)$$

- T_{avr} : An average of polling time
- T_{poll} : Time to be able to poll a unit
- N : The number of units for polling
- M_r : Size of data per a time unit
- T_D : Transmit time for the average size of data

As can be seen, if N is not large like the proposed network structure, can be reduced. When operating separated groups (e.g. U_3 and U_4 in the example described above) in the network, each station can poll different group of units respectively at the same time. It increases utilization of communication infrastructure.

2.3 Communication Algorithm and Protocol

To implement the proposed system, the communication software program is installed in each unit with the following subroutines:

- Searching subroutine to search adjacent units. It is activated when the system is initially installed or updated.
- Data transmission subroutine to send/receive data between adjacent units. This subroutine is installed in the RF modem and the recharging controller.
- Checking/controlling subroutine to check each unit's status and to control it. This is for units to play the relay role.
- Searching the shortest route subroutine to find the shortest route. This is installed in the host PC.
- Checking/correction subroutine to check transmission error and to correct it. This is installed in the host PC.

The ID of each unit is defined in the hardware system. The current network status is regularly checked by the control programme. If there are any changes in the network (e.g. adding new units or removing existing units), the host PC is informed to update the DB of units. After updating the DB, the host PC re-identifies the shortest route to each unit. Bellman-Ford Algorithm was adopted for the shortest route identification. In addition to RF modems, CDMA modems were also installed to secure communication in case of disconnection of the RF-based link. If a RF modem does not response or the responding time is over, the shortest route is lost. In the case of malfunction of the RF modem, the CDMA modem is automatically activated to send the alert message of the RF modem error to the host PC. And then, the CDMA modem takes over the role of the RF modem sending data and checking the network. In this way, the communication can be secured at any incidents in the network. In this study, Binary Synchronous Control (BSC) was adopted for such semi-double data communication.

3 Results of the Field Test

Data communication between the hybrid RE systems at the trial site was confirmed by measuring the waveforms with an oscilloscope. Fig. 3 shows waveforms between U_0 and U_3 when U_1 searches for its adjacent unit (i.e. reconnecting algorithm). In Fig. 3, the oscilloscope channel 1 (Ch1) and Ch2 represent Transmit (TX) and Receive (RX) terminal in U_1. Ch3 and Ch4 in Fig. 3 (a) represent TX and RX terminal in U_0 respectively. Ch3 and Ch4 in Fig. 3 (b) represent TX and RX terminal in U_3 respectively.

As can be seen in Fig. 3 (a), when sending a signal from TX in U_1(i.e. Ch1), the signal is received at RX in U_0 after some delay which is caused by using RF modem speed rate with 2400 bps. In Fig. 3 (a), the waveform displayed at the left part in Ch2 is the responding signal from U_1 while the waveform at the right part in Ch2 is the responding signal from U_3. As shown in Fig. 3, the waveform of Ch3 in Fig. 3 (a) (i.e. TX in U_0) can be seen in Ch4 in Fig. 3 (b). It is, therefore, confirmed that the responding signal from U_3 arrives in U_0 as well as U_1. As the waveforms are displayed with exactly same shape in other units, it is also confirmed that there are no signs of faulty data transmission.

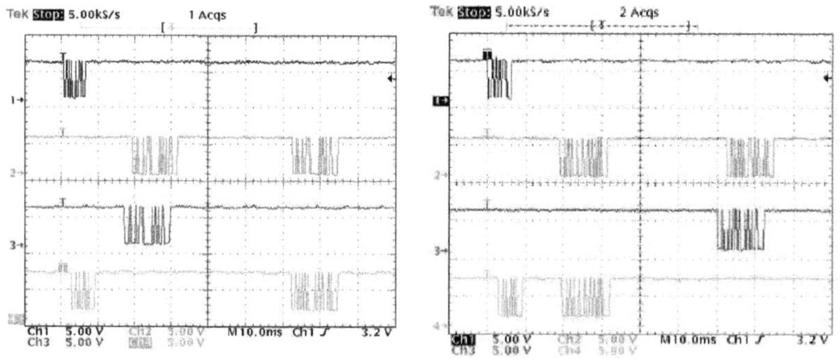

Fig. 3. Waveforms of Reconnecting Algorithm with Command Protocol (Left : between U_1 and U_0, Right : between U_1 and U_3)

4 Integrated Management System at the Host PC

To According to all test results, it proved that it is possible to make long distance communication using the multi-point link technique with RF modems. In order to manage condition monitoring of all units, it is required to establish a central database system at the host PC. The integrated real-time monitoring system was, therefore, designed using the LabVIEW program with Graphical User Interface (GUI) as shown in Fig. 4 (a). The data from all units accommodated in the host PC can be accessed via the Internet or exported into text files for further analysis. Fig. 4 (b) shows a graph of profiles (i.e. wave voltage, solar voltage, battery voltage, charging power) monitored from a hybrid PV+Wave power generation system.

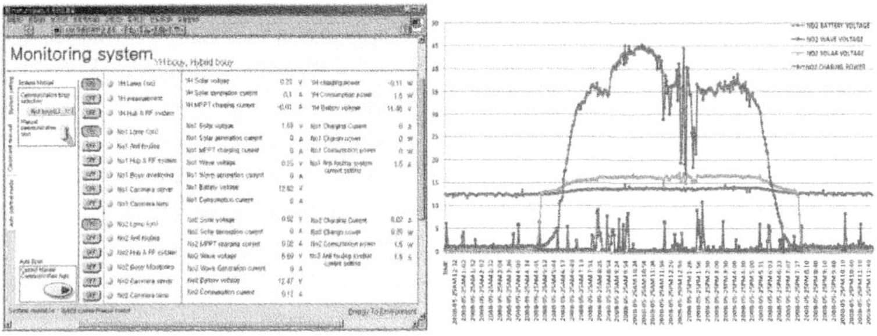

Fig. 4. Monitoring screen and graphs of hybrid power system 25th May 2008

5 Conclusions

Following the body in this study, a remote condition monitoring system was proposed on the basis of RF modems. Communication control algorithms and multi-point link architecture were adopted to enable long range communication. The proposed system was tested at field trials with hybrid RE systems installed on roofs of buildings land and a buoy at sea. Data transmission betweens units in the multi-point link network were confirmed successfully. In conclusion, it was verified that the remote condition monitoring can be implemented with low cost devices (i.e. RF). The communication in the network is secured by adopting CDMA technology.

The proposed system can be applied for RE plants at both large and small scale as long as adjacent units are located within transmission range of RF modems. To establish more flexible network infrastructure, however, various communication protocols should be integrated in a standard system.

Acknowledgment

This study was sponsored by British Council PMI 2 Connect – Research Co-operation Award.

References

[1] Koutroulis, E., Kalaitzakis, K.: Renewable Energy 28, 139–152 (2003)
[2] Kalaitzakis, K., Koutroulis, E., Vlachos, V.: Measurement 34, 75–83 (2003)
[3] World Energy Council, 2007 Survey of Energy Resources (2007)
[4] Song, S.: Journal of Air-Conditioningand Refrigeration 20(3), 155–166 (2008)
[5] Lessing, P.A.: OCEANS 2003, vol. 4, pp. 1997–2001 (2003)
[6] Halsall, F.: Data communications, computer networks and OSI. Addison-wesley publishing company, Reading (1988)
[7] Stallings, W.: Data and computer communications. Prentice-hall, Inc., Englewood Cliffs (2000)
[8] Oh, J.-S., Kwak, J.-H.: Journal of the Korean Society of Marine Engineering 29(5), 579–586 (2005)
[9] Rashid, H.: Wireless Application Protocol Programming. Hungry Minds (2001)
[10] Weisman, C.J.: The Essential Guide to RF and Wireless. Prentice-Hall, Inc., Englewood Cliffs (2002)

Level Set Method for Reconstruction of Thin Electromagnetic Inclusions

Won-Kwang Park[1] and Dominique Lesselier[2,*]

Abstract. In this paper, we consider the recently developed level set evolution technique in order to reconstruct two-dimensional thin electromagnetic inclusions with dielectric or magnetic contrast with respect to the embedding homogeneous medium. For a successful reconstruction, two level set functions are employed; the first one describes the location and shape, and the other one the connectivity and length. Speeds of evolution of level set functions are calculated via Fréchet derivatives by means of an adjoint technique. Several numerical experiments illustrate how the proposed method behaves.

1 Introduction

Recent progress towards inverse problem is highly remarkable due to the developments of the computing system environment. One of the many interests in inverse problem is the non-destructive reconstruction of inhomogeneities embedded in known media from measured boundary data, refer to [2], which is a topic arising in physics, medical science, material engineering, and so on, all domains highly related with human life. Recent promising techniques which use a level set representation of shapes for solving inverse scattering problems have been successfully developed. Starting out from the original binary approach for solving the shape reconstruction problem, [14], a number of generalizations have been developed, refer to [5] for an in-depth review of the topic of level set evolutions in the field of electromagnetic and elastic imaging, and to a set of classical references [7,8]. So far, all these techniques are mostly focusing on the retrieval of volumetric objects or regions with certain properties.

In this paper, we consider a technique of electromagnetic imaging which uses the level set method to efficiently reconstruct a thin inclusion with either dielectric or magnetic contrast with respect to the embedding homogeneous medium.

[1] Département de Recherche en Electromagnètisme, Laboratoire des Signaux et Systèmes (CNRS-Supélec-Univ. Paris Sud 11), 3 rue Joliot-Curie, 91192 Gif-sur-Yvette cedex, France
Won-Kwang.Park@lss.supelec.fr

[2] Département de Recherche en Electromagnètisme, Laboratoire des Signaux et Systèmes (CNRS-Supélec-Univ. Paris Sud 11), 3 rue Joliot-Curie, 91192 Gif-sur-Yvette cedex, France
Dominique.Lesselier@lss.supelec.fr

* Corresponding author.

J.H. Lee et al. (Eds.): EU-Korea Conference on Science and Technology, SPPHY 135, pp. 99–108.
springerlink.com

Opposite to the classical reconstruction of volumetric objects, a thin inclusion cannot be represented by a single level set function. To that effect, we adopt two level set functions to do the work: the first one describes the location and shape of the thin inclusion and the second one its connectivity and length, based on the investigations of [1,5,12].

The paper is organized as follows. In section 2, the direct scattering problem and the modeling of the thin inclusion via two level set functions are presented. In section 3, we introduce the basic concepts of shape deformation and the needed Fréchet derivatives by means of an adjoint technique. In section 4, we explain how to evolve such two level set functions and briefly sketch the reconstruction algorithm. In section 5, several numerical examples are presented for demonstrating the performance of the proposed algorithm. A short conclusion follows.

Before proceeding, we would like to emphasize that various approaches have been suggested in order to reconstruct thin inclusions or perfectly conducting cracks. Recently, a MUSIC (MUltiple SIgnal Classification)-type non-iterative algorithm has been developed, [10,11]. This is a very fast, effective algorithm and it can be easily applied to multiple inclusions. We also wish to introduce the first author's PhD thesis, [9], wherein various techniques for reconstructing not only penetrable thin inclusions but also perfectly conducting cracks are proposed and investigated.

2 Mathematical Formulation and Two Level Set Functions

Let Ω be a two-dimensional homogeneous domain with smooth boundary $\partial\Omega$. The thin inclusion is curve-like, i.e., it is in the neighborhood of a curve:

$$\Gamma=\{\mathbf{x}+\eta\mathbf{n}(\mathbf{x}) : \mathbf{x}\in\sigma,\ \eta\in(-h,h)\} \tag{1}$$

where σ is a simple, smooth curve of finite length, and h is a positive constant which gives the thickness of the inclusion, refer to Fig. 1.

All materials involved are characterized by their dielectric permittivity at frequency of operation ω; we define the piecewise constant permittivity $\varepsilon(\mathbf{x})$ and permeability $\mu(\mathbf{x})$ as:

$$\varepsilon(\mathbf{x})=\begin{cases}\varepsilon_0 \text{ for } \mathbf{x}\in\Omega\setminus\Gamma\\ \varepsilon \text{ for } \mathbf{x}\in\overline{\Gamma}\end{cases} \quad\text{and}\quad \mu(\mathbf{x})=\begin{cases}\mu_0 \text{ for } \mathbf{x}\in\Omega\setminus\Gamma\\ \mu \text{ for } \mathbf{x}\in\overline{\Gamma}.\end{cases}$$

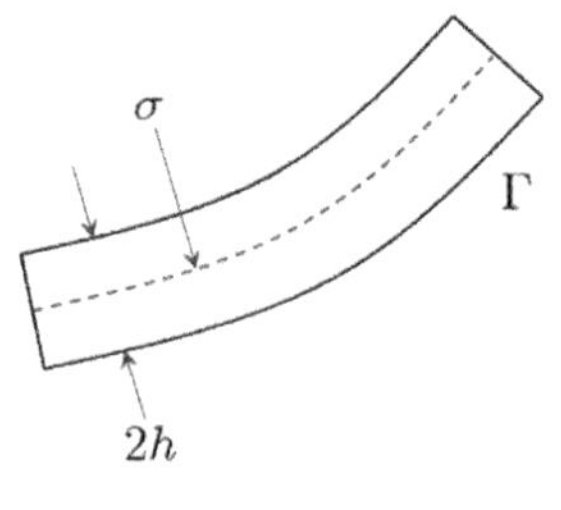

Fig. 1. Illustration of a two-dimensional thin penetrable inclusion Γ

Note that in the absence of inclusion, $\mu(\mathbf{x})$ and $\varepsilon(\mathbf{x})$ are equal to μ_0 and ε_0, respectively. Throughout the present contribution, ones assumes that the electric permittivity and magnetic permeability of the inclusion at a single, non-zero frequency of operation are known; they are finite-valued and differ (either one or both of them) from those of the homogeneous embedding medium. The unknown information on the inclusion consists in its location, shape, and connectivity.
At strictly positive operation frequency ω(wavenumber $k_0 = \omega\sqrt{\varepsilon_0\mu_0}$), let $u(\mathbf{x})$ be the time-harmonic total field which satisfies the Helmholtz equation

$$\nabla\cdot\left(\frac{1}{\mu(\mathbf{x})}\nabla u(\mathbf{x})\right)+\omega^2\varepsilon(\mathbf{x})u(\mathbf{x})=0 \text{ in } \Omega \tag{2}$$

with boundary condition

$$\frac{1}{\mu_0}\frac{\partial u(\mathbf{x})}{\partial \nu(\mathbf{x})}=\frac{1}{\mu_0}\frac{\partial \exp(ik_0\theta\cdot\mathbf{x})}{\partial \nu(\mathbf{x})}=g(\mathbf{x}) \text{ in } \partial\Omega \tag{3}$$

where $\upsilon(\mathbf{x})$ represents the unit outward normal to $\mathbf{x}\in\partial\Omega$ and $\theta=(\theta_x,\theta_y)$ is an incident direction defined on the unit circle S^1, i.e., θ satisfies $\theta\cdot\theta=1$. Notice that if there were no inclusion, one would have the incident field $u(\mathbf{x})=\exp(ik_0\theta\cdot\mathbf{x})$ once replaced in the above $\varepsilon(\mathbf{x})$ and $\mu(\mathbf{x})$ by ε_0 and μ_0 respectively.

Now, let us represent the thin inclusion by two different level set functions. For a more detailed description, we suggest a set of references [1,5,12]. Throughout the section, we assume that the level set function φ is a continuously differentiable function. With this feature in mind, we define a thin inclusion represented by φ in the following way. Let us denote $\Omega_0=\{\mathbf{x} : \varphi(\mathbf{x})\leq 0\}$ and $\partial\Omega_0=\{\mathbf{x} : \varphi(\mathbf{x})=0\}$. Notice that generally, $\partial\Omega_0$ is called the zero level set. With this representation, we are able to define the thin region Ω_T of thickness $2h$ as follows (see Fig. 2):

$$\Omega_T=\Omega_0\cap\{\mathbf{y}:\mathbf{y}=\mathbf{x}-\eta\mathbf{n}(\mathbf{x}) \text{ for } \mathbf{x}\in\partial\Omega_0,\ 0\leq\eta\leq 2h\}. \tag{4}$$

In order to describe the thin region of finite length (which is connected or disconnected) in Ω, we assume that another level set function φ is also a continuously differentiable function. Let us define a band structure Ω_B by

$$\Omega_B=\{\mathbf{x}=(x_1,x_2) : \psi(\mathbf{x})<0\}. \tag{5}$$

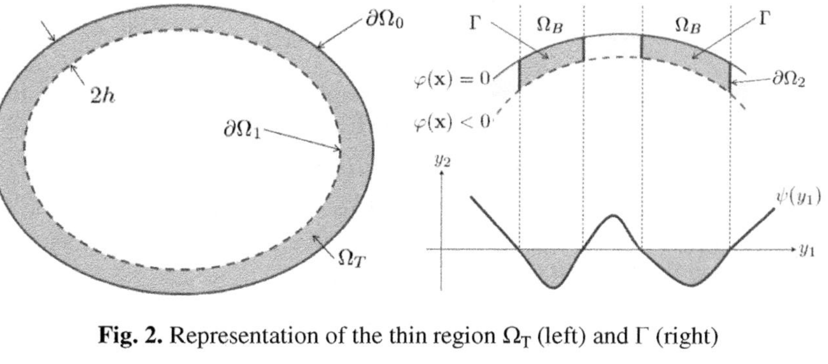

Fig. 2. Representation of the thin region Ω_T (left) and Γ (right)

With these representations (4) and (5), the thin inclusion Γ is $\Gamma=(\Omega_T \cap \Omega_B) \cap \Omega$ and boundary $\partial\Omega_2=\{\mathbf{y}=(y_1,y_2)\in\Gamma : \psi(y_1)=0\}$ of Γ for evolving at crack tips.

3 Shape Deformation by Calculus of Variations

In this section, we introduce a basic concept of shape deformation and of its representation via a level set function. The method is about the deformation of an already existing thin inclusion, say $\Gamma^{(n)}$ at the n-th time step of evolution, into the normal direction with an optimal velocity. In order to obtain the optimal velocity of level set functions, we have to introduce proper residual operators. For convenience, let us introduce the parameter distribution $b(\mathbf{x})$ as

$$b(\mathbf{x}) = \begin{cases} b_0 \text{ for } \mathbf{x}\in \Omega\setminus\overline{\Gamma} \\ b \text{ for } \mathbf{x}\in \overline{\Gamma}. \end{cases}$$

For a given thin inclusion Γ, we denote the measured (or 'true') data as u_{true} and similarly, for a given existing thin inclusion (or appropriate initial guess) $\Gamma^{(n)}$, we denote the measured data as u_{comp}. Due to the discrepancy of shapes between Γ and $\Gamma^{(n)}$, the parameter distributions are also different and this is causing the discrepancy between u_{true} and u_{comp}. Hence, for a given parameter distribution b, we can define the residual operator $R(b)$ on $\partial\Omega$: operator $R(b)$ on $\partial\Omega$:

$$R(b)= u_{\text{true}} - u_{\text{comp}}. \tag{6}$$

With this residual operator, we get the least-square cost functional

$$T(b) = \frac{1}{2}\| R(b)\|^2_{\partial\Omega} = \frac{1}{2}\langle R(b), R(b)\rangle_{\partial\Omega} = \frac{1}{2}\int_{\partial\Omega} R(b)R(b)^* dS \tag{7}$$

where * is the mark of complex conjugate. We henceforth assume that $R(b)$ admits the expansion

$$R(b+\delta b) = R(b) + R'(b)\delta b + O(\| \delta b \|^2_{\Omega})$$

for a sufficiently small perturbation δb. If the linear operator $R'(b)$ exists, it is denoted as the Fréchet derivative of $R(b)$. Then, the least-square cost functional also satisfies

$$T(b+\delta b) = T(b) + \mathrm{Re}\langle R'(b)^* R(b), \delta b\rangle_{\Omega} + O(\| \delta b \|^2_{\Omega})$$

where Re is the real part of corresponding quantity, the operator $R'(b)^*$ is the adjoint of $R'(b)$ with respect to spaces $\partial\Omega$ and Ω:

$$\langle R'(b)^* R(b), \delta b\rangle_{\Omega} = \langle R(b), R'(b)\delta b\rangle_{\partial\Omega}$$

and $R'(b)^* R(b)$ is the *gradient direction* of T in b. Thus in order to find the variation of $T(b)$ for obtaining the optimal velocity of level set functions, we have to compute δb and the gradient direction $R'(b)^* R(b)$.

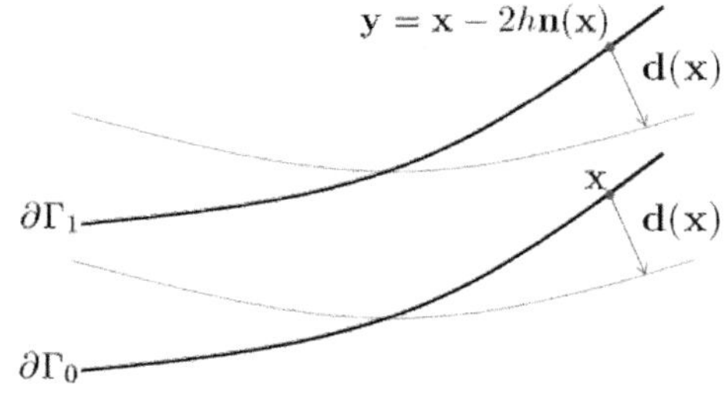

Fig. 3. Deformation of a thin inclusion

First, let us explain the parameter change δb due to the evolution of level set function. When every point $\mathbf{x}\in\partial\Gamma_0$ moves by a small distance $\mathbf{d}(\mathbf{x})$, Γ is deformed, refer to Fig. 3. Let us specify that the distance can be represented as $\mathbf{d}(\mathbf{x})=F(\mathbf{x})\mathbf{n}(\mathbf{x})\tau$ for some scalar function $F(\mathbf{x})$ and time step τ.

By adopting the interpretation of δb in [1,12], we can immediately obtain the parameter change δb due to the first and second level set functions as:

$$\text{(8)}\quad \begin{aligned} \delta b\big|_{\partial\Omega_0} &= (b-b_0)\mathbf{n}(\mathbf{x})\cdot d(\mathbf{x})[\delta_{\partial\Omega_0}(\mathbf{x})-\delta_{\partial\Omega_0}(\mathbf{x}+2h\mathbf{n}(\mathbf{x})]\aleph_{\Omega_B}(\mathbf{x}) \\ \delta b\big|_{\partial\Omega_2} &= (b-b_0)d(y_1)\delta_{\partial\Omega_2}(\mathbf{y})\aleph_{\Omega_T}(\mathbf{y}). \end{aligned}$$

Next, let us consider the gradient direction $R'(b)^*R(b)$. It is derived for the following two cases of interest (a rigorous derivation can be found in [9,12]):

(i) Permittivity contrast case ($\varepsilon\neq\varepsilon_0$ and $\mu=\mu_0$): Let $R(\varepsilon)$ be defined as (6). Then

$$R'(\varepsilon)^*R(\varepsilon)=u^*z.$$

Here, u satisfies (2) and (3), and z solves the following adjoint equation:

$$\frac{1}{\mu_0}\Delta z(\mathbf{x})+\omega^2\varepsilon(\mathbf{x})z(\mathbf{x})=0 \text{ in } \Omega \text{ with } \frac{1}{\mu_0}\frac{\partial z(\mathbf{x})}{\partial\nu(\mathbf{x})}=R(\varepsilon) \text{ on } \partial\Omega.$$

(ii) Permeability contrast case ($\varepsilon=\varepsilon_0$ and $\mu\neq\mu_0$): Let $R(\mu)$ be defined as (6). Then

$$R'(\mu)^*R(\mu)=\nabla u^*\cdot\nabla z$$

Here, u satisfies (2) and (3), and z solves the following adjoint equation

$$\nabla\cdot\left(\frac{1}{\mu(\mathbf{x})}\nabla z(\mathbf{x})\right)+\omega^2\varepsilon_0\, z(\mathbf{x})=0 \text{ in } \Omega \text{ with } \frac{1}{\mu_0}\frac{\partial z(\mathbf{x})}{\partial\nu(\mathbf{x})}=-R(\mu) \text{ on } \partial\Omega.$$

4 Evolution of Level Functions and Reconstruction Algorithm

The evolution of level set function φ (or ψ) satisfies the Hamilton-Jacobi type equation

$$\frac{\partial\varphi(\mathbf{x})}{\partial t}+F_j\,|\nabla\varphi(\mathbf{x})|=0.$$

From the formula (8) and the gradient directions in (i), (ii), we can choose the descent directions of level set functions φ and ψ as follows

1. The descent direction F_1 for the first level set goes as

$$F_1(\mathbf{x}) = -\mathrm{Re}[R'(b)^* R(b)](b-b_0)[\delta_{\partial\Omega_0}(\mathbf{x}) - \delta_{\partial\Omega_0}(\mathbf{x}+2h\mathbf{n}(\mathbf{x}))],\ \mathbf{x}\in\partial\Omega_0 \tag{9}$$

and the iterative procedure for φ reads as $\varphi^{(n+1)} = \varphi^{(n)} + \tau_T^{(n)} F_1^{(n)} |\nabla\varphi^{(n)}|$.

2. The descent direction F_2 for the first level set reads as

$$F_2(x_1) = -\mathrm{Re}\int_{\partial\Omega_2}[R'(b)^* R(b)](b-b_0)\delta_{\partial\Omega_2}(\mathbf{x})\aleph_D dx_2,\ \mathbf{x}=(x_1,x_2)\in\partial\Omega_2 \tag{10}$$

and the iterative procedure for ψ can be $\psi^{(n+1)} = \psi^{(n)} + \tau_B^{(n)} F_2^{(n)} |\nabla\psi^{(n)}|$.

In the previous section, every essential factor for evolving the two level set functions has been introduced. Now, we sketch the reconstruction algorithm.

Before starting, let us denote the initial guesses (functions) as $\varphi^{(0)}$ and $\psi^{(0)}$ and assume that the n-th level set functions $\varphi^{(n)}$ and $\psi^{(n)}$ have been obtained in the previous step. The iteration procedure can be summarized as follows

1. For a thin inclusion Γ and $\Gamma^{(n)}$, calculate the residual $R(b)$ by solving the forward problem (1) and (2).
2. With $R(b)$, solve the adjoint problem mentioned in (i) and (ii) and obtain the gradient direction $R'(b)^*R(b)$.
3. Calculate the descent direction F_1 and F_2 by (9) and (10), respectively.
4. Evolve the level set functions $\varphi^{(n)}$ and $\psi^{(n)}$. In this step, applying suitable regularization (for smoothing the descent directions F_1 and F_2) is required.
5. Update the parameter distribution $b(\mathbf{x})$.
6. When the residual $R(b)$ converges or oscillates around some suitable value, stop the process. If not, go to the step 1.

5 Numerical Examples

In the numerical simulations, the homogeneous domain Ω is chosen as a unit disk centered at (0,0), the value of thickness h is set to 0.015 and parameters μ, ε, μ_0 and ε_0 are chosen as 5, 5, 1 and 1, respectively. Since ε_0 and μ_0 are unity, applied frequency reads as $\omega=2\pi/\lambda$, at wavelength λ. For a successful evolution of level sets, we have chosen the time step as $\Delta t(s)=3\times10^{-s}$ for some positive integer s. We suggest [13] for more detailed discussion and examples about the influence of time step.

Two single inclusions $\Gamma_j=\{\mathbf{x}+\eta\mathbf{n}(\mathbf{x}) : \mathbf{x}\in\sigma,\ \eta\in(-h,h)\}$ for $j=1,2$ are chosen with following support curves:

$\sigma_1=\{(s,0.5s^2+0.1): s\in(-0.5,0.5)\}$

$\sigma_4=\{(s,2s^3-0.1s^2+0.1s): s\in(-0.5,-0.15)\cup(0.15,0.5)\}$.

The test configuration, e.g., wavelength λ, incident direction θ and time step $\Delta t(s)$ for reconstructing the thin inclusions are arranged in Table 1.

Table 1. Test configuration for thin inclusions Γ_j for j=1 and 2

Item	Γ_1	Γ_2
Wavelength λ	0.7	0.7
Incident direction θ	(cos π/2, sin π/2)	(cos 3π/4, sin 3π/4)
Time step $\Delta t(s)$	$\Delta t(4)$	$\Delta t(5)/3$

It is worth mentioning that for a suitable initial guess, we can estimate the location of the end points of thin inclusion, refer to [3,4,6] or the informal shape via a MUSIC-type algorithm, refer to [10,11]. Throughout this paper, we assume that the location of every end points of thin inclusions has been identified. By connecting them by a straight line, we can choose it as an initial support curve σ. Notice that without a good initial guess, we might suffer from large computational costs. For example, let the supporting curve σ of initial guess be the straight line $\sigma=\{(s,0.425): s\in(-0.5,0.5)\}$. Although we have performed 500 iterations, further iterations would be required for a successful reconstruction, refer to Fig.4.

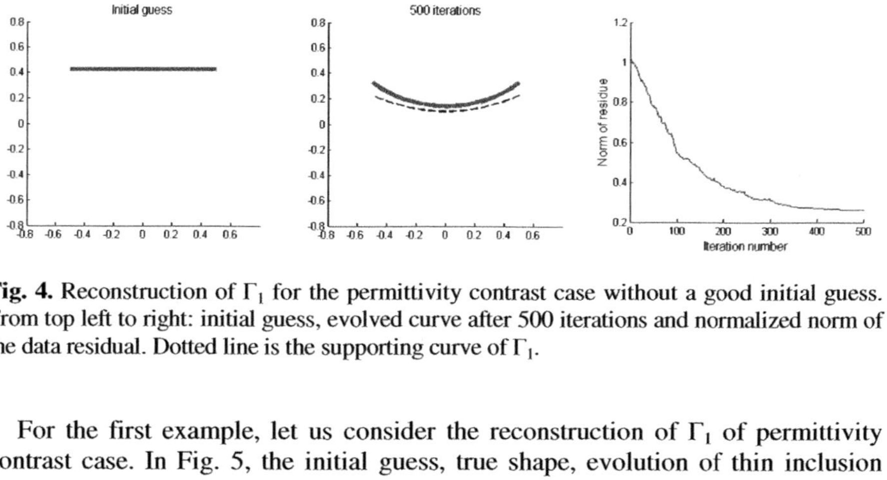

Fig. 4. Reconstruction of Γ_1 for the permittivity contrast case without a good initial guess. From top left to right: initial guess, evolved curve after 500 iterations and normalized norm of the data residual. Dotted line is the supporting curve of Γ_1.

For the first example, let us consider the reconstruction of Γ_1 of permittivity contrast case. In Fig. 5, the initial guess, true shape, evolution of thin inclusion and normalized norm of residual $R(\varepsilon)$ are displayed. From this result, we can say that Γ_1 is successfully reconstructed.

Now, let us consider the reconstruction of Γ_1 when the measured data was polluted by 30% noise. By comparing to the result exhibited in Fig. 6, we can say that although the normalized norm of residue is more oscillating than the noiseless one, Γ_1 is well retrieved.

From now, we consider the reconstruction of multiple thin inclusions. In Fig. 7, the reconstruction of thin inclusion Γ_2 of permeability contrast case is described. Similarly with the single inclusion case, Γ_2 is well retrieved.

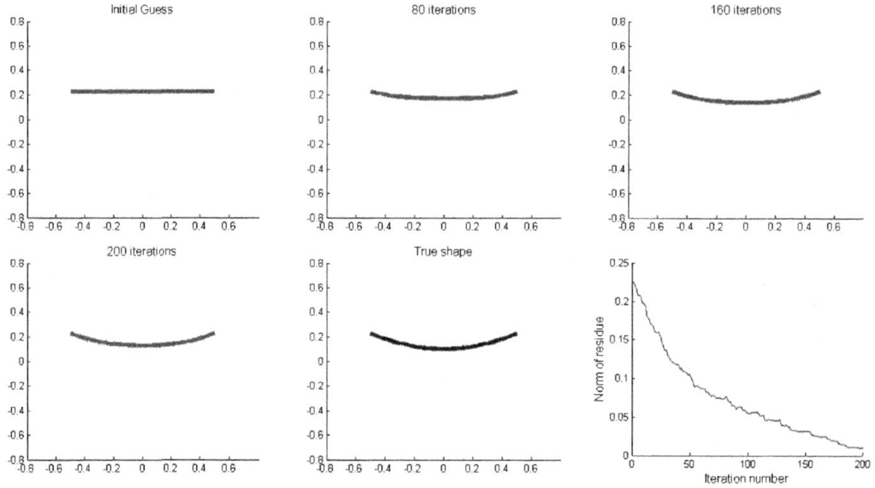

Fig. 5. Reconstruction of Γ_1 for the permittivity contrast case. From top to bottom, left to right: initial guess, evolved curve after 80, 160 and 200 iterations, true shape and normalized norm of the data residual.

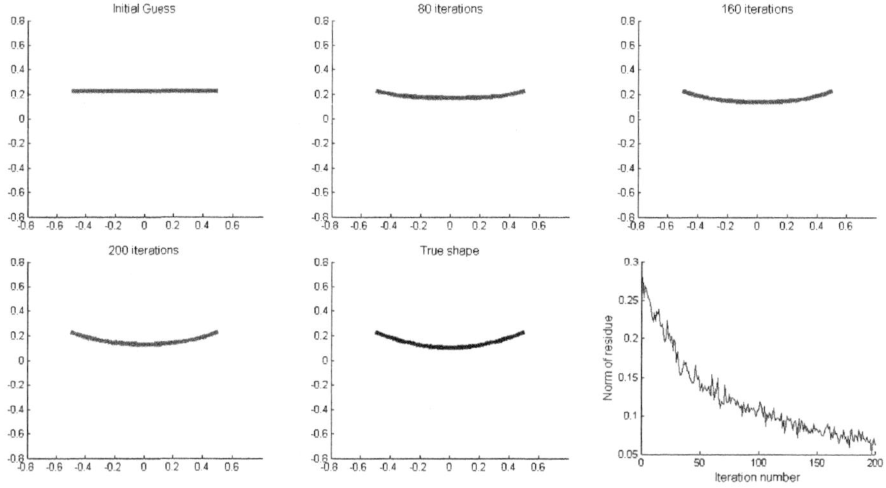

Fig. 6. Same as Fig. 5 with 30% noise

In Fig. 8, the reconstruction of Γ_2 with 30% noisy data is presented. From these results, the multiple thin inclusions are also successfully retrieved and it appears that the proposed reconstruction algorithm does not significantly suffer from noise.

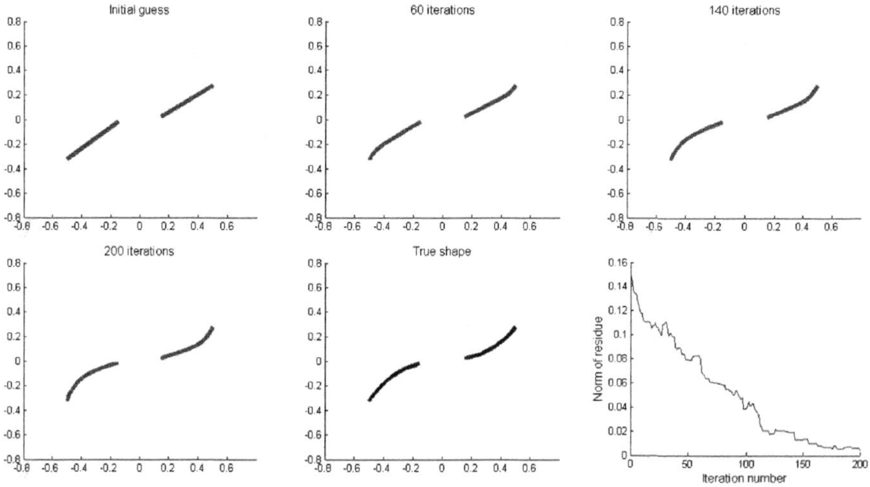

Fig. 7. Reconstruction of Γ_2 for the permeability contrast case. From top to bottom, left to right: initial guess, evolved curve after 60, 140 and 200 iterations, true shape and normalized norm of the data residual.

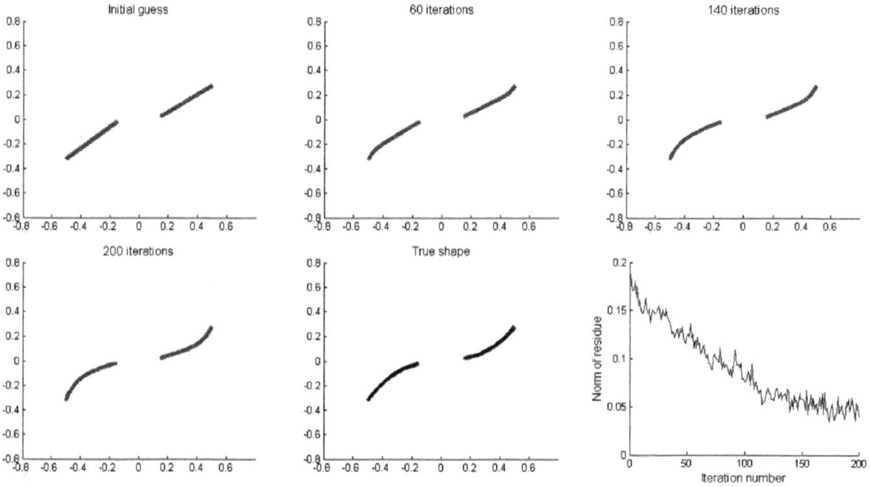

Fig. 8. Same as Fig. 7 with 30% noise

6 Conclusion

In this paper, a level set method of evolution aimed at the electromagnetic retrieval of a thin inclusion, with purely dielectric or magnetic contrast with respect to the embedding homogeneous medium, has been proposed. Employing two level set functions, the first one to describe its location and shape and the second one to describe its connectivity and length, enables us to retrieve thin inclusions by using a rigorous derivation of gradient directions via Fréchet derivatives (by solving an adjoint problem each time) in order to access to suitable velocities of evolution. Numerical simulations show that the proposed technique is stable and efficient, even in the influence of noise. The main challenge would be the three-dimensional, vector scattering cases, which is requiring both the availability of a proper computational tool of the electromagnetic fields that will be due to a non-necessarily planar inclusion and the ability of handling level sets of higher dimension.

References

[1] Àlvarez, D., Dorn, O., Moscoso, M.: Reconstructing thin shapes from boundary electrical measurements with level sets. Int. J. Inf. Syst. Sci 2, 498–511 (2006)
[2] Ammari, H.: An Introduction to Mathematics of Emerging Biomedical Imaging. Mathematics and Applications Series. Mathematics and Applications Series, vol. 62. Springer, Berlin (2009)
[3] Ammari, H., Beretta, E., Francini, E.: Reconstruction of thin conductivity imperfections. Applicable Analysis 83, 63–76 (2004)
[4] Ammari, H., Beretta, E., Francini, E.: Reconstruction of thin conductivity imperfections, II. The case of multiple segments. Applicable Analysis 85, 87–105 (2006)
[5] Dorn, O., Lesselier, D.: Level set methods for inverse scattering. Inverse Problems 22, R67–R131 (2006)
[6] Lee, H., Park, W.K.: Location search algorithm of thin conductivity inclusions via boundary measurements. In: ESAIM: proc., vol. 26, pp. 217–229 (2009)
[7] Osher, S., Fedkiw, R.: Level set methods and dynamic implicit surfaces. Springer, New York (2003)
[8] Osher, S., Sethian, J.A.: Fronts propagating with curvature-dependent speed: algorithms based on Hamilton-Jacobi formulations. J. Comput. Phys. 79, 12–49 (1988)
[9] Park, W.K.: Inverse scattering from two-dimensional thin inclusions and cracks. Thèse dedoctorat, Ecole Polytechnique (available via ParisTech) (2009), http://paristech.bib.rilk.com/4834
[10] Park, W.K., Ammari, H., Lesselier, D.: On the imaging of two-dimensional thin inclusions by a MUSIC-type algorithm from boundary measurements. Electromagnetic Nondestructive Evaluation (XII), Studies in Applied Electromagnetics and Mechanics 32, 297–304 (2009)
[11] Park, W.K., Lesselier, D.: MUSIC-type imaging of a thin penetrable inclusion from its multi-static response matrix. Inverse Problems 25, 075002 (2009)
[12] Park, W.K., Lesselier, D.: Reconstruction of thin electromagnetic inclusions by a level set method. Inverse Problems 25, 085010 (2009)
[13] Ramananjaona, C., Lambert, M., Lesselier, D., Zolésio, J.-P.: Shape reconstruction of buried obstacles by controlled evolution of a level set: from a min-max formulation to numerical experimentation. Inverse Problems 17, 1087–1111 (2001)
[14] Santosa, F.: A level set approach for inverse problems involving obstacles. ESAIM Control Optim. Calc. Var. 1, 17–33 (1996)

Uniaxial Compressive Behaviour of Carbon Fibre-Epoxy Laminates – Part 1: Unnotched

Joung Hwan Lee[1]

Abstract. An extensive experimental investigation was performed to understand the uniaxial compressive behavior of unnotched composite laminates using the T300/924C and T800/924C composite systems. The unnotched unidirectional ($[0_4]_{2s}$ for the T300/924C and T800/924C material) and multidirectional (T300/924C $[45/-45/0/90]_{3s}$ and T800/924C $[45/0/-45/90]_{3s}$) laminates were used for the mechanical tests. From the microstructure investigation, major factors (fibre distributions, voids, resin rich regions and fibre waviness) which influence compressive strengths were identified. Compressive failure behavior of unidirectional and multidirectional specimens was characterised through the SEM analysis and post failure investigations. Finally failure strength predictions were compared and explained for unidirectional and multidirectional composites with various commonly used models.

1 Introduction

The choice of material used in an application is probably the most crucial decision to be made since it affects the component's manufacture, design approach and the ultimate performance of the finished product. For this reason many industries put much time and effort into researching and developing materials which are effectively tailor made for a specific type of application. Probably the most important class of these innovative materials is advanced composites, such as those containing continuous carbon fibres in an epoxy resin. These materials have become very attractive materials in structural applications where strength and stiffness to weight ratios as well as good resistance to fatigue and corrosion are the important factors of design, for example in aircraft and automotive constructions. However, their properties are not as well characterized as those of more conventional materials (primarily metals), which they are designed to replace.

Imperfections, such as misalignment of fibres, microvoids in the matrix and clustering of fibres, could originate in a composite during the manufacturing process. Localized deformation due to these imperfections leads to premature failure and often dictates the dominating failure mechanisms of fibre reinforced composites. It is commonly accepted that fibre misalignment among the imperfections is the most important parameter in determining the axial compressive strength of the composite. The presence of the fibres misoriented with respect to the loading axis

[1] University of Sheffield, The AMRC with Boeing, Sheffield, United Kingdom
j.h.lee@sheffield.ac.uk

J.H. Lee et al. (Eds.): EU-Korea Conference on Science and Technology, SPPHY 135, pp. 109–129.
springerlink.com

leads to final failure via formation of a kink band initiated by the rotation of initially misaligned fibres within a certain band (plastic shear deformation) or via 0° fibre microbuckling triggered by the local instability of fibres embedded in the matrix.

In this study, in order to measure the uniaxial compressive strength and understand the behaviour, a broad investigation was performed using the data obtained from the current work and the published data on the T300/924C and T800/924C composite systems. Composite microstructure of the unidirectional composite laminates is examined first as the microstructure can affect the mechanical behaviour of the composite considerably. Through the mechanical tests of the unnotched unidirectional ($[0_4]_{2s}$ for the T300/924C and T800/924C material) and multidirectional (T300/924C $[45/\text{-}45/0/90]_{3s}$ and T800/924C $[45/0/\text{-}45/90]_{3s}$) laminates, the compressive properties are presented. Then tensile and compressive strengths of the unidirectional laminates are compared. The progression of compressive damage was also monitored using optical microscopy and scanning electron microscopy (SEM). Finally unnotched compressive failure strength predictions are carried out for unidirectional and multidirectional composites with various commonly used models.

2 Composite Microstructure

In order to investigate the microstructure of unidirectional composite laminates, two different reinforcing materials (T300 and T800 Toray fibres) are used with the same matrix (epoxy resin 924C), i.e. the T300/924C and T800/924C carbon/epoxy systems.

2.1 Fibre Distribution

Figure 1 (a) and 1 (b) show the optical micrographs of a polished transverse section of T300/924C and T800/924C unidirectional laminate, respectively. It is clear that the fibres are non-uniformly distributed, with regions of high and regions of low packing fraction and in different types of cross sectional fibre shapes, i.e. circular, elliptical and 'kidney' shaped. It can also be seen that in some locations the fibres are in contact with each other. These irregularities in fibre shapes and distribution are not desirable because a considerable spread in strengths about a mean value is caused in that the variation could occur as a result of variations in the fibre diameters. The contact of fibres is also not desirable since cracks pass from one fibre to another, resulting in premature failure of the composite

Figure 1 exhibits that the cross section fibre density in the T800/924C composite is higher than the T300/924C. From the resin burn off test it was identified that the fibre volume fraction of the T800/924C (61%) was higher than that of the T300/924C (56%). The fibre volume fraction can be obtained from the following expression

$$V_f = \frac{\rho_{ce}}{\rho_f} W_f \tag{1}$$

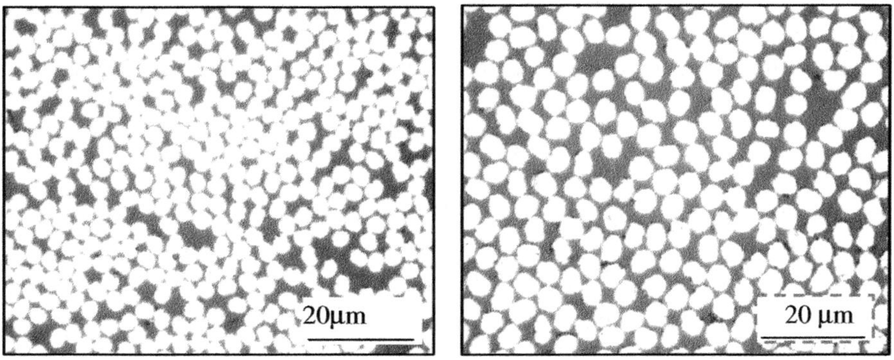

Fig. 1. Optical micrographs of cross sections of (a) T300/924C and (b) T800/924C unidirectional composite $[0_4]_{2s}$

Table 1. Material Properties (Manufacturer's Data)

Property	T300	T800	924C
Tensile Strength, MPa	3650	5490	64.6
Modulus, GPa	231	294	3.79
Failure Strain, %	1.4	1.9	2.45
Density, g/cm^3	1.76	1.81	1.31
Fibre Diameter, μm	7	5.1	-

where ρ_{ce} = composite density, ρ_f = fibre density and W_f = fibre weight fraction. It was measured: $\rho_{ce}^{T300} = 1.5719 g/cm^3$ and $\rho_{ce}^{T800} = 1.6092 g/cm^3$, $\rho_f^{T300} = 1.8093 g/cm^3$ and $\rho_f^{T800} = 1.8493 g/cm^3$, and $W_f^{T300} = 66.27\%$ and $W_f^{T800} = 71.12\%$. The density values and other material properties for the fibres and the resin are presented in Table 1.

2.2 Voids and Resin Rich Regions

The presence of voids in the composite reduces its strength and elastic modulus since they reduce the volume of solid, within which the stresses are distributed. Also, voids can act as sites of local stress concentration that can initiate failure. The compressive strength is affected most, because the fibres have less side support by the resin and buckle at lower applied loads [1]. The knowledge of void content is desirable for estimation of the quality of the composite. A good composite should have less than 1% voids, whereas a poorly manufactured composite can have up to 5% void content.

Figure 1(a) and 1 (b) illustrate void content in the T300/924C and T800/924C system, respectively. It can be seen that the T300/924C has higher void content than the T800/924C in the sections examined. It is also clear that the voids generally exist at the boundary of fibres. From the resin acid digestion it was observed that the void volume fraction of T300/924C (2.0%) was greater than that of the T800/924C (1.6%).

The volume fraction of voids can be calculated from the following expression

$$V_v = \frac{\rho_{ct} - \rho_{ce}}{\rho_{ct}} \tag{2}$$

ρ_{ct} is the theoretical composite density obtained from the rule of mixture, i.e.

$$\rho_{ct} = V_f \rho_f + (1 - V_f)\rho_m \tag{3}$$

where ρ_m is the matrix density. The matrix density (ρ_m) of the 924C epoxy resin used in this study is 1.31 g/cm^3.

2.3 Fibre Waviness

Fibre waviness is a type of manufacturing defect occurring especially during the filament winding process and curing process. This can cause a dramatic reduction in the compressive strength/stiffness properties of the composite. Wisnom [2] studied the effect of specimen misalignment of a composite with a non-linear stress-strain matrix response. The study has shown that a misalignment angle between the fibres and the loading axis of only 0.25° is sufficient to reduce the compressive strength of the XAS/914 carbon-epoxy system from 2720 MPa to 1800 MPa (i.e. more than 40% reduction). At 3° this is reduced to 700 MPa. The actual compressive strength of the XAS/914 system is generally considered to be about 1400 MPa.

Figure 2 (a) and 2 (b) summarize the fibre waviness measurement results of the unidirectional laminates $[0_4]_{2s}$ for the T300/924C and T800/924C obtained in accordance with the Yugratis [3] method. From both figures it can be seen that about

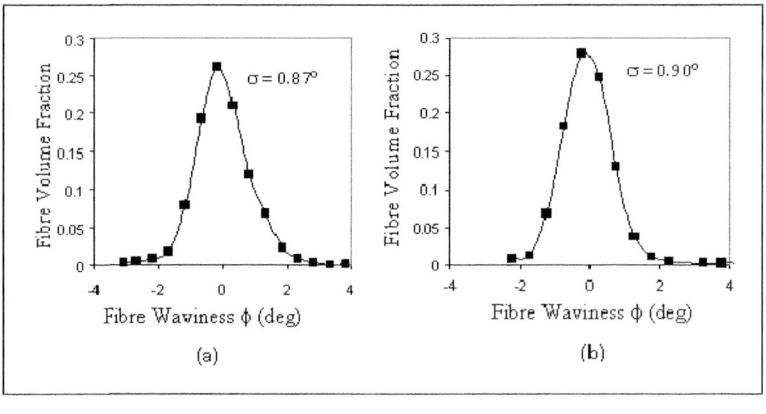

Fig. 2. Fibre angle distributions for the unidirectional T300/924C (a) and T800/924C (b) laminates (σ = Standard deviation)

94% of the fibre volume lie within ±1.5° of the main fibre direction. This indicates that the fibres are very well aligned in the laminates. Both laminates also have quite similar fibre angle distribution. The standard deviation is calculated to be 0.87° and 0.90° for the T300/924C and T800/924C laminates, respectively.

3 Compressive Test Results

The results obtained from the uniaxial compressive test of unnotched unidirectional ($[0_4]_{2s}$ for T300/924C and T800/924C) and multidirectional ($[45/-45/0/90]_{3s}$ for T300/924C and $[45/0/-45/90]_{3s}$ for T800/924C) specimens are presented. The ICSTM and modified ICSTM test fixture are used for the unidirectional and multidirectional specimens, respectively. The dimensions for the unidirectional and multidirectional specimens are 10mm × 10mm and 30mm × 30mm in the gauge section and specimen width, respectively.

Five specimens are used to determine mechanical strengths, each derived from a minimum of five satisfactory tests. Multidirectional specimens that failed near the end tabs are discarded.

3.1 Unnotched Unidirectional Specimens

3.1.1 Stress-Strain Behaviour

Figure 3 (a) and 3 (b) illustrate a typical compressive strain-stress plot for unidirectional T300/924C and T800/924C laminates, respectively. The specimens were instrumented with two normally back-to-back strain gauges at the centre of the specimen gauge section. The direct strain measurements are plotted, along with the mean of the two readings, from which the elastic modulus and tangent modulus are determined. The strain curves for both T300/924C and T800/924C unidirectional laminates intersect at 1.08% and 0.72% strain, respectively and increasingly diverge, indicating that the specimen is bending. This is due to a slight lack of flatness of the specimen and slight misalignment of the test piece in the test rig.

Figure 3 (a) shows that non-linearity in the stress-strain curve commences at a strain of approximately 0.3% - 0.4 % with the ultimate compressive strength of 1592 MPa and the fracture strain of 1.12 %, whereas Figure 3 (b) illustrates non-linearity initiating around 0.7 % with the ultimate strength of 1564 MPa and the average fracture strain of 0.97 %. This non-linearity for both materials seems to be caused by non-linear behaviour of the fibres because the matrix is generally observed to remain elastic at these strain values [11]. The longitudinal modulus for the T300/924C and T800/924C materials measured at 0.25 % applied axial strain are 157 GPa and 167 GPa, respectively.

Table 2 summarises the mean ultimate compressive strength for both materials measured in this study and presents the ultimate tensile failure strength[4] for

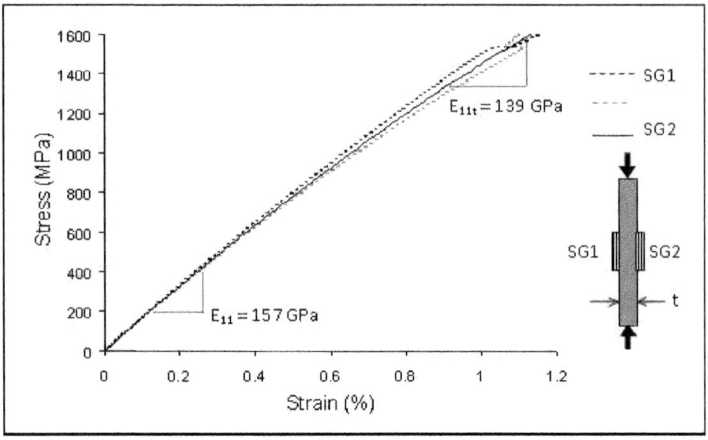

(a) T300/924C unidirectional specimen $[0_4]_{2s}$

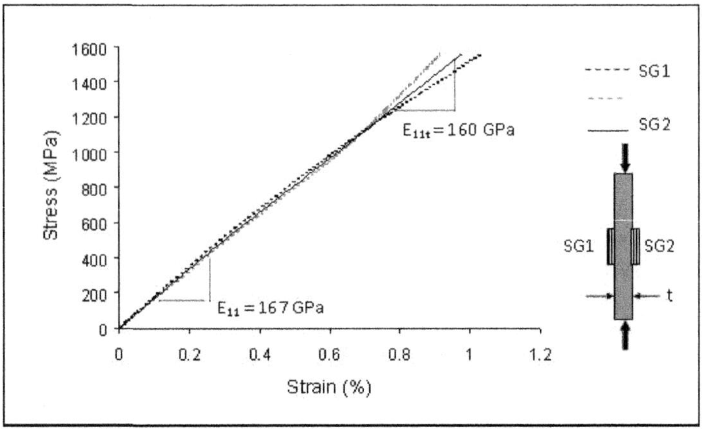

(b) T800/924C unidirectional specimen $[0_4]_{2s}$

Fig. 3. Typical compressive stress-strain curves for the T300/924C (a) and T800/924C (b) unidirectional laminates

comparison purposes. It is evident from the table that the tensile strength of the T800/924C laminate is higher than that of the T300/924C laminate. This is because T300 fibres (7 μm) have larger diameter than T800 fibre (5.1 μm) i.e., higher possibility of defects within the fibre that affects tensile strength. However, bigger diameter is beneficial in compression since it provides higher resistance to bending than the thinner T800 fibres. The T800/924C laminate shows slightly higher compressive strength since the fibre volume fraction is higher (61%) compared to the volume fraction (56%) of the T300/924C laminate.

Table 2. Compressive and tensile strength data of unidirectional T300/924C and T800/924C

	Compression Data				Tension Data	
	T300/924C	C.V. (%)	T800/924C	C.V. (%)	T300/924C	T800/924C
Failure Strength (MPa)	1612[b]	4.74	1625[b]	4.79	2000[a]	2320[4]
Young's Modulus (GPa)	162[5]	6.1	168[6]	-	130[a]	157[4]

([a]: Supplied by the prepreg manufacturer, Hexcel Composites, Duxford and [b]: Measured in the present study, C.V.: coefficient variation).

3.1.2 Characteristics of Fractured Specimens

All unidirectional specimens of the T300/924C and T800/924C materials failed suddenly and catastrophically near the junction of the end tab and gauge section, accompanied by the spontaneous release of sound. When failure occurred, the test piece was partitioned into two pieces with fracture surfaces inclined at typical angles of between $\beta = 10° \sim 30°$ (β: kinkband inclination angle) in either the width or thickness direction for both materials (see Figure 4). In addition the fracture surfaces usually exhibited some longitudinal splitting and interlaminar cracking, a fact that is attributed to post-failure damage rather than being a failure cause as shown in Figure 4. It was found from Figure 4 that the failure behaviour for both materials was similar. Figure 4 (a) and (b) show the overall compressive failure of the unidirectional T300/924C and T800/924C specimens, respectively.

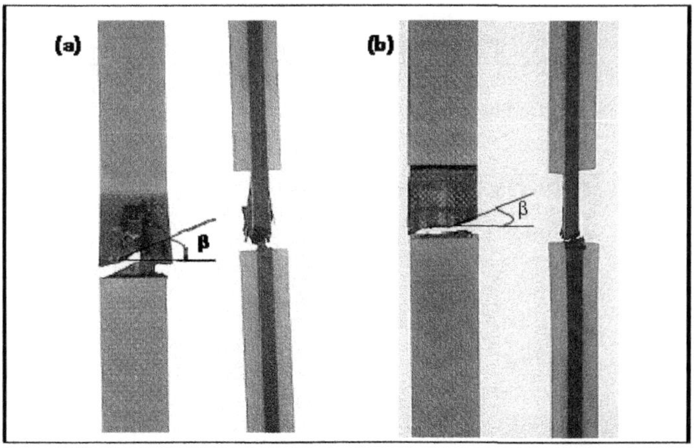

Fig. 4. Overall failure mode of unidirectional compression specimen for (a) the T300/924C and (b) T800/924C material

3.1.2.1 Scanning-Electron-Microscope (SEM) Examination

After compression tests, some of the broken T300/924C specimens were selected and examined in the SEM. Figure 5 depicts the microscopic feature of a typical fracture surface, in which the characteristic step features associated with micro-buckling of the fibre forming a kink band are clearly evident as described in most publications. There is also little reason to doubt that failure in general is by micro-buckling and kinking of fibres because the characteristic kink band angle is similar to the fracture surface angle of test piece fragments (see Figure 4), though usually slightly larger. It is observed that the fibres break at two points, which create a band inclined at $\beta = \sim 23°$ to the horizontal axis, where β is defined in Figure 5. Kink band starts at the free edge or locations of stress concentration such as a pre-existing material defect and load introduction, i.e. at the end of end tabs where the specimen emerges from the clamps. Then the band is propagated, keeping its direction. The higher magnification image, Figure 6, shows fractured fibre ends embedded in the matrix. The group of fibres illustrated appears to have failed due to bending.

Fig. 5. SEM micrograph of fibre kinking in a unidirectional T300/924C laminate

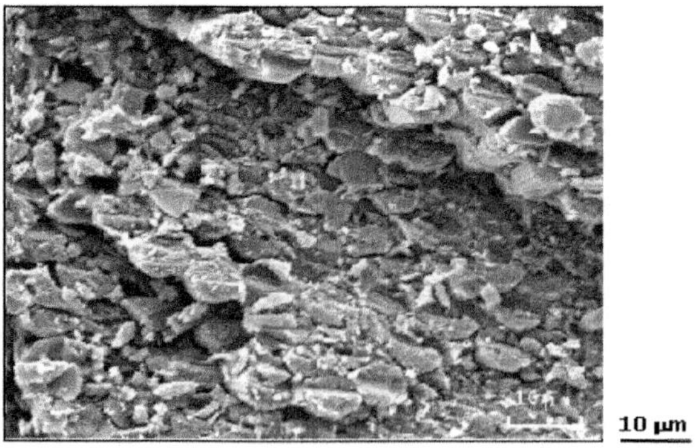

Fig. 6. SEM micrograph [5] of a group of fibres failed in bending

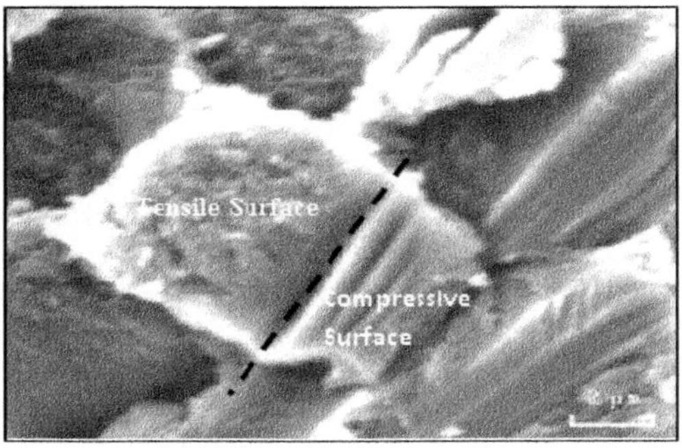

Fig. 7. Micrograph [5] illustrating tensile and compressive surfaces on an individual failed fibre due to bending

Looking more closely at the fibre surface, Figure 7, it is possible to identify the areas of local tension and compression that exist during fibre buckling. The tensile surface is relatively planar and perpendicular to the loading and fibre direction. The compressive surface, however, appears to be distinctly non-planar, irregular and serrated in appearance. In addition, the surface is inclined to about 45° to the fibre axis on the compressive side. It can be confirmed from the figure that all the fibres failed due to a bending type of deformation.

3.2 Unnotched Mutidirectional Specimens

3.2.1 Stress-Strain Behaviour

Uniaxial compressive tests were performed for the multidirectional T300/924C ($[45/\text{-}45/0/90]_{3s}$) and T800/924C ($[45/0/\text{-}45/90]_{3s}$) laminates with an antibuckling device. A typical compressive stress-strain response of a specimen selected from both T300/924C and T800/924C materials is shown in Figures 8 (a) and 8 (b), respectively. The longitudinal strain on the two faces of the test piece is initially the same but as the applied load is increased the strains diverge indicating bending similar to the stress-strain curves of the unidirectional specimen (see Figure 3).

The mean line plotted through the strain points exhibited an initial straight portion with an elastic modulus of 51 GPa and 56 GPa for T300/924C and T800/924C laminate, respectively, followed by a continuously curved portion with a tangential modulus at failure approximately 14 % for the T300/924C laminate and 11 % for the T800/924C laminate less than that of the linear part. The measured elastic moduli for both materials are in reasonable agreement when compared with those estimated by the laminate plate theory (62 GPa for the multidirectional T300/924C laminate and 64 GPa for the multidirectional T800/924C laminate).

The mean failure strains of T300/924C and T800/924C are 1.5 % and 1.27 %, respectively. These strains are larger than those achieved in the unidirectional laminates (see Figure 3). This suggests that the mechanism of failure in the axial

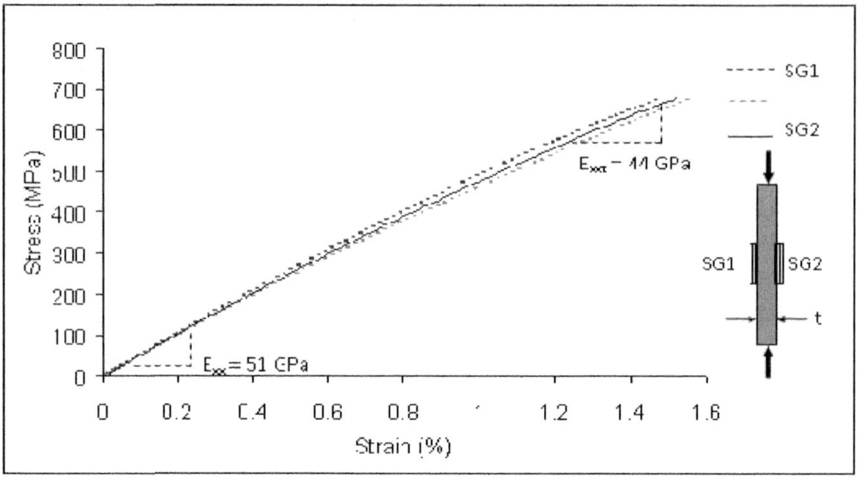

(a) T300/924C multidirectional specimen ([45/-45/0/90]$_{3s}$)

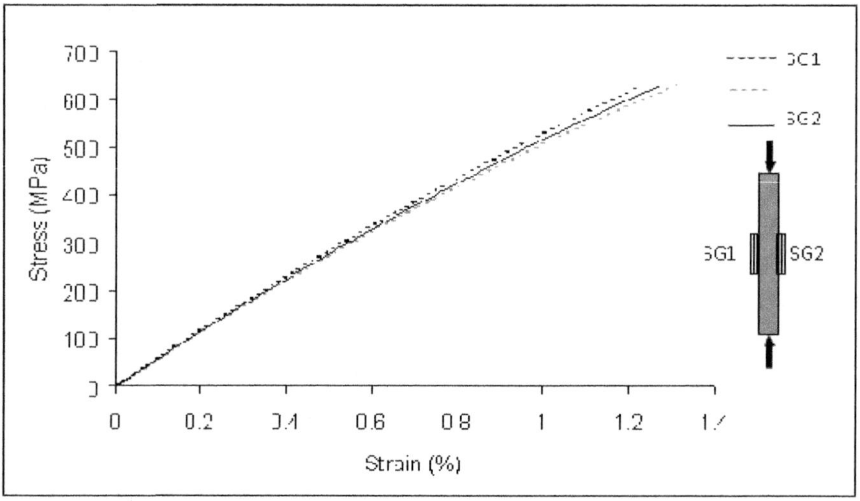

(b) T800/924C multidirectional specimen ([45/0/-45/90]$_{3s}$)

Fig. 8. Typical compressive strain-stress curves for the (a) T300/924C and (b) T800/924C system

plies is affected by the constraining effect of adjacent off-axis plies. The off-axis layers (±45° plies) provide lateral support to the 0° axial plies and delay the initiation of fibre microbuckling. These results could also be confirmed from the Berbinau *et al.* experimental study [6].

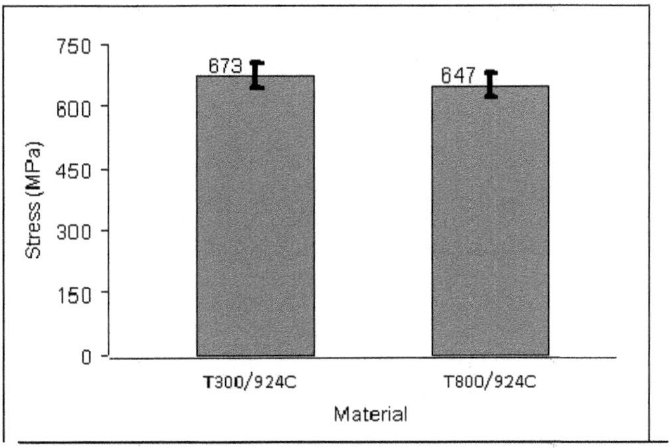

Fig. 9. Measured mean compressive strength for the multidirectional T300/924C ($[45/-45/0/90]_{3s}$) and T800/924C ($[45/0/-45/90]_{3s}$) laminates

The measured mean compressive strengths are presented in the form of a bar chart with error bars in Figure 9. The error bars indicate a maximum value and a minimum value. It is immediately evident that the mean strengths for both materials are quite similar like the unidirectional mean compressive strengths (1612 MPa for T300/924C and 1625 MPa for T800/924C). The difference is only within 3.8 % of one another as shown in Figure 9.

The scatter in strength is satisfactorily low (4.04 % for the T300/924C and 3.85 % for T800/924C). The scatter is probably due to imperfections introduced during manufacture of the laminates resulting in fabrication defects and non-uniform laminate thickness. Imperfections in the specimen geometry can also produce misalignment of the specimen in the testing fixture that causes bending and reduction in the measured compressive strength.

3.2.2 Characteristics of Fractured Specimens

All the multidirectional specimens of the T300/924C and T800/924C materials broke within the specimen's gauge section emitting a cracking sound just prior to failure (see Figure 10). The failure occurred suddenly and catastrophically.

Post-failure examination suggests that in-plane microbuckling in the 0° plies, which are the primary load-bearing members of the multidirectional laminate, is the critical damage mechanism, which causes the catastrophic fracture. Longitudinal splits, fibre/matrix de-bonding and delamination between neighbouring plies take place suddenly and concurrently with the final failure (see Figure 4.10). This is supported by failure strain measurement (see Figure 8). In general, the 0° plies fail on a plane $\beta = 10° \sim 30°$ to the horizontal axis, Figure 10, which is similar to the fracture line observed in the unidirectional laminate although this may be affected also by the fibre direction in the off-axis plies.

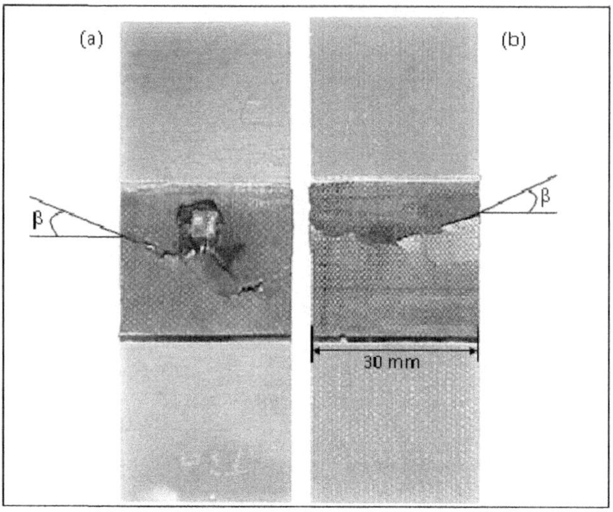

Fig. 10. Overall failure mode of multidirectional compression specimen for the (a) T300/924C $[45/\text{-}45/0/90]_{3s}$ and (b) T800/924C $[45/0/\text{-}45/90]_{3s}$ material

4 Prediction of Composite Compressive Strength

The theoretical analysis used during the duration of the current project is explained in this section. Compressive strengths for the unnotched unidirectional laminates of T300/924C and T800/924C materials are predicted with various commonly used models. Compressive strength predictions are also performed for the unnotched multidirectional laminates of T300/924C and T800/924C materials using a fibre microbuckling model together with stiffness-ratio method. All the predictions are compared to each other and to the experimental data. The models are discussed and their limitations are pointed out.

4.1 Unnotched Unidirectional Laminates

A number of theories for the compressive strength of unidirectional composite laminates have been developed and these have been reviewed by several researchers [7-10]. The theories are divided into two major models for the strength predictions, namely, the fibre microbuckling model and the kink band model. These models are based on the compressive failure due to fibre microbuckling or kinking. These failure behaviours were also identified in the current work (see Section 3.1.2).

To predict the compressive strengths in the current work, both microbuckling models developed by Rosen [11] and Berbinau *et al* [6, 12] and kink band models developed by Argon [9] and Budiansky [13] are applied to the T300/924C and T800/924C system.

4.1.1 Fibre Microbuckling Model

In the microbuckling model, compressive failure is assumed to be triggered by the local instability of fibres embedded in the matrix. Rosen [11] proposed a 2-D model treating a unidirectional fibre composite as parallel fibre layers. He assumed that the fibre layers were perfectly straight and evenly spaced. The matrix material and fibre material were also assumed to be linearly elastic. Under compressive loading, two possible fibre microbuckling modes, i.e. extension mode and shear mode, were considered for the bifurcation buckling analysis. The energy method was utilized to derive the elastic microbuckling stress for these two modes. For the shear mode, the compressive strength was obtained as

$$\sigma_c = \frac{G_m}{1 - V_f} = G_{12} \tag{4}$$

where G_m is the matrix shear modulus, V_f is the fibre volume fraction and G_{12} is the elastic shear modulus of the composite.

According to Rosen's bifurcation buckling analysis, the compressive strength of a unidirectional composite was approximately equal to the elastic shear modulus of the composite. The result is quite impressive since the material strength generally has nothing to do with its own material constants. The assumptions are, however, not acceptable for the case of modern carbon fibre/epoxy systems. Most crucially, fibres cannot be treated as straight, parallel layers. It is known that a major factor contributing to the reduced compressive strength is the fibre waviness (fibre misalignment), which is introduced into the composite during the manufacturing stages. Furthermore, the more modern matrix materials cannot be treated as linearly elastic, in fact these materials exhibit a nonlinear behaviour.

As expected, Rosen's prediction was found to be much greater than the experimental results when applied to the unidirectional laminates of the T300/924C and T800/924C material (see Figure 12).

Berbinau *et al.* [12] obtained a general microbuckling equation where 0° fibre is modelled as an Euler slender column (equation (5)), considering fibre misalignment and matrix nonlinearity effects on the compressive strength of unidirectional fibre composites. They assumed that a fibre had an initial sinusoidal shape of amplitude, V_0. As the compressive load was applied the fibre deformed further such that the amplitude increased to V.

$$E_f I \frac{d^4(V - V_0)}{dx^4} + \frac{A_f \sigma_0}{V_f} \frac{d^2 V}{dx^2} + A_f G(\gamma) \frac{d^2(V - V_0)}{dx^2} = 0 \tag{5}$$

where E_f, A_f, I are the fibre elastic modulus, the cross-section area, and the second moment of area; V_f is the fibre volume fraction, and $G(\gamma)$ is the composite shear modulus (as a function of the shear strain γ).

From equation (5) a relationship between the applied compressive stress, σ_0 and the maximum amplitude, V of the 0° buckled fibre during uniaxial compression is obtained as described in equation (6). Failure of the unidirectional material occurs when the fibre amplitude V starts to increase asymptotically.

$$\frac{V}{V_0} = \left[1 - \frac{\left(A_f / V_f\right)\sigma_0}{E_f I\left(\pi/\lambda\right)^2 - A_f G}\right]^{-1} \tag{6}$$

Following the above concept, Berbinau *et al* [6] further studied the effect of the off-axis ply orientation on fibre microbuckling in the 0° plies of multidirectional laminates. The study is based on the concept that interlaminar shear stresses may influence the laminate failure since 0° fibre microbuckling may initiate from the specimen free edge, where the fibre support is substantially reduced. They also showed that among the interlaminar stresses, σ_z, τ_{zy} and τ_{zx}, the shear stress components τ_{zy} and τ_{zx} are more likely to influence the in-plane movement of the 0° fibres. The theoretical approach is to calculate first the interlaminar shear stresses τ_{zy} and τ_{zx} using the Puppo and Evensen model[14]. Then the shear stresses are incorporated into a general microbuckling equation described in equation (5). The modified equation including the interlaminar shear stresses is shown as follows:

$$E_f I \frac{d^4(V - V_0)}{dx^4} + \frac{A_f \sigma_0}{V_f} \cdot \frac{d^2 V}{dx^2} - 2d_f \left\{ \left[\frac{d\tau_{zy}}{dy}\right]_{\frac{W}{2}} \right\} \cdot V - A_f G(\gamma) \frac{d^2(V - V_0)}{dx^2} = 0 \tag{7}$$

Equation (7) again gives a relationship for the compressive stress σ_0 developed in the 0° ply in terms of the maximum amplitude V of the 0° buckled fibre and interlaminar shear stress τ_{zy} during uniaxial compression. This relationship can be written in the following closed form:

$$\frac{V}{V_0} = \left(1 - \frac{\frac{A_f}{V_f}\sigma_0 + 4r_f \left\{ \left[\frac{d\tau_{zy}}{dy}\right]_{\frac{W}{2}} \right\} \cdot \left(\frac{\lambda}{\pi}\right)^2}{E_f I \left(\frac{\pi}{\lambda}\right)^2 + A_f G} \right)^{-1} \tag{8}$$

In general, the Berbinau *et al.* model yields a conservative prediction. It contains all the effects, which are known to influence fibre microbuckling, namely the initial fibre misalignment and the effects due to the nonlinear shear behaviour. However, it is noted that the assumed shear response does not match up exactly with the experimentally determined one. Furthermore, because the model is microbuckling based, stress redistributrion, which may occur at the onset of microbuckling is not captured. It is clear that if stress redistribution does occur then a higher strength will result. These effects explain why the model gives a relatively large underestimation of the compressive strength. For industry purposes this model is of great value because of the conservative prediction that it yields.

Rosen and Berbinau microbuckling predictions of compressive failure stress for the unidirectional laminate of the T300/924C and T800/924C material used in the current study are compared with the measured experimental results in Figure 12.

The predictions of Berbinau *et al.* model were solved numerically by using a Fortran programme.

4.1.2 Fibre Kink Band Model

In the kink band model, the compressive failure is assumed to be initiated by plastic shear deformation associated with the rotation of initially misaligned fibre within a certain band. Argon [9] investigated the fibre misalignment effect on compressive failure and suggested that, once the shear stress in the region of fibre misalignments reaches the composite shear yield stress, compressive failure would take place. Based on his argument, the compressive strength was expressed as

$$\sigma_c = \frac{\tau_y}{\phi} \tag{9}$$

where τ_y is the shear yield stress of the composite and ϕ is the initial fibre misalignment. Argon also assumed that the matrix material was perfectly plastic in shear. In actual fact, epoxy type matrix material initially undergoes elastic deformation. Again, for carbon epoxies, this model is expected to over predict actual compressive strength (see Figure 12).

Busiansky [13] extended Argon's concept and proposed an elastic-perfectly plastic kink band model, considering the plastic collapse mechanism to be more likely for carbon fibre composites. The principle here is that the elastic shear limit is reached (i.e. the shear yield point) before further plastic deformation is required to cause failure. This resulted in the equation

$$\sigma = \frac{\tau_y^*}{\phi_0 + \gamma_y} \tag{10}$$

where ϕ_0 is the initial fibre misalignment angle in the kink band, γ_y is the yield shear strain and

$$\tau_y^* = \tau_y \left[1 + \left(\frac{\sigma_{Ty}}{\tau_y} \right)^2 \tan^2 \beta \right]^{\frac{1}{2}} \tag{11}$$

where τ_y,σ_{Ty} and β are the in-plane shear yield strength, transverse yield strength and kink band inclination angle of the composite, respectively.

A model based on the fibre and matrix properties like the Budiansky's model would be preferred since, such a model would allow compressive strength predictions of concept materials to be made.

4.1.3 Comparison of Experimental and Analytical Results

The input data for the compressive failure strength predictions calculated for the current study are shown in Table 3. The shear yield stress (τ_y) and strain (γ_y) of the T300/924C material were obtained from Mackinley's experimental work[5]. He determined a shear yield point graphically by plotting a tangent to the shear strain-stress curve at the origin and another tangent to the curve at 6% shear strain.

Table 3. Input Data for the Compressive Failure Predictions of the Unidirectional T300/924C and T800/924C Laminates

Model	Input Data	
	T300/924C	T800/924C
Rosen [11][1]	G_{12}= 5.7 GPa	G_{12}= 6.0 GPa
Berbinau et al [6,12]	V_f= 0.56, r_f= 3.5μm, E_f= 231 GPa, λ_0= 70μm, ϕ_0= 1.15°	V_f= 0.61, r_f= 2.55μm, E_f= 294 GPa, λ_0= 51μm, ϕ_0= 1.15°
Argon [9]	τ_y= 64.6 MPa, γ_y= 0.019	
Budiansky [13]	τ_y= 64.6 MPa, γ_y= 0.019, ϕ_0= 1.15°, β= 20°	

(G_{12}: In-plane Shear Modulus, V_f: Fibre Volume Fraction, r_f: Fibre Radius (Table 1), E_f: Fibre Young Modulus (See Table 1), λ_0: Half-wavelength (Berbinau et al [6,12]), ϕ_0: Initial Fibre Misalignment (Measured in the current study), τ_y: Shear Yield Stress (Mackinley [5]), γ_y: Shear Yield Strain (Mackinley [5]) and β: Kinkband Inclination Angle (See Section 3.1.2))

A vertical line, which passed through the point of intersection of the tangents was then plotted. The yield point was taken to be where this line intersected with the response curve. The shear yield stress (τ_y) and strain (γ_y) of the T800/924C material [13] were very similar to those of the T300/924C material when measured in accordance with the method used by Mackinley [5].

The analytical models of compressive failure require the knowledge of some 'maximum' fibre misalignment in the composite. However, none give an experimental definition of this quantity. In fact, the extent of fibre waviness in the laminates may vary from 0° to 5°, even bigger as shown in Figure 2. This maximum fibre waviness will trigger fibre microbuckling at a certain load, then the microbuckling will be propagated quickly with increasing applied load, depending on the degree of neighbouring fibre waviness or defects and finally catastrophic failure will occur at an ultimate load. Therefore average fibre misalignments have to be used for the analytical models rather than a maximum single fibre misalignment. There are two ways to measure the fibre waviness i.e., graphical method by Bardorf and Ko [9] and statistical method by Yurgartis [3].

The Yurgartis method requires a number of major axis lengths of sectioned fibres for reliable statistical results. The results can be easily distorted by human error even though morphological features can be identified and appropriate measurement made within a reasonable time using computer aided image analysis techniques upon digitized images. In addition, since the fibres are not uniformly misaligned but rather in a statistical manner the definition of a typical fibre misalignment angle is another main problem. In general, two methods to define a typical angle from the distribution of fibre angles have been suggested [16]: the standard deviation of the fibre angle distribution and a 'linearised' standard deviation, i.e. the weighted mean of the absolute fibre angle distribution.

The fibre waviness results measured by Yurgartis [3] method will be used only to investigate the sensitivity of the fibre misalignment related to increasing laminate thickness and the standard deviation of the fibre angle distribution will be adapted as a value of a typical fibre misalignment angle.

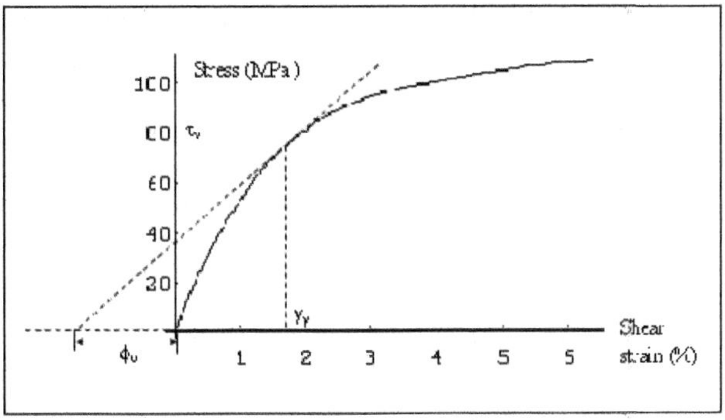

Fig. 11. Typical shear stress-strain curve [12] for the T800/924C composite and graphical determination of initial fibre misalignment, ϕ

In the present section, the average fibre misalignment was determined graphically by using a method similar to that described by Bardorf and Ko [9]. Effectively, a tangent passing through the yield point on the shear response curve measured from both materials used in the current study was drawn as shown in Figure 11. The value of the strain at the point of intersection of the tangent and the shear strain axis was taken to be the initial fibre misalignment. Using this method it was found that the average fibre misalignments for both materials were equal to 1.15°.

Figure 12 presents the predicted unidirectional compressive strengths according to the fibre microbuckling models and kink band models. The analytical results are compared with the measured experimental results in the figure. As explained in the above sections, Rosen and Argon's models significantly overestimate the actual compressive strengths because of the unrealistic assumptions in their

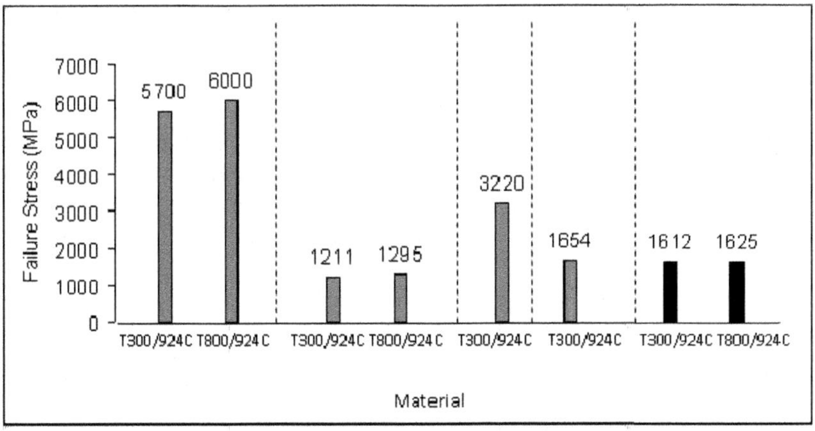

Fig. 12. Comparison of predicted and measured compressive failure strengths of the unidirectional T300/924C and T800/924C laminate

model. The model suggested by Berbinau *et al.* yields a conservative prediction. It should be noted that the predicted strength is strictly the critical stress at which fibre microbuckling initiates and not final failure stress. Finally Budiansky's kink band model predicts the strengths within 3% when compared with the measured experimental values. The model gives the most accurate values among the above models due to the assumptions similar to the stress-strain response of the modern composite epoxies.

4.2 Unnotched Multidirectional Laminates

The fibre microbuckling failures of unnotched multidirectional laminates in compression have received very limited attention and there is no theory for analysing microbuckling failure in these laminates. Researchers have used two general methods to predict failure strength of unnotched multidirectional laminates to date, i.e. the classical laminate plate theory[15] (maximum stress theory, maximum strain theory, Tsai-Hill theory, Hoffman theory, Tsai-Wu stress theory and Tsai-Wu strain theory) and the stiffness-ratio method. In the classical laminate plate theory, to estimate the failure strength of a multidirectional laminate, it is assumed that the strain in the composite is uniform, that the response is linear elastic to failure, and that the failure strain equals that of the unidirectional material.

In the present study, the maximum stress theory with the first ply failure approach and the stiffness-ratio method were used to predict the compressive failure strength of the multidirectional T300/924C ($[45/-45/0/90]_{3s}$) and T800/924C ($[45/0/-45/90]_{3s}$) laminates.

In the maximum stress criterion, stresses are resolved in the fibre directions for each ply. A ply is then considered to fail when the stress that it experiences reaches the associated strength. This is a very simplistic approach since it does not take into account any interaction between stresses or strains, compared with other failure criteria (Hoffman theory, Tsai-Wu stress theory and Tsai-Wu strain theory). In order to apply the criterion lamina ultimate stress is required.

Failure strength predictions using the stiffness-ratio method are governed by 0° ply failure strength. If failure strength of the unidirectional laminate is successfully predicted, the multidirectional failure strength could be calculated by integrating the strength of 0° laminate and the elastic stiffnesses of the 0° and ±θ° plies, i.e.

$$\sigma_{lam} = \frac{\sigma_0}{NE_{11}} \sum_{i=1}^{N} n_\theta^i E_{x\theta}^{(i)} \tag{12}$$

where σ_{lam} is the laminate strength, σ_0 is the strength of the 0° laminate, N is the total number of the laminae in the laminate, E_{11} is the 0-ply stiffness in the fibre direction, n is the number of plies of a given orientation θ and $E_{x\theta}$ is the modulus of a θ°-ply in the loading direction (x).

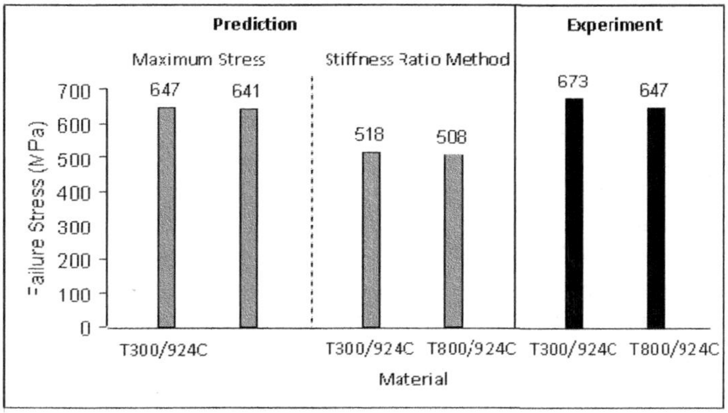

Fig. 13. Comparison of predicted and measured multidirectional compressive failure strengths of the T300/924C ($[45/-45/0/90]_{3s}$) and T800/924C ($[45/0/-45/90]_{3s}$) laminates

In order to calculate the failure strengths of the multidirectional laminates using the maximum stress criterion and the stiffness ratio method, the unidirectional failure strength (1654 MPa) successfully estimated by the Budiansky's model was used. The moduli of θ°-plies in the loading direction (x) were calculated by the laminate plate theory [15].

Figure 13 shows the comparison of the measured experimental results and the estimated results of the multidirectional laminates. The analytical strengths obtained by the maximum stress criterion are in good agreement with the experimental results. On the other hand, the failure strengths predicted by the stiffness ratio model are about 23 % lower than the experimental values, indicating that the model gives a conservative strength value.

5 Conclusion

An extensive experimental investigation from the composite microstructures to the composite laminate tests was carried out to identify uniaxial compressive behaviour. In addition, the strengths of the unidirectional and multidirectional laminates fabricated from the T300/924C and T800/924C material were calculated analytically using the commonly used models.

In the examination of the composite microstructure of both T300/924C and T800/924C material, the fibres were non-uniformly distributed and the cross sectional fibre shapes were of different types, i.e. circular, elliptical and 'kidney' shape. These irregularities in fibre shapes and distribution can be notified as a factor causing a considerable spread in compressive strengths about a mean value. Fibre volume, void contents and fibre waviness for the unidirectional laminates of both materials were measured using the resin acid digestion and Yagurtis technique, respectively. The T800/924C laminates had slightly higher fibre volume fraction (61%) and lower void volume fraction (1.6%) than those of T300/924C laminates (56% for fibre volume fraction and 2.0% for void volume fraction). The extent of the fibre waviness for both laminates was quite close. The standard deviation of the fibre waviness was 0.87° and 0.90° for the T300/924C and T800/924C laminates respectively.

From the uniaxial compressive strength test, the unidirectional failure strengths (1612 MPa for T300/924C and 1625 MPa for T800/924C) between both materials were almost equal even though the measured microstructure properties were different and the fibre tensile strength of T800 is much higher that that of T300 (see Table 1), while the unidirectional tensile strength of the T800/924C laminate was higher than that of T300/924C laminate (see Table 2), indicating that the tensile strength depends very much on the fibre tensile strength. From this experimental result, it can be inferred that the compressive failure strength is less sensitive to the microstructural properties (such as fibre volume fraction, void contents and fibre strength) than the tensile strength and is dominated by the fibre diameter, the fibre waviness and the matrix (resin) in the laminate. The tensile strength, however, seems to be proportional to the microstructural properties' values, especially to the fibre strength when compared to the laminate strengths of two different carbon fibres reinforced in the same matrix (resin). The unnotched (673 MPa for T300/924C and 647 MPa for T800/924C) multidirectional compressive failure strengths for both T300/924C and T800/924C laminate were also almost equal to each other.

The unnotched unidirectional specimens failed near the grip presumably because of a stress concentration there; the test piece was partitioned into two pieces with fracture surfaces inclined at typical angles of between 10° and 30° in either the width or thickness direction for both materials. Scanning electron micrographs (SEM) showed that failure is by microbuckling of the fibre forming a kink band. The characteristic kink band angle, β measured in SEM was similar to the fracture surface angle of test piece fragments.

The unnotched multidirectional specimens also failed by microbuckling of the 0° fibres. The failure strains for both T300/924C and T800/924C laminate are larger than those obtained in the unidirectional laminates indicating that the off-axis plies provide lateral support to the 0° axial plies and delay the initiation of fibre microbuckling.

Finally, the strengths for both T300/924C and T800/924C laminates were predicted analytically for unnotched unidirectional and muntidirectional specimens. The unidirectional compressive strength was successfully predicted by Budiansky [14] model and a conservative prediction was made by the model suggested by Berbinau *et al.* [13]. For the unnotched multidirectional laminates compressive failure strengths, the maximum stress criterion and stiffness ratio method was used with the unidirectional laminate compressive strength estimated by the Budiansky's model. The analytical strength from the maximum stress criterion is in good agreement with the experimental results within the difference of about 5%.

References

[1] Soutis, C.: Compression Testing of Pultruded Carbon Fibre-Epoxy Cylindrical rods. Journal of Materials Science 34, 3441–3446 (2000)

[2] Wisnom, M.R.: The Effect of Fibre Misalignment on the Compressive Strength of Unidirectional Carbon Fibre/Epoxy. Composites 21, 403–407 (1990)

[3] Yugartis, S.W.: Measurement of Small Angle Fibre Misalignments in Continuous Fibre Composites. Composite Science and Technology 30(4), 279–293 (1987)
[4] Smith, F.C.: The Effect of Constituents Properties on the Mechanical Performance of Fibre-Reinforced Plastics, PhD Thesis, University of London (May 1998)
[5] Mackinley, C.P.: Compressive failure of CFRP laminates containing pin-loaded holes, PhD thesis, Imperial College of Science Technology and Medicine, UK (2000)
[6] Berbinau, P., Soutis, C., Goutas, P., Curtis, P.T.: Effect of Off-axis Ply Orientation on 0° Fibre Microbuckling. Composite: Part A~30(10), 1197--1207 (1999)
[7] Camponeschi Jr., E.T.: Compression of Composite Materials: A Review. In: O'Brien, T.K. (ed.) Composite Materials: Fatigue and Fracture (3rd Volume), American Society for Testing and Materials, Philadelphia. ASTM STP, vol. 1110, pp. 550–578 (1991)
[8] Schultheisz, C.R., Wass, A.M.: Compressive Failure of Composites, Part I: Testing and Micromechanical Theories. Prog. Aerospace Sci. 32, 1–42 (1996)
[9] Soutis, C.: Compressive Behaviour of Composites. Rapra Review Reports, Report 94 8(10) (1997)
[10] Creighton, C.J.: The Role of Fibre Alignment in The Axial Compressive Failure of Carbon-Fibre/Polymer Composite, PhD thesis, University of Cambridge, UK (2000)
[11] Kulkarni, S.V., Rice, J.S., Rosen, B.W.: An Investigation of the Compressive Strength of Kevlar 49/Epoxy Composites. Composites, 217–225 (September 1975)
[12] Berbinau, P., Soutis, C., Guz, I.A.: Compressive Failure of 0° Fibre Microbuckling. Composite: Part A~30(10), 1197--1207 (1999)
[13] Budiansky, B.: Micromechanics. Computers and Structures 16(1-4), 3–12 (1983)
[14] Puppo, A.H., Evensen, H.A.: Interlaminar Shear in Laminated Composites under Generalised Plane Stress. Journal of Composite Materials 4, 204 (1970)
[15] Datoo, M.H.: Mechanics of fibrous Composites. Elsevier Science Publishers Ltd., Amsterdam (1991)
[16] Haberle, D.C.: Strength and Failure Mechanisms of Unidirectional Carbon Fibre-Reinforced Plastics Under Axial Compression, PhD Thesis, University of London (December 1991)

Fabrication of Three-Dimensional Magnetic Microcomponents

Jung-Sik Kim[1,*] and Miha Zakotnik[2]

Abstract. This paper presents a new fabrication process for producing magnetic microcomponents. The process is based on micro powder injection moulding technology, while the micro moulds are produced using microelectromechanical systems (MEMS) technology. The process involves (1) fabrication of polydimethylsiloxane (PDMS) micro moulds from SU-8 masters, which are produced using UV photolithography process; (2) mould filling with magnetic powder and demoulding; and (3) sintering of moulded components in vacuum. The proposed process has been used to sinter Nd-Fe-B microcomponents successfully without using binder system. This research proposes a new approach to fabricate 3-D magnetic microcomponents to meet the needs in applications where magnetic field required.

1 Introduction

Permanent magnets experienced a rapid development throughout the 20th century, with the maximum energy product rising from < 10 kJm^{-3} to > 400 kJm^{-3}. Advances in processing and the discovery of new materials have led this development. The most recent, and highest maximum energy product, permanent magnet material was discovered in 1984[1-2] and is based on a ternary alloy: NdFeB. A higher maximum energy product means that a smaller permanent magnet can be used to perform a certain function, such as generating torque in a motor. Thus the introduction of NdFeB based permanent magnets has allowed, for example, the miniaturisation of hard disk drives where permanent magnet motors are used to spin the disk and scan the head across the surface of the disk.

Conventionally sintered NdFeB permanent magnets are produced by powder metallurgy. The alloy is exposed to hydrogen; causing hydrogen decrepitation (HD) of the alloy and the formation of a powder with ~100mm particle size, which is further milled to ~5mm particle size. The particles are single crystals, which have a preferred direction of magnetisation, that can be aligned in a magnetic field and pressed to give a green compact that is sintered under vacuum at ~1080°C for 1 hour. The alignment process is essential for producing the optimum

[1] Department of Aeronautical & Automotive Eng., Loughborough University, UK. LE11 3TU, Telephone: +44 (0)1509 227219
e-mail: j.kim@lboro.ac.uk

[2] Metallurgy and Materials, University of Birmingham, UK. B15 2TT

* Corresponding author.

J.H. Lee et al. (Eds.): EU-Korea Conference on Science and Technology, SPPHY 135, pp. 131–139.
springerlink.com

magnetic properties; the maximum energy product of a fully aligned magnet can be up to four times greater than an isotropic, unaligned magnet.

There have been many reports (a recent review can be found in reference 3) of potential applications of permanent magnets in MicroElectroMechanical Systems (MEMS), sometimes referred to as MAGnetic Micro-Actuators and Systems (MAGMAS). These applications include micro-motors, micro-switches and micro-generators. Almost all of these reports use magnets produced by wire electro-discharge machining from large blocks of NdFeB or SmCo magnets. This is not suitable for producing a large number of components so alternate routes are being sought, including screen printing[4], sputtering[5] and pulse laser deposition.[6] As far as the authors are aware the work presented here is the only work to attempt the sintering of near net shape microcomponents from NdFeB powder.

2 Result and Characterisation

Figure 1 shows a sintered magnetic microgear. It can be seen that a relatively good shape retention of the microgear has been achieved after sintering. Figure 2 shows an SEM image of the sintered microstructure.

Magnetic characterisation of the HD sintered microcomponents was carried out using vibration sample magnetometer (VSM) which is shown in figure 3.

The second quadrant curve used to describe the quality of the magnetic sintered microcomponents have a few important parameters: remanence Br (T), intrinsic coercivity iHc (kA/m), inductive coercivity bHc (kA/m), and maximum energy product BHmax (kJ/m^3). Squareness factor (SF) was use to determinate the

Fig. 1. Sintered microgear

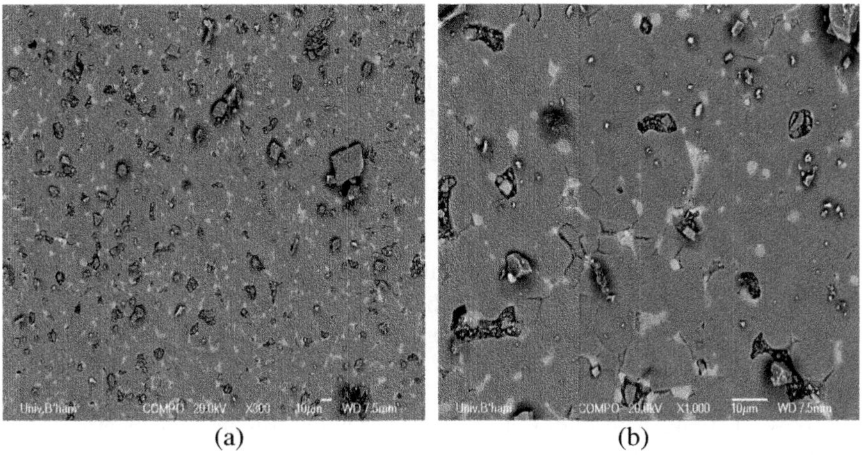

Fig. 2. The microstructure of the sintered component at lower mag.(a) and higher mag.(b)

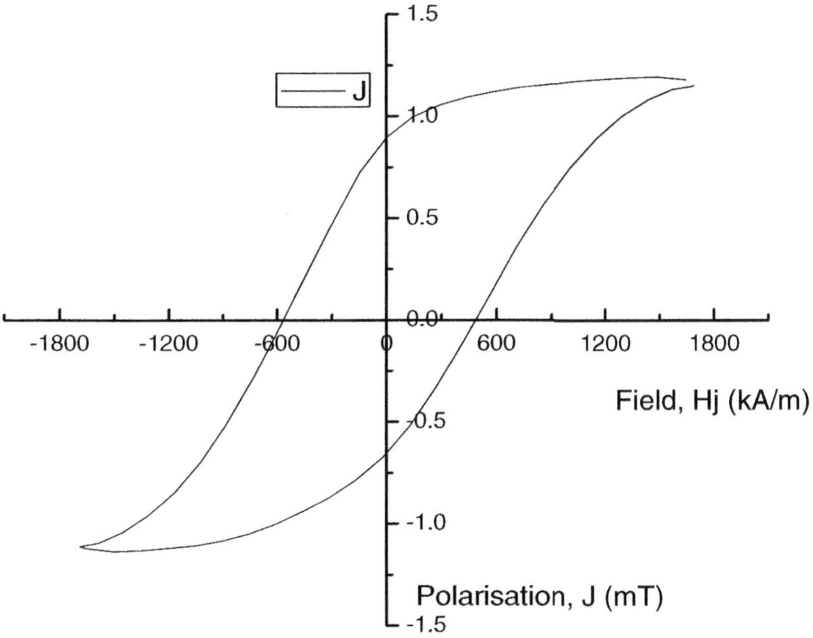

Fig. 3. Typical hysteresis loop of sintered microcomponents, determined by VSM.

stability of the magnets while undergoing demagentisation and defined as the applied field at 90% of the remanence Hs divided by the intrinsic coercivity. Measurements were performed after saturation in a pulsed field of 4.5T. A summery of magnetic properties is given in table 1. The density of the sample was calculated using the Archimede's principle where by a body submerged in a fluid experiences a buoyant force equal to the weight of fluid displaced.

Table 1. Summarised magnetic properties of sintered microcomponents

Milling Time (h)	Br (±10) (mT)	Hjc(±10) (kA/m)	bHc(±10) (kA/m)	BHmax (±10) (kJ/m^3)	Hs (kA/m)	SF=Hs/iHc (ratio)
22	892	564	439	140	66	0.12

Conclusions

A new fabrication process is presented for producing magnetic microcomponents from Nd-Fe-B powder via sintering. The technology is based on μPIM but extended with the help of MEMS technology. The process involves (1) fabrication of micro master moulds in SU-8 using UV lithographical process and PDMS negative moulds are made from the SU-8 masters; (2) filling the moulds with the magnetic powder along with compaction using isostatic pressure and peeling off the PDMS moulds when the patterns are consolidated; (3) sintering patterned components under vacuum. Nd-Fe-B microcomponents have been successfully produced following this process.

The use of soft PDMS moulds makes the demoulding easier in avoiding the damage of the green patterns by the moulds. The investigation has also achieved progress in identifying the best mould filling method. After repeated experiments, using isostatic pressure method stands out. It basically avoids the forming of air bubbles and increases the density and mechanical strength of the components. The most advantage of using isostatic pressure for shape patterning is by-passing using of binder system. Applying isostatic pressure for powder compaction is eliminating the worries regarding chemical reaction between powder particles and binder during sintering and requiring no additional debinding process. Also, it decreases shrinkage ratio during sintering process. Although isostatic pressure method was most secure method among introduced mould filling methods, it caused a great shrinkage to both elastic moulds and green patterns during powder compaction. This shrinkage ratio will be calculated and enlarged moulds will be designed to fabricate net-shape magnetic microcomponents.

The HD powder was employed successfully produce fully dense, sintered magnetic micro components and the magnetic properties were determined by means of a vibrating sample magnetometer. Square loop was obtained and it was found that the remanence value was significantly reduced to those of the original magnets, probably due to poor alignment of the powder. The powder flowability is an important factor affecting the alignment degree, remanence and energy product. The main factor affecting powder flowability was mould filling and magnetic aggregation. The iHc value was significantly smaller than that of the expected NdFeB magnet (~ 1272kA/m) and this can be associated with the oxide surface of the micro components. These results have been discussed in terms of a possible method of producing magnetic microcomponents using an HD-type process through the use of MEMS technologies. The proposed process shows a new way of fabricating three-dimensional magnetic microcomponents.

Experimental Procedure

Mould Fabrication

The fabrication process to produce magnetic microcomponents using Nd-Fe-B alloy powder starts from making master moulds. The master mould is relatively rigid, of high precision and usually fabricated using MEMS technology, such as deep reactive ion etching (DRIE) and X-ray exposure on SU-8 photoresist. SU-8 UV lithography has gained much progress in recent years. 1 mm thick micropatterned master mould was fabricated through UTSP. More details can be found in references.[7-8] Figure 4 show images of 1mm thick SU-8 microgears.

Fig. 4. SU-8 master moulds(top) and soft PDMS moulds(bottom) replicated from Su-8 master

The gear is 1 mm in height and 2.9 mm in diameter, with two through holes in the middle. The SU-8 master moulds have very smooth surfaces that can be replicated to the negative moulds.[9] The microgear was used as the master mould and a negative soft mould was produced from it. A widely adopted soft moulding technique is using elastomer PDMS to pattern the micrometer and sub-micrometer sized structures. Further details about the PDMS moulds can be found in literature.[8, 10-14]

Powder Preparation

The starting material was, in the form of sintered magnets with commercial composition given in table 2. Their composition was obtained by inductive coupled plasma (ICP). Magnets were placed into stainless steel hydrogenation vessel, which was then evacuated using rotary pump to a pressure of 1×10^{-1} bar. The hydrogen decrepitation (HD) process was then carried out by admitting hydrogen at

Table 2. ICP measurements of the typical composition of the magnets. (±0.1at%)

Nd at%	Pr at%	Dy at%	B at%	Nb at%	Al at%	Fe at%
13.98	0.06	0.63	6.44	0.36	0.69	77.81

Table 3. Particle sizes and sources of the powders used

Powder	Mean size	Shape	Source
Nd-Fe-B	4.5 micrometer	irregular	Philips, UK

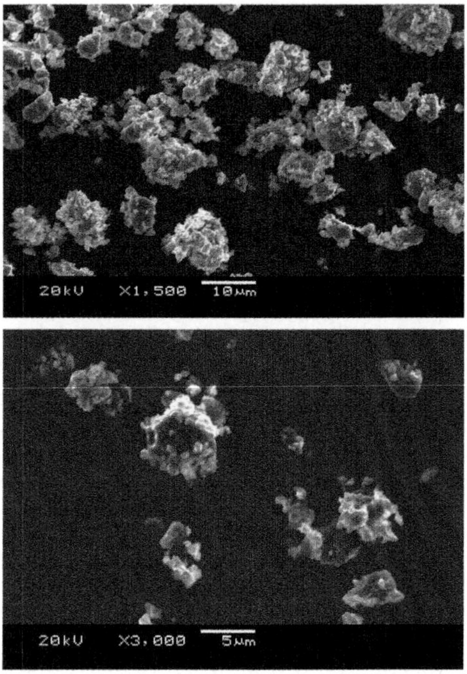

Fig. 5. SEM images of Nd-Fe-B magnetic powder (top) ×1500 (bottom) ×3000.

25°C to a pressure of 2 bar, resulting in decrepitation of the sintered material. Decrepitated material was then transferred to a roller ball mill under argon atmosphere and milled in a cyclohexane medium for 22 hrs. The resultant fine powder was dried for 30 min in vacuum [15-17].

The magnetic powder used in the experiments is micron sized Nd-Fe-B powder, supplied by Philips, UK. Table 3 shows the particle sizes, shapes and sources of the powder. Figure 5(a) and (b) show the Scanning Electron Microscope (SEM) images of milled powder at different magnifications.

Table 4. Mould filling methods used in this experiments

Name of method	Process
A*. Pouring method	Pour powder slurry on PDMS mould and dry, peel off mould
B**. Immersing method	Sink PDMS mould into the thin slurry and dry, peel off mould
C**. Magnetic field method	Repeat method B on a bulk magnet
D. Mould loosing method	Tamp powder into SU-8 mould and sinter together with mould
E. Isostatic pressure method	Fill up rubber tube with powder and PDMS mould, close tube, and compacted using isostatic pressure, peel off mould

* Magnetic powder-15wt%adhesive+3ml of acetone
** Magnetic powder-15wt%adhesive+30ml of acetone

Micro Mould Filling Methods

The magnetic powders are then prepared in two types, i.e. with and without binder, to fill up the micro moulds. Table 4 shows the different types of mould filling methods with simple explanation.

For methods A ~ C, the powders and binder were mixed with acetone to form slurry. The mixing process was carried out for 2 hours using a magnetic stirrer(Hanna Instruments HI180H/D). When the adhesive binder is mixed with the powders in acetone, it is diluted and its adhesion is lost temporarily. When the mixture is dry, the adhesive binder regains its bonding characteristics so the moulded components can maintain their patterns. No additional debinding process was required, as the binder evaporates in-situ during the sintering process in the furnace.

Then the cavity of patterned micro mould was filled up with the powder. The following methods were experimented in the mould filling process: A, pouring the powder premix on to the moulds under gravity; B, immersing PDMS mould into the mixture; C, process method B on a bulk of magnet, so the powder will compacted into the patterned mould; D, fill up dry powder into SU-8 mould and tamp down gently; E, fill up rubber tube with powder and place mould on top of them and compact tube under isostatic pressure. Extensive experiments have been carried out to fill the micro moulds using the five methods and the results are shown in Fig. 6. It is found that methods A and B tend to leave some holes unfilled in the moulds and cracked during sintering, referring to Figs 6a and 6b. Cracks are avoidable when green structures have thick substrate like E. Method C results lined up powder to vertical direction all over rather then filling up the mould. Handling magnetic powder within magnetic field was not handy. Also, produced green structure has rough surface profile. Method D is unlikely using SU-8 mould rather than soft PDMS mould. In this method, when powder is compacted into the mould, both mould and powder are undergone sintering process. This process consists of two steps. The first step is pyrolysis process which burn SU-8 mould out at the temperature below 480°C followed by the second step which is sintering of the remained powder. As the result of this process is shown in the Fig. 6c the SU-8 mould was not only completely burned out, also caused great distortion, deformation, and cracks on green structures. More details regarding SU-8 pyrolysis process can be found in references.[13, 18] Fig. 6d shows the micropillars formed using isostatic pressure method, i.e. method E. The results are obviously better than the other four moulding methods. In isostatic pressure method, the powder and soft PDMS mould are pressed isostatically at a pressure of 60 MPa for 10 min. Then the PDMS mould was carefully removed to achieve the powder components with a good shape retention and low porosity level, as shown in Fig. 6d. Since the fabricated green structure retains its patterned shape without any assistant of binder, less shrinkage ratio compare with others which using binder system is expected after sintering.

Demoulding and Sintering

Once the powder and mould are placed into the rubber tube, the fine powder was then aligned by pulsing two times in a 4.5T magnetic field and compacted. The

Fig. 6. Sintered components prepared from various mould filling methods: (a)~(d) correspond to filling methods A,B,C, and E respectively

green component was achieved by peeling off the soft PDMS mould. Next, the patterned component was placed inside a furnace under vacuum and heated to 1080°C, 10°C/min. The sample was held for 1 hour at designated temperature and taken out after the furnace was cooled down to room temperature.

References

[1] Sagawa, M., Fujimura, S., Togawa, M., Yamamoto, H., Matsuura, Y.: J. Appl. Phys. 55, 2083–2087 (1984)
[2] Croat, J.J., Herbst, J.F., Lee, R.W., Pinkerton, F.E.: J. Appl. Phys. 55, 2078–2082 (1984)
[3] Cugat, O., Reyne, G., Delamare, J., Rostaing, H.: Sensors and Actuators A 129, 265–269 (2006)
[4] Yuan, Z.C., Williams, A.J., Shields, T.C., Blackburn, S., Ponton, C.B., Abell, J.S., Harris, I.R.: Jouranal of Magnetism and Magnetic Materials 247(3), 257–269 (2002)
[5] Kapitanov, B.A., Kornilov, N.V., Linetsky, Y.L., Tsvetkov, V.Y.: J. Magn. Magn. Mater. 127, 289 (1993)
[6] Nakano, M., Katoh, R., Fukunaga, H., Tutumi, S., Yamashita, F.: IEEE Transactions on Magnetics 39(5), 2863 (2003)
[7] Jin, P., Jiang, K.C., Sun, N.: Journal of Microlithography, Microfabrication, and Microsystems. 3(4), 569–573 (2004)
[8] Kim, J.-S., Jiang, K., Chang, I.: J. Micromech. Microeng. 16, 48 (2006)
[9] Lee, C.H., Jiang, K., Davies, G.J.: Proceedings of Lamdamap 2005 International Conference, Cranfield, UK, pp. 402–411 (2005)
[10] Biebuyck, H.A., Larsen, N.B., Delamarche, E., Michel, B.: IBM J. Res. & Dev. 41, 159 (1997)
[11] Kim, J.-S., Jiang, K., Chang, I.: Advanced Engineering Materials 8(1-2), 38–41 (2006)
[12] Kim, J.-S., Jiang, K.C., Falticeanu, L., Davies, G.J., Chang, I.: Materials Science Forum. 534-36(2), 1041–1044 (2007)
[13] Kim, J.-S., Chang, I.T., Falticeanu, C.L., Davies, G.J., Jiang, K.C.: Materials Science Forum 534-36(1), 769–772 (2007)
[14] Kim, J.-S., Jiang, K., Chang, I.: Advanced Engineering Materials 11(1-2), 106 (2009)
[15] Zakotnik, M., Harris, I.R., Williams, A.J.: Journal of Alloys and Compounds 450 (1-2), 525–531 (2008)
[16] Zakotnik, M., Devlin, E., Harris, I.R., Williams, A.J.: Journal of Iron and Steel Research, International 13(suppl.1), 289–295 (2006)
[17] Zakotnik, M., Harris, I.R., Williams, A.J.: Journal of Alloys and Compounds 469 (1-2), 314–321 (2009)
[18] Kim, J.-S.: Design and Fabrication of a MEMS Reciprocating Engine Using High Temperature Resistant Materials. PhD thesis, Mechanical Engineering Dept., The University of Birmingham (2006)

A Multi-Agent Emotion Generating System for Mobile Robots System

Shivashankar B. Nair[1,*] and Dong Hwa Kim[2]

Abstract. Emotions are a complex phenomenon and are in constant research for so long. Emotions can be presupposed as a dynamic association of cells on elicitation. The use of agents for emotions generation and emotion modeling are on the rise in the recent years. In this work, we propose a dynamic emotion generation system based on multiple agents. A Multi-Agent System (MAS) to generate emotions targeted towards a mobile robot has been implemented. The agent-robot communication is still being investigated. An illustration of the use of this emotion generation module is explained in the context of the emotion system. Some ancillary investigations performed in using mobile robots for localization and terrain mapping have also been described.

1 Introduction

Emotions form a complex phenomenon exhibited by living beings and seem to be generated as an attempt at self-preservation. They are partly based on the being's past experiences. Emotions have always been hard to understand as they are dynamic processes with a degree of uncertainty attached to them [3]. Though many theories have been propounded on these complex emotional processes every since Plato [15], a concrete understanding still seems elusive. Computational models of emotion provide an important tool for understanding the complex mechanisms involved in emotion generation. These tools play a central role in the applications developed for Human-Computer Interaction.

In the present age of automation, machines take the roles of humans in many areas. Replacing humans with machines increases the degree of interaction that these machines have with humans. In order for the communication to be more

[1] Department of Computer Science & Engineering, Indian Institute of Technology Guwahati, Guwahati, India
{w.godfrey,sbnair}@iitg.ernet.in

[2] Department of Instrumentation and Control Engineerong, Hanbat National University, 16-1 San Duckmyong-Dong Yuseong-Gu, Daejeon City, South Korea, 305-719,
kimdh@hanbat.ac.kr

* Corresponding author.

J.H. Lee et al. (Eds.): EU-Korea Conference on Science and Technology, SPPHY 135, pp. 141–153.
springerlink.com

meaningful and interactive, the machines should not just provide mere answers to queries but respond with an understanding of the emotions of the human being interacting with it. Failing to do so makes it feel more like a passive entity. Replacement of humans by machines can only happen when the machines are able to capture the feelings of human and adjust themselves accordingly. This improves the *feel-good* factor of the human who communicates with the machine.

The communication between a human being and a machine may involve four basic modules viz. –

1) Perception of human emotion,
2) Conversion of Perceived emotion into Machine understandable terms,
3) Generation of New emotion in response to the understood input and
4) Expression of emotion in response to the perceived emotion in human intelligible form.

In this work, we propose a method for the conversion of the perceived emotion into an understandable form and also propose a new algorithm for generation of response emotions. The conversion of perceived emotion is based on the FACS coding method for facial images [6] and the algorithm for dynamic emotion generation is based on the concept of mobile agents [19]. Since it may not be possible for us to implement the complete system, considering the magnanimity of the problem, we decided to test the emotion generating system on a mobile robot based on its cognition of the world around it. We bear in mind yet that once the FACs to intensity converter is in place, the same emotion generating system can be coupled to it to yield similar results.

2 Emotions and Facial Expressions

Cognition and emotion are intertwined. Through processes involved with cognitive appraisal, cognition plays an important role in emotion generation. Both cognition and emotion are related to behavior [20]. Together they form the first level of motivation for an entity in behaving intelligently. Understanding the mechanisms of these interactions give a foundational grasp in building computational tools for artificial emotion generation.

3 Theories of Emotion

There are several theories of emotion based on what causes it. These theories [15] are broadly classified into two types - Non-Cognitive and Cognitive. Non-cognitive theories of emotion maintain that Emotion and cognition are independent. According to cognitive theory, cognition plays an integral role in emotions, the decision for emotion comes after interpretation of an event that has happened [22]. The cause of the event and the positive and negative impact of that event form the basis for feelings or emotion. The main theories [13] under the former kind include the James-Lange theory and Facial Feedback theory while the

Canon-Bard theory and Schachter-Singer experiment fall under the latter type. According to the James-Lange theory, the brain interprets physiological responses to a situation, and only then do we use these responses to formulate an experience of emotion. In the Facial feedback theory the muscles of the face move to form an expression. The brain in turn interprets these muscle movements which are felt as an emotion and this culminates into the demonstration of an observable behavior.

Basically, all theories on the complex phenomenon like emotion attempt to explain the mechanism, meaning, impact, origin, evolution and function of emotion. These theories play a important role in emotion modeling and emotion recognition since they help in modularizing and interpreting the complex emotion process.

4 Computational Models of Emotion

Computational models of emotion are based on the theories or descriptions of emotion. They aim to formalize the emotion or describe emotions in modular terms. These models could be either Deep or Shallow [21]. Deep models consider the state, the factors constraining the interpretation of an emotion like memory, feelings etc. and predict the emotion of the individual. Shallow models on the other hand, consider only the external appearances or the physical cues of an individual to comprehend and predict the resultant emotion of the individual.

A significant amount of research has been carried out in computational models for generating emotions using Deep models. These models consider the state and the factors influencing emotion through numerous variables. One of the early models was proposed by Fumio Hara and Shusuke Mogi [1]. They propose a computational model for artificial emotion using a networked architecture, which they call harmony theory closely resembling neural network which takes as input some facial cues and produce emotion output. The input consists of the facial components, eye, eyebrow and mouth. The output comprises one of the 6 basic emotions viz joy, anger, sadness, fear, surprise and disgust. Duy Bui et al [2] describe a model for generating facial expressions using a fuzzy rule based system. They introduce the concept of fuzziness to generate life-like facial expressions on a 3-dimensional face. They define fuzzy rules mapping emotional representations onto contraction levels of facial muscles. The fuzzy system enables the expressions to constitute a blend of emotions in addition to the basic emotions. Arun Chandra [3] models the emotional states and the state transitions based on the Markov modeling theory. However for such a complex phenomenon like emotion, state transitions may still prove to be incomplete depending on the number of the states.

Shallow models of Emotion models are shallow and too simple to exhibit an intricate process like emotion that they compute emotion outputs in terms of simplistic input output relations [23]. Depending on the purpose of usage a decision can be made to use either shallow or deep models. Shallow models are more suited for modeling response emotions on service machines since they are supposed to respond to the prompting of what those machines perceive through their cameras or audio inputs.

5 Somatic Model for Emotion Generation

Computational models for emotion generation try to simplify the theoretical models propounded on emotion to create an emotion milieu. The somatic theory [13] claims that bodily responses rather than judgments are essential to emotions. This seems specifically suited to service robots. The justification for this could be twofold. Firstly in robots, the internal state remains indifferent and they need to respond to the visible cues of the human. Secondly from the first person perspective, the somatic theory gives a simpler and better understanding of an individual's emotions than other theories. The proposed emotion generation system described in this paper is based on Somatic theory of Emotions.

6 Dynamic Facial Expressions

Emotions elicit behavioral responses which communicate to others. Facial expressions are common and form a powerful mode of communication of emotion. Experimental studies [14] have shown an improved facilitation of processing of emotions while using dynamic presentation of facial expressions their static equivalents. The dynamics in emotion has thus been exploited in the area of facial expression recognition [15][16][17].Yongmian Zhang [18] discusses a dynamic model for facial expression generation using Dynamic Bayesian Networks. Though dynamic facial expressions provide for efficient means of communicating strong emotions and eliciting natural emotional responses, capturing of the temporal and intermediate state information of these expressions pose potential problems. The dynamic nature of the emotion responses pose the need for identifying the temporal information associated with the emotions to be integrated along with the computational models of emotion modeling.

7 Agents in Computational models for Emotion

Agents have been widely used in a vast gamut of application ranging from weaning and searching specific items in the World Wide Web, learning habits and generating profiles of users, understanding market trends, etc. [23]. Auctions have been another prominent area for agent interaction and decision making [24]. Agents have been used in emotion generation models as in Cathexis [8], a distributed computational model that features an approach to model dynamic affective phenomenon like moods and temperaments. Jan Allbeck and Norman Badler [25] present a new scheme called Parameterized Action Representation that can be used for action, planning, reasoning, behaviors and animation. Elliott and Ortony [26] use logical case based reasoning for their emotional model which is made up of two types of personalities called interpretive personality and manifestive personality. Jonathan Gratch [27] proposes a plan based approach for emotional modeling. STRIPS [28] planning representation is used in this system.

8 Role of Multi-Agent Paradigm in Emotion Generation

A dynamic emotion generation system derived on the principles from facial expression feedback theory [15] and based on the philosophy of multiple agents can be used to embed the much needed dynamism in artificial emoting. Some basic motivations that prompted the use of mobile agents include;

- Multiple agents can play the role of biological cells or even secretions that can elicit emotions.
- The abilities to function independently and in parallel make a distributed implementation simple and straightforward.
- Controlling the flow of inter-agent signals can aid in emulating the dynamics of the emotion.
- The decay of emotions can be emulated using signals – stimulation and suppression from each agent.

All these aspects provide a framework for the generation of affective and dynamic behaviors. In the following sections we describe the architecture and working of an emotion emulation engine that makes use of a multi-agent scenario to embed emotion dynamism.

9 Multi-Agent Dynamic Emotion Emulator

We discuss our efforts to model a Multi Agent based Dynamic Emotion Emulator (MADEE) to be used in conjunction with mobile robots. The fact that emotions can play a key role in optimizing parameters, which in the current scenario is tantamount to feeling happy or contented with the rate of progress towards a goal, forms the rationale behind using them in such robots. The emotion emulator comprises an agent for each emotion which functions to quite an extent like their biological counterparts, and a referee agent that forms the bandmaster.

10 Base Agents

The Base agents, referred to as B-agents, are static and form the main source of emotion related activations. They secrete exciters and suppressors of other positive and negative emotions and can be loosely treated as metaphors of their biological equivalents responsible for emotion generation such as the amygdala and the prefrontal cortex. Dynamism is provided by the secretions and their sub sequent activation or inhibition of other glands to secrete hormones. In this computational world these Base agents when stimulated tend to generate neurotransmitters that either stimulate or suppress other B-agent activity. In this paper we have tried to model the system using three emotions viz. BHp-Agent and BSd-Agent, where the subscripts indicate each of the three emotions – happiness, sadness and anger respectively. However, the system is scalable and one needs to only attach more such emotion agents to support a more complex emotion generating system.

Each of these B-agents houses a variable called Basic Emotion Intensity for that corresponding basic emotion. Periodically the current intensities of all these B-agents are given as output of the emotion emulator.

11 Suppression and Stimulation

When the emotional intensity of a B-agent crosses a threshold it gets excited and transmits messages that suppress or stimulate other agent activity. The messages on reaching other B-agent territories stimulate or inhibit them, creating a positive or negative effect on the current emotion of the B-agent. These messages form metaphors for biological neurotransmitters. They also have a life-time stamped on them by their creators at the time when they emanate, which is proportional to the change in the associated emotion intensity. This facilitates for an emotional recovery time, generally attributed to the prefrontal cortex. These messages thus provide the much necessary dynamism in stimulation and suppression.

The messages can be either stimulate or suppress the B-agent onto whose territory they are sent. This behavior is embedded a priori in the system. Table 1 depicts the B-agents, the emotion they represent and the type of messages they generate. For instance, the message from the BHp-Agent stimulate the BSr and also the BHp Agents. The clones of BHp-Agent suppress the BSd-Agent, BFr-Agent, BAn-Agent and the BDg-Agent on reaching their respective territories.

Emotion (B-agent)	Message Destinations & their Positive/Negative behaviours
Happiness (Hp)	Hp+,Sr+, Sd-, Fr-, An-
Sadness (Sd)	Sd+,An+, Fr+, Hp-, Sr-,
Angry (An)	Fr+, Dg+, An+, Sr-, Hp-, Nt-
Fear (Fr)	Dg+, Sd+, Fr+, Hp-, Nt-
Disgust (Dg)	Fr+, An+, Sd+, Dg+, Hp-, Nt-, Sr-
Surprise (Sr)	Hp+, Sr+, Sd-, Fr-, An-
Neutral (Nt)	Hp+, Sr+, Nt+, Sd-, Fr-, An-, Dg-

The decision as to which B-Agent to stimulate and which to suppress has been made by intuitively classifying the affection of other emotions on a basic emotion, as positive or negative. For e.g. if a predominant happy emotion is created, then it has to suppress (act negatively on) the emotions like sadness and fear. On the contrary, it has to stimulate (act positively on) the emotions like happiness and surprise. Table 1 has been formulated based on these aspects. The table depicts the messages generated by each B-Agent and their stimulating (positive) or suppressive (negative) behavior and also to which B-Agent they migrate.

The lifetimes of the messages that migrate and occupy other B-agent territories decide the amount of suppression or stimulation of the emotional intensity generated by the invaded B-agent. The intensity of the emotion exhibited by each B-agent is thus proportional to the lifetimes of stimulations and suppressions that have been sent by other B-Agents. It is also proportional to the input intensity to that B-agent among other factors discussed later.

12 Referee Agent

The referee agent plays the bandmaster more like the pituitary gland. It collects the emotional intensities of each B-agent and passes the information onto other B-agents to facilitate in further iterative dynamic emotional intensity calculations. It accumulates all the current emotional intensities to provide the output intensity vector [EISd, EIAn, EIFr, EISr, EIHp, EIDg, EINt] comprising the intensities of all the basic emotions. The Referee thus acts more like a go-in-between the B-agents, monitors and provides the concerned information and formulates the final output.

13 Emotion Dynamics

The emotion dynamics here refers to the dynamic change of emotions from one state to another by the emotion generation system instead of an instantaneous transition from current state to an output state. The equations that govern the intensity I of an emotion e are governed by:

14 Emotional Input

Emotional change due to a change in input emotional intensity. By input emotional intensity we mean the intensity that is perceived by the emoting system. While normally this would depend on the facial coding and intensity converting systems, in the mobile robot's world these could be caused by the information rendered by its respective sensors.

$$Ie(t+1) = Ie(t)\ OP\ \{(Iei(t) - Ie(t))*(1-we.\sum_j^N Stje(t) + we' \sum_j^N Suje(t)\}\ (Re\,(t)/Re\ max) \quad \text{.........(1)}$$

$j \neq e$, where e belongs to the set of all N emotions under consideration and OP is an operator such that –OP is + (plus) if $Iei(t) >= Ie(t)$); and – (minus) otherwise.

Iei(t) is the input intensity obtained by processing the on-board sensors, St and Su are the stimulations and suppressions from other emotions respectively, we and we' are weights and Re is the resource of the emotion and Re max is the maximum resource of emotion e.

15 Concept of Resource

The resource of the emotion could be compared to the water in a freshly soaked piece of cloth at time t. Assuming that A ml of water has been squeezed out of it at time t, the amount of effort required to be put in to squeeze the same amount of water again at time (t+1) will be greater than that in the previous attempt. The resource forms a metaphor of the emotion related secretion capability within a B-agent. Thus more input intensity is required to extract the same output emotion intensity as time progresses. Repeated increase in emotion inputs will not result in similar rise in the emotion generated. This does not mean that emotions will diminish with time. The emotion resource of the B-agent is periodically refreshed so that R attains its maximum value. Resource is calculated iteratively using the equation (2) below.

$$R(t+1) = R(t) - Ie(t+1) \qquad \text{.........(2)}$$

Resource corrections are also applied so that emotions do not out-squeeze the resource.

Stimulations and suppressions from other agents:

The stimulations and suppressions from other B-agents k have a life time and thus are added up with the older ones till they die out.

$$Stje(t+1) = \sum\_j\hat{}N\ StjeOld(t) + Stje(t) \qquad \text{.........(3)}$$

Individual stimulations die out exponentially as

$$St(t) = k1e\text{-}L \qquad \text{.........(4)}$$

where L is the lifetime and k1 is a constant that could be selected to suit the application.

Stimulations and suppressions for other agents:

The stimulations and suppressions to be sent to other B-agents are computed using the following equation.

$$\begin{aligned} te(t+1) = Sue(t+1) &= [\{Iei(t) - Ie(t)\}/Ie(t)]\{R(t)/Rmax\} \\ &\quad \text{if } Iei(t) > Ie(t) \\ &= 0 \text{ otherwise.} \qquad \text{.........(5)} \end{aligned}$$

Lifetime L of the stimulation or suppression is proportional to the change in current and input intensity and the current resource.

$$\begin{aligned} L &= k1.\{Iei(t) - Ie(t)\}/R(t) \text{ if } Iei(t) > Ie(t) \\ &= 0 \text{ otherwise.} \qquad \text{.........(6)} \end{aligned}$$

where k1 is a positive constant and could be taken as unity.

The Emotional Intensity I of a specific emotion is calculated locally by the associated B-Agent. Every B-Agent uses its associated component of intensity from the incoming intensity vector, its current intensity, the stimulations and suppressions of other B-Agents to iterate into the next I. Emotional states cannot last forever and have to die out over a period of time. Modelling their decay can rake up a hot debate. Nevertheless this modeling needs to be addressed and done. We have modelled emotion to decay based on what happened to it in the immediate past. We gather the state of what happened to the emotion by defining a simple term –

$$De(TN) = k3\{\sum\_(i=1)\hat{}N\ (Ie(i+1) - Iei)\}/(TN - T) \qquad \text{.........(7)}$$

where k3 is a constant.

As can be seen the numerator is the summation of the difference of all N intensities computed in a period of time (TN - T). It is proportional to the rises and falls the emotion has made in the given time period. N is the look-back or the depth to which we inspect the past record of the emotion.

De is used to model the decay the emotions as

$$Ie(t) = k4e\text{-}De(t) \quad \text{.........(8)}$$

16 Emotions and the Mobile Robot

The scenario used for testing our model comprised a mobile Lego robot. It has a bumper on the front and rear sides, a light sensor that can differentiate between green, red and other colours and an IR sensor that can detect moving obstacles. We assume that the robot gets angry when it encounters repeated bumps, is sad when it is in a red coloured area and happy when it is on a green coloured area. A conceptual sketch is shown in Fig. 1.

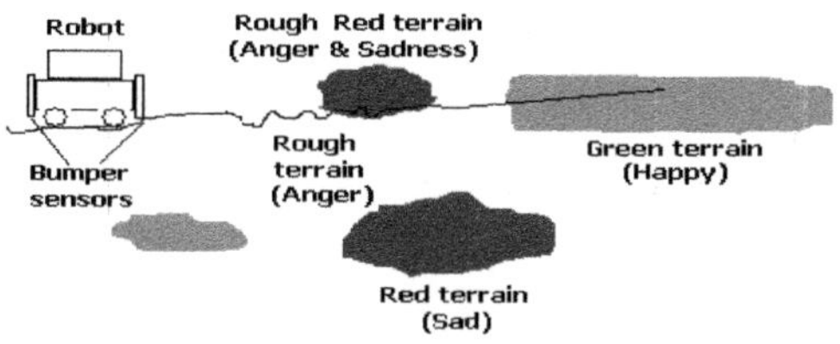

Fig. 1. Mobile robot terrain model

These signals are processed to yield corresponding emotion signals which constitute the individual components Iei(t), of the input emotion vector. It is this vector that is finally provided to the emotion generating system. Some elementary sensor signal processing mechanisms to yield the components of the emotion vector are shown in Fig. 2.

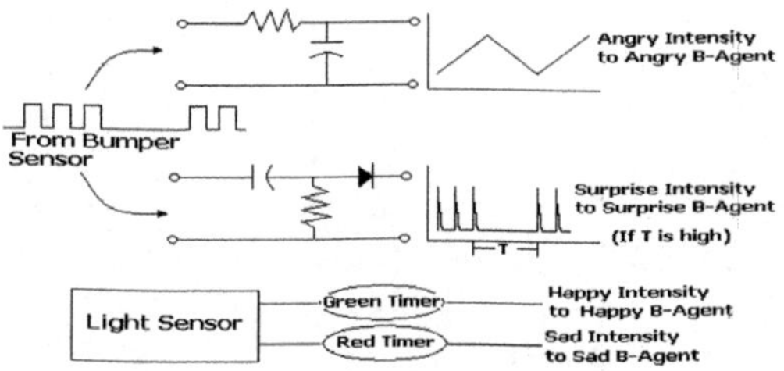

Fig. 2. Some processing mechanisms for generating inputs for the emotion generating system

As can be seen from Fig. 2 the on-off type bumper sensor will relay a series of pulses. These can be passed through an RC low pass filter and integrated to form the angry input intensity to the emotion generator. More specifically its values for the angry component of the input emotion vector. The same signal from the bumper sensor could be injected through a high pass filter, rectified to find the maximum width Tmax between pulses in a given period of time. If the value is greater than a pre-decided threshold it may be treated as a bump that occurred suddenly as a surprise. The light sensor in turn finds number of times the robot is on a green or red area and corresponding generates the happy or sad signal.

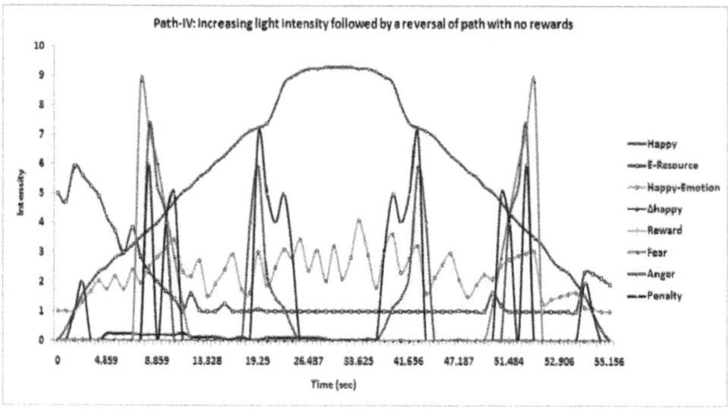

Fig. 3. Emotion recognition to situation by using emotion equation suggested.

17 Implementation and Conclusion

We have used a combination of technologies to embed emotions into a mobile robot. The emotion generation model was implemented using Prolog (Fig. 1-3). Prolog was used as it is known to be a versatile language for rapid prototyping.

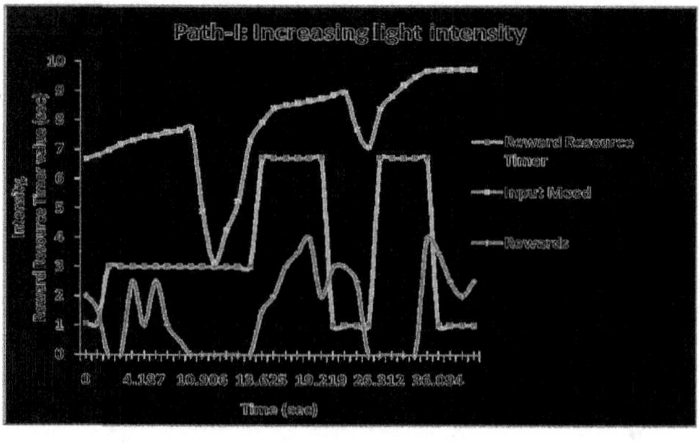

Fig. 4. Reaction to simple emotion situation by the suggested dyanamic equation.

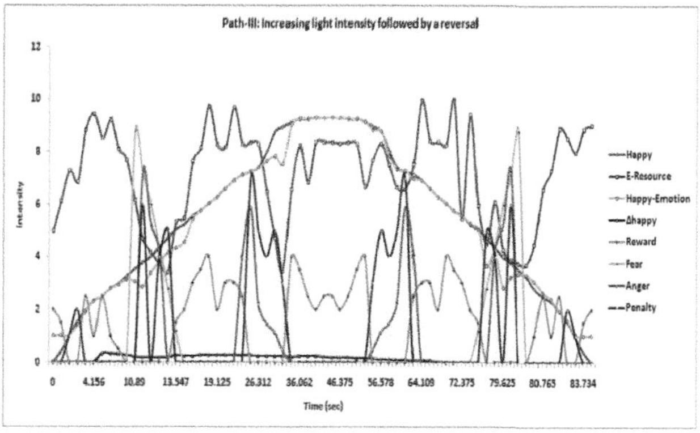

Fig. 5. Emotion recognition to different situation by using emotion equation suggested.

The LPA Prolog compiler used also supported the construction of multiple agents. We have used LPA Prolog's Chimera Agent System to realize three separate agents that for generating the emotions – Happiness, Sadness and Anger. The Chimera system uses WinSock for communication amongst the agents and provides a well regulated protocol for inter-agent communication. The agents can be scattered anywhere in a network. A Lego robot configured to be a mobile robot was used to test the system. We have used the Lego Java Operating System (LeJOS) on-board the robot to facilitate a real-time-like control of the robot.

Through this work, we can develop new idea and promising new idea and products.

References

[1] Hara, F., Mogi, S.: A computational model of artificial emotion by using harmony theory and genetic algorithm. In: IEEE International Workshop on Robot and Human Communication, pp. 414–419 (1993)

[2] Bui, T.D., Heylen, D., Poel, M., Nijholt, A.: Generation of facial expression from emotion using a fuzzy rule based system. In: Proceedings of the 14th Australian Joint Conference on Artificial Intelligence: Advances in Artificial Intelligence, pp. 83–94 (2001)

[3] Chandra, A.: A computational Architecture to model Human Emotions. In: Proceedings of the 1997 IASTED International Conference on Intelligent Information Systems, pp. 86–89 (1997)

[4] Poel, M., et al.: Learning Emotions in Virtual Environments Recognizing Action units for Facial Expression Analysis. IEEE Transactions on Pattern Analysis and Machine Intelligence 23(2) (February 2001)

[5] Ekman, P., Friesen, W.V.: The facial Action coding system:A technique for the measurement of Facial Movement. Consulting Psychologists Press, San Francisco (1978)

[6] Scherer, K., Ekman, P.: Handbook of methods in Nonverbal Behavior Research. Cambridge Univ. press, Cambridge (1982)
[7] Velasquez, J.D., Maes, P.: Cathexis: A computational Model of Emotion. In: Proceedings of the first international conference on Autonomous agents, pp. 518–519 (1997)
[8] Davis, D.N., Lewis, S.C.: Computational models of Emotion for autonomy and reasoning. Informatica (Special Edition on Perception and Emotion Based Models of Reasoning) 27(2), 159–165 (2003)
[9] Marinier, R.P., Laird, J.E.: Toward a comprehensive computational model of emotions and feelings. In: International Conference on Cognitive Modeling 2004 (2004)
[10] Dias, J., Paiva, A.: Feeling and Reasoning_A computational model for emotional characters. In: Bento, C., Cardoso, A., Dias, G. (eds.) EPIA 2005. LNCS (LNAI), vol. 3808, pp. 127–140. Springer, Heidelberg (2005)
[11] Marsella, S., Gratch, J.: EMA: a computational model for appraisal dynamics. In: European Meeting on Cybernetics and Systems Research (2006)
[12] Strongman, K.T.: The Psychology of Emotion: Theories of Emotion in Perspective, 4th edn. John Wiley & Sons, New York (1996)
[13] Sato, W., Yoshikawa, S.: Emotional elicitation by dynamic facial expressions. In: Proceedings of the 4th International Conference on Development and Learning, pp. 170–174 (2005)
[14] Essa, I.A., Pentland, A.P.: Facial expression recognition using a dynamic model and motion energy. In: Proceedings of the Fifth International Conference on Computer Vision, pp. 360–367 (1995)
[15] Shan, C.F., Gong, S.G., McOwan, P.W.: Dynamic facial expression recognition using a Bayesian temporal Manifold Model. In: Proceedings of British Machine Vision Conference, Edinburgh (September 2006)
[16] Kätsyri, J., Klucharev, V., Frydrych, M., Sams, M.: Identification of synthetic and natural emotional facial expressions. In: Proceedings of Audio-Visual Speech Processing Conference (2003)
[17] Zhang, Y., Ji, Q.: Active and Dynamic Information fusion for Facial Expression understanding from Image sequences. IEEE transactions on pattern analysis and machine intelligence 27(5) (May 2005)
[18] David, C., Harrison, C., Kirshenbaum, A.: Mobile agents: Are they a good idea? In: Second International Workshop on Mobile Object Systems - Towards the Programmable Internet (1996)
[19] Sloman, A.: Deep and Shallow models of Motivation and Emotion, School of Computer Science. The University of Birmingham, Birmingham
[20] Jain, L.C., Chen, Z., Ichalkaranje, N.: Intelligent agents and their applications. Physica-Verlag GmbH, Heidelberg (2002)
[21] Allbeck, J., Badler, N.: Toward representing agent behaviors modified by personality and emotion. In: Embodied Conversational Agents at AAMAS 2002, Bologna, Italy. ACM Press, New York (2002)
[22] Elliott, C., Ortony, A.: Point of View: Modeling the emotions of others. In: Proceedings of the 14th Annual Meeting of the Cognitive Science Society, pp. 809–814 (1992)
[23] Gratch, J.: Emile: Marshalling passions in training and education. In: Proceedings of the fourth International conference on Autonomous Agents, pp. 325–332. ACM Press, New York (2000)

[24] Harrison, C.G., et al.: Mobile Agents: Are they a good idea? Research Report, IBM T.J. Watson Research Center (1995)
[25] Bylander, T.: Complexity results for planning. In: Proceedings of the Twelfth International Joint Conference on Artificial Intelligence (IJCAI 1991), pp. 274–279 (1991)
[26] RPC: Remote Procedure Call Protocol Specification Version 2, Request for Comments: Network Working Group (1831)
[27] Kim, H.-R., et al.: Emotional interaction model for a service robot. In: Proceedings of International workshop on Robot and Human Interactive Communication, pp. 672–678 (2005)

Quality Prediction for a Fed-Batch Fermentation Process Using Multi-Block PLS

Jeong Jin Hong[1,*] and Jie Zhang[2]

Abstract. Quality prediction is usually required for product quality monitoring and setting up control strategy can reduce operating cost and improve production efficiency. Partial least square (PLS) regression is a popular statistical method for predictive modelling. The amount of data measured and stored in a typical industrial process is dramatically increased due to the fast development of computer and measuring system. It is hard to analyse all measured data using one matrix for its complexity. Multi-Block PLS model allows the data to be separated into sub-blocks and the sub-blocks can be analysed independently. Data from the fed-batch fermentation process is used to build models. Data is divided by different modes and different phases and model parameters are used to select variables that can be used as good predictors. The new set of data after variable selections is used to build a new model again. In most cases, new models show improved prediction performances compared with results from the conventional method.

1 Introduction

Analysing data measured from different process stages/phases or different types of measurements is difficult using the conventional statistical methods such as principal component analysis (PCA) and PLS where all data (or all predictor data) go into one matrix as it is hard to find out the relationships among the variables or stages within one big matrix. Multi-block methods enable the data to be separated into sub-blocks by appropriate knowledge of the systems. For example, different types of measured data, different process stages or phases. Sub-blocks can be analysed individually. Therefore, it can provide information about within sub-block as well as between sub-blocks. And it is useful to understand the process better. However, many literatures stated that there is no advantage of using the multi-block methods for predictive modelling only as prediction performances of PLS model and multi-block PLS (MBPLS) model are the same. Thus, multi-block methods are usually used for process fault detection and diagnosis.

[1] Newcastle University, School of Chemical Engineering and Advance Materials, Newcastle upon Tyne, United Kingdom
`j.j.hong@ncl.ac.uk`
[2] Newcastle University, School of Chemical Engineering and Advance Materials, Newcastle upon Tyne, United Kingdom
`jie.zhang@ncl.ac.uk`
[*] Corresponding author.

J.H. Lee et al. (Eds.): EU-Korea Conference on Science and Technology, SPPHY 135, pp. 155–162.
springerlink.com

The hypothesis that prediction performance can be improved if appropriate variable selection in each sub-block is obtained is considered in this research. To obtain the appropriate variable selection, model parameters from MBPLS models are used as high magnitudes of model parameters generally represent high contributions of the corresponding variables. In other words, variables in each sub-block showing higher contributions according to model parameters are selected to build models. Data is produced using a fed-batch penicillin fermentation simulator, and data is divided into two sub-blocks by different process modes, batch mode and fed-batch mode. 10 cases with different sets of training and validation data are investigated. Each training and validation data set is used to build a model and to select appropriate number of latent variables. Data is not selected as training and validation data is used as unseen testing data to evaluate the models. Models are built by PLS, MBPLS and MBPLS with selected variables using the proposed method. All cases using the proposed variable selection method show improved prediction performances on validation data and 8 cases out of 10 cases show better results on testing data than those from conventional PLS and MBPLS models.

2 Background

The conventional PLS model has two matrices of X and Y. X usually represents process variables measured during process operation, and Y is a output variable measured once at the end of process / batch. It mainly represents product quality such as impurity or concentration and so on. So, X is used to build a model to predict Y. MBPLS model is normally used to build a model for a process where data are from different stages or different phases because it is possible to put variables having the same importance or similar relations together into one space to classify or to break the process into smaller groups by using the MBPLS algorithm [1].

2.1 Multi-Block PLS

MBPLS model separates the whole data into several blocks. Each separated box calculates different scores called block scores that are combined into one matrix and then from that matrix, big scores called super scores can be calculated by block scores [2]. It is possible to see relations between a certain predictor block and predicted block using the obtained block scores. It also allows seeing relations between all predictor blocks and predicted block using the super scores [3, 4]. Unlike a PLS model, an MBPLS model can not only give overall information of an entire process but also give information of each stage or between stages [5]. It might allow providing a better understanding about the whole process. MBPLS algorithm used for this research is presented below.

MBPLS Algorithm

1. Take one column from Y, and let the taken column be u
   (If there is only one column in Y then Y=u)
2. Calculate weight, w, using u and each X block:
   $w_1^T=u^TX_1/u^Tu$, $w_2^T=u^TX_2/u^Tu$
3. Normalize w_1 and w_2 to unit length
4. Calculate block scores t using each block and weights:
   $t_1=X_1w_1/w_1^Tw_1$, $t_2=X_2w_2/w_2^Tw_2$
5. Put obtained score of each block, t_1 and t_2 together into one block, T
6. Calculate weight, w of block T: $w_3^T=u^TT/u^Tu$ (Super weight)
7. Normalize w_3 to unit length
8. Calculate super score, t_3: $t_3=Tw_3/w_3^Tw_3$
9. Obtain q using t and Y: $q^T=t_3^TY/t_3^Tt_3$
10. Compute new u using Y and q: $u=Yq/q^Tq$
11. Check convergence of u
    if yes move to Step 12 otherwise go back to Step 2
12. Determine loadings of each block:
    $p_1=X_1^Tt_1/t_1^Tt_1$, $p_2=X_2^Tt_2/t_2^Tt_2$ (Obtain loadings using block scores)
13. Determine regression parameter of inner relation: $b=u^Tt_3/t_3^Tt_3$
14. Compute residuals of both block matrices:
    $E_1=X1-t_3p_1^T$, $E_2=X_2-t_3p_2^T$ (Deflation by using super scores)
15. Calculate new regression coefficient, b: $b=u^Tt_3/t_3^Tt_3$
16. Compute residual of Y: $F=Y-bt_3q^T$
17. Replace X_1, X_2 and Y by E_1, E_2 and F to calculate next latent variable
18. Stop calculation when required number of latent variables is obtained, otherwise go back to Step 2

Even though an MBPLS model can provide better interpretation of process data and relationships among them, there is no advantage for using this method for the purpose of only quality prediction as the MBPLS model shows same weights and scores as the ordinary PLS model. Thus, the prediction performance by the MBPLS model shows the same prediction ability as the ordinary PLS model [6-8]. However, using information from each sub block obtained by MBPLS model can provide a criterion of appropriate variable selection for enhanced predictive modelling.

2.2 *Variable Selections for Enhanced Predictive Modelling*

The method proposed here is based on a hypothesis that an enhanced predictive model could be achieved if appropriate variables are selected, and the appropriate variables can be obtained by separating data into sub-blocks by time which can provide relation profile by time series, because variables measured obviously are time-dependant. If all data are treated with one big matrix, some interactions at a certain time containing valuable information could be offset by other interactions at different time. But if data is divided by time, it would be able to prevent the interactions being offset and enable the interactions being used for modelling.

After dividing data, MBPLS model is computed using obtained sub-blocks of X. Model parameters of the MBPLS model then can be determined for each variable in each block. For example, if a process has 11 variables and MBPLS model is obtained with two sub-blocks, model parameters can be divided into 22 groups which mean that each group of parameters represents a variable in one sub-block. . The sum of absolute model parameter values for each variable in each block and mean of those sum values is calculated. This leads to a total of 22 mean values of the sum of absolute model parameter values. These values are arranged in ascending order and the 25th percentile and the 75th percentile are obtained. If the mean value of a variable shows lower than 25th percentile value, the variable is decided as not being used for modelling. In a contrast, if the mean value of a variable is higher than 75th percentile, it is used for modelling. If it is located in a region between 25th and 75th percentile, mean values of a variable in both sub-blocks are compared and the one showing a higher value is used for model building. Fig. 1

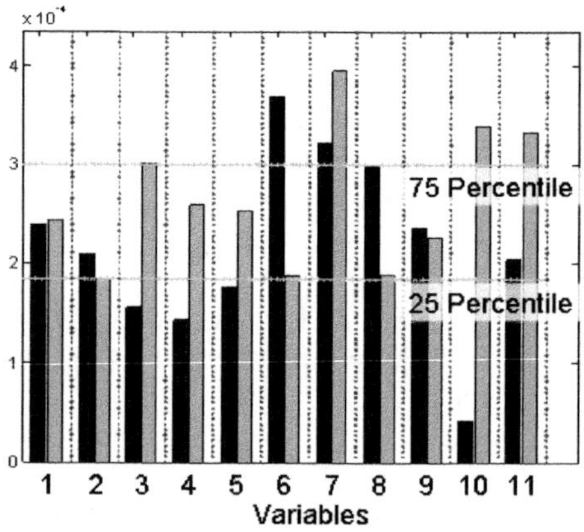

Fig. 1. The mean values for the model parameters for variables in sub-blocks

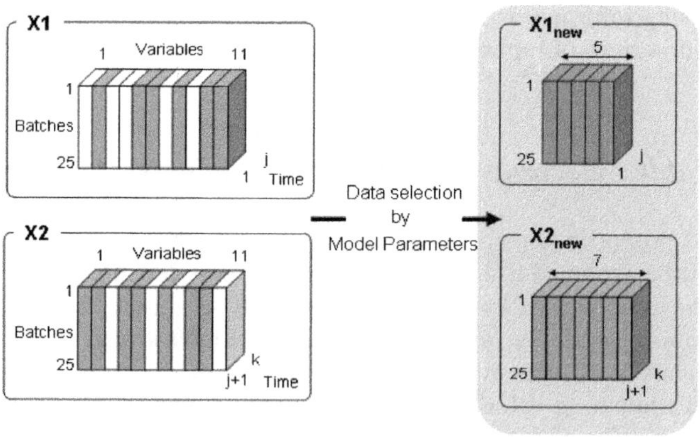

Fig. 2. Change of sub-block structures after variable selection

shows an example plot mean values of parameters for variables in sub-blocks. Black bars represent sub-block 1 and the greys represent sub-block 2. In case of 4 sub-block MBPLS model, each variable will have four mean values.

By using this procedure, different data partitioning structure is obtained as $X1_{new}$ and $X2_{new}$ described in Fig. 2 and a new MBPLS model is developed with these data blocks.

The reason of selecting variables showing high means values calculated from the model parameters is that model parameters of a certain variable having high magnitudes can be considered as the variable has a big impact on the response variable. Therefore, in this variable procedure, it is tried to select variables which can be best predictors for output variable by using model parameters.

3 Case Study: Fed-Batch Penicillin Fermentation Process

The data of the fed-batch fermentation process is generated by PenSim v2.0, simulation software developed by Undey et al. in Illinois Institute of Technology [8] based on the mechanistic model of Bajpai and Reuss [9]. Process flow sheet is shown in Fig. 3. The process has two modes, one is batch mode and another is fed-batch mode. In penicillin production process, necessary cell mass is mainly generated during the initial pre-culture stage which runs in batch mode, and the process then is operating in fed-batch mode for penicillin production by adding glucose continuously. The process is usually operating in batch mode for the first 45hrs, and then it switches to fed-batch mode [10]. Process run for 400hrs and 11 process variables are measured at 30min interval. 45 batches were produced and 10 batches were left and used as unseen testing data to evaluate model performance. The other 35 batches are used to make 10 different groups of training and validation data and build modes in order to show the effectiveness of the proposed method. Training data of 25 batches is used to build models and validation data of 10 batches is used to choose the best model among the models built with different number of latent variables. Models were built using the conventional PLS and MBPLS algorithms and a new MBPLS model with selected variables. Mean squared prediction error (MSE) is used to evaluate prediction performances.

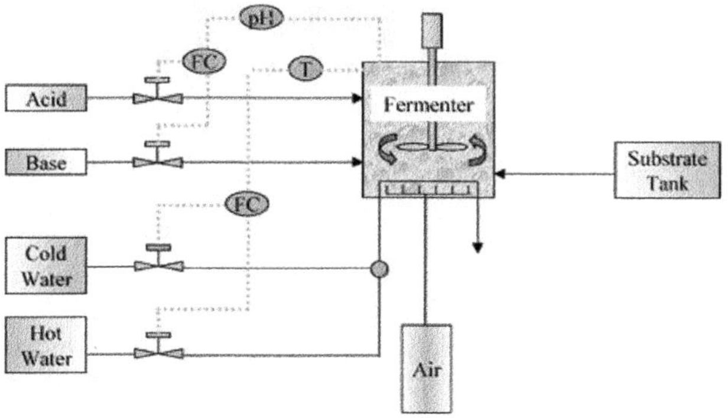

Fig. 3. Flow Sheet of Penicillin Fed-Batch Fermentation Process [10]

Model built at number of latent variables showing the lowest MSE on the validation data is selected as a good model. This good model is used with the unseen testing data to see how good prediction performance the model can deliver.

3.1 Results and Discussions

PLS models were built using one X matrix consisting data from the whole process whereas MBPLS models were built using two blocks of X. 1st sub-block has data collected for batch processing time and 2nd sub-block is formed with data measured for fed-batch processing time. Prediction results on unseen testing data from PLS, MBPLS and a MBPLS model are compared. Fig. 4 shows the MSE values on the unseen testing data for PLS/MBPLS models and Time-MBPLS models in the 10 cases (experiments). It can be seen that the conventional PLS models and MBPLS models represented by blue (solid) bars show the same prediction performances. This is because the conventional PLS and the conventional MBPLS model compute exactly same scores which used for determine model parameters from the same data as mentioned before. In the 9 cases out of 10 cases, a new MBPLS models with selected variables represented by white bars show improved prediction performances than the conventional MBPLS models. Only experiment 10 shows a worse prediction performance from the new MBPLS model, but the differences between conventional MBPLS and Time-MBPLS models are insignificant whereas other cases show that the proposed MBPLS model produces bigger improvements as shown in Fig. 4.

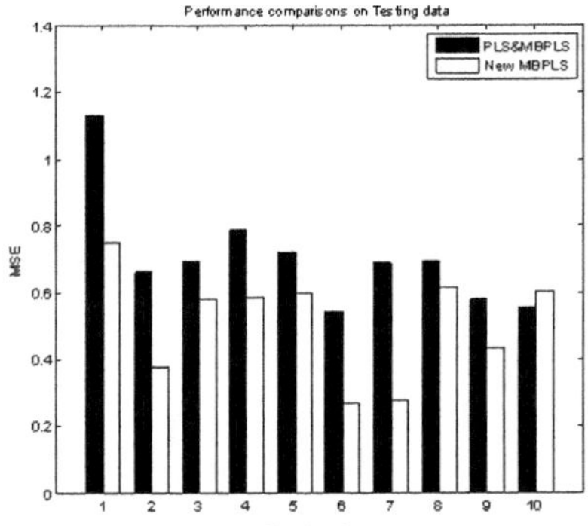

Fig. 4. Prediction Results on Unseen Testing Data

The result from on experiment is displayed in Fig. 5 which is results of experiment 4. The real measured testing data and the predictions from Time-MBPLS models and conventional PLS and MBPLS are represented by "O", "X", and "+" respectively. It can be said that data blocking by time for different process mode based on process knowledge and sum of the absolute model parameters can provide more information related to dynamics among the process variables than

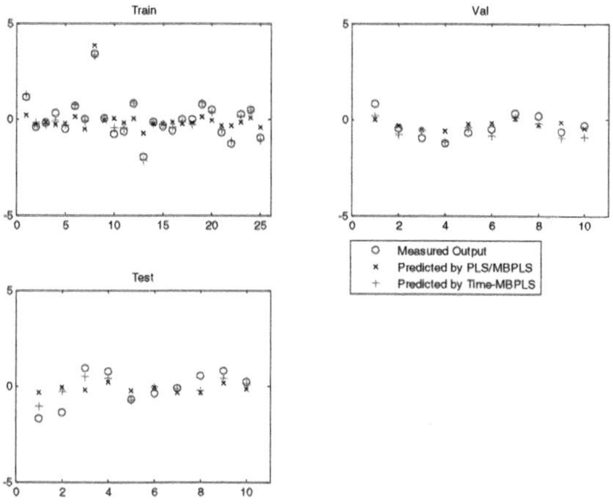

Fig. 5. Prediction Results of Experiment 4

the conventional model having one big matrix of data on these cases, because it prevents interactions between two different modes to involve in modelling and helps to see actual relationship within each mode. Thus, it can be said that the proposed variable selection method using MBPLS model can generally achieve improvement in prediction capability.

4 Conclusions

A variable selection method using MBPLS modelling is proposed for enhancing predictive modelling and the proposed method is applied to inferential estimation of product quality in a fed-batch penicillin fermentation process. In the proposed method, data is divided into multiple blocks by time based on process knowledge and means of the sum of the absolute model parameters for individual variables in each of the blocks are used to determine the variable blocking structures as higher mean value indicates high contribution to the process. Results from the proposed methods and the conventional methods are compared and it shows improved prediction performance in 9 experiments out of 10 regarding mean squared prediction error on unseen testing data. However, it is still unclear of how to divide data into sub-block structure if there is no process knowledge available or process knowledge does not provide appropriate information. In case of batch processes, they usually have multiple phases during a whole process time, if it is possible to detect time when each phase start, it would be useful to form appropriate sub-block structures. Therefore further studies to obtaining appropriate blocking structure are required in order to further improve the performance.

Acknowledgments. The first author wishes to acknowledge the financial support of Syngenta for his PhD study.

References

[1] Kourti, T.: Multivariate dynamic data modeling for analysis and statistical process control of batch processes, start-ups and grade transitions. Journal of Chemometrics 17, 93–109 (2003)
[2] Lopes, J.A., Menezes, J.C., Westerhuis, J.A., Smilde, A.K.: Multiblock PLS analysis of an industrial pharmaceutical process. Biotechnology and Bioengineering 80, 419–427 (2002)
[3] Macgregor, J.F., Jaeckle, C., Kiparissides, C., Koutoudi, M.: Process Monitoring and Diagnosis by Multiblock PLS Methods. AIChE Journal 40 (1994)
[4] Westerhuis, J.A., Smilde, A.K.: Deflation in multiblock PLS. Journal of Chemometrics 15, 485–493 (2001)
[5] Bras, L.P., Bernardino, S.A., Lopes, J.A., Menezes, J.C.: Multiblock PLS as an approach to compare and combine NIR and MIR spectra in calibrations of soybean flour. Chemometrics and Intelligent Laboratory Systems 75, 91–99 (2005)
[6] Westerhuis, J.A., Kourti, T., Macgregor, J.F.: Analysis of multiblock and hierarchical PCA and PLS models. Journal of Chemometrics 12, 301–321 (1998)
[7] Qin, S.J., Valle, S., Piovoso, M.J.: On unifying multiblock analysis with application to decentralized process monitoring. J. Chemometrics 15, 715 (2001)
[8] Birol, G., Ündey, C., Çinar, A.: A modular simulation package for fed-batch fermentation: penicillin production. Comput. Chem. Eng. 26 (2002)
[9] Bajpai, R.K., Reuss, M.: A mechanistic model for penicillin production. J. Chemical Technology and Biotechnology 30, 330 (1980)
[10] Lee, J.-M., Yoo, C.K., Lee, I.-B.: Enhanced process monitoring of fed-batch penicillin cultivation using time-varying and multivariate statistical analysis. J. Biotechnology 110, 119 (2004)

Constrained Sintering Stress-Review

Samuel Taub[1] and Jung-Sik Kim[2,*]

Abstract. When a thin film is sintered on a rigid substrate, the film will become constrained in its plane. Densification can therefore only occur in one direction, that being perpendicular to the plane of the substrate. The constraint will lead to the development of an in-plane tensile stress in the film, which exactly opposes the tendency for shrinkage in the plane. The magnitude of these stresses has been of much technological interest.

This paper reviews a methodologies to calculate the sintering stresses generated in films constrained on rigid substrates.

1 Introduction

The study of constrained film sintering is a research area that has been of interest for some twenty five years. Although much of the work has focused on general constrained film sintering, most of the results and discussion are applicable to the field of sintering constrained electrolyte films.

This area of research can be divided quite equally into theoretical and experimental studies, in many cases the latter being exploited to check the predictions of the former.

As will be shown, research calculating the sintering stresses in constrained films is quite sparse; however, many studies have been conducted into the related area of constrained film sintering kinetics.

2 Methodologies

2.1 The Model of Scherer and Garino

The work of George W. Scherer and Terry Garino[1] set the precedence for researching the sintering of constrained thin films on rigid substrates in their paper "Viscous Sintering on a Rigid Substrate".

The authors assumed that the densification rate and magnitude of stress caused by a constraint could be analysed using constitutive equations appropriate for a

1 Department of Materials, Imperial College London
2 Department of Aeronautical & Automotive Eng., Loughborough University, United Kingdom, LE11 3TU
j.kim@lboro.ac.uk
* Corresponding author.

J.H. Lee et al. (Eds.): EU-Korea Conference on Science and Technology, SPPHY 135, pp. 163–173.
springerlink.com

porous, viscous body together with measurements of the free strain rate. When reference is made to sintering, the term *viscous* is normally associated with glassy materials, however in the context of this model, *viscous* refers to a strain rate that is proportional to the magnitude of the stress to the first power. With this definition in place, viscous sintering can apply to both viscous flow and solid-state diffusional processes[2], and therefore should be applicable to the sintering of yttria-stabilised zirconia.

Using equations originally formulated by Scherer[3] the sintering rate was denoted as[4]:

$$\dot{\varepsilon}_u = \dot{\varepsilon}_f + \frac{1}{E_p}\left[\sigma_u - \upsilon_p(\sigma_v + \sigma_w)\right] \tag{3.01a}$$

$$\dot{\varepsilon}_v = \dot{\varepsilon}_f + \frac{1}{E_p}\left[\sigma_v - \upsilon_p(\sigma_u + \sigma_w)\right] \tag{3.01b}$$

$$\dot{\varepsilon}_w = \dot{\varepsilon}_f + \frac{1}{E_p}\left[\sigma_w - \upsilon_p(\sigma_u + \sigma_v)\right] \tag{3.01c}$$

Where $\dot{\varepsilon}_u, \dot{\varepsilon}_v, \dot{\varepsilon}_w$ and $\sigma_u, \sigma_v, \sigma_w$ are the strain rates and stresses in the *u*, *v* (in-plane of film) and *w* (perpendicular to plane of film) directions, $\dot{\varepsilon}_f$ is the densification rate of the unconstrained film, E_p is the uniaxial viscosity and υ_p is the Poisson's ration of the porous film.

Since the film is constrained along the *u* and *v* directions ($\varepsilon_u = \varepsilon_v = 0$) i.e. those directions in the plane of the film, and there is no resistance to densification in the *w* direction ($\sigma_w = 0$), equation 3.01 can be written as[4]:

$$\sigma = -\frac{E_p\dot{\varepsilon}_f}{1-\upsilon_p} \quad \text{and} \quad \dot{\varepsilon}_w = \left(\frac{1+\upsilon_p}{1-\upsilon_p}\right)\dot{\varepsilon}_f \qquad \begin{matrix}(3.02a)\\(3.02b)\end{matrix}$$

given that $\sigma_u = \sigma_v = \sigma$

The rate of volume change can also be defined as[4]:

$$-\frac{\dot{\rho}}{\rho} = \frac{\dot{V}}{V} = \dot{\varepsilon}_u + \dot{\varepsilon}_v + \dot{\varepsilon}_w \tag{3.03}$$

Where ρ is the relative density, V is the volume.

Once again by putting $\varepsilon_u = \varepsilon_v = 0$ and substituting equation 3.03 into 3.02b, the shrinkage kinetics (i.e. the densification rate) of a constrained film $(\dot{\rho}/\rho)_c$, can be calculated from the densification rate of the unconstrained film, $(\dot{\rho}/\rho)_u = -3\dot{\varepsilon}_f$, i.e. by measuring the free strain rate[4]:

$$\left(\frac{\dot{\rho}}{\rho}\right)_c = -\left[\frac{1+\upsilon_p}{3(1-\upsilon_p)}\right]3\dot{\varepsilon}_f \tag{3.04}$$

Using equation 3.02b, Scherer and Garino were then able to predict the stress in the sintering layer, σ, using equations derived[3] for E_p, $\dot{\varepsilon}_f$ and υ_p[4]:

$$\sigma = \frac{\gamma_S}{l_0 \rho_0^{1/3}} f_1(\rho) \tag{3.05}$$

Where γ_S is the surface tension, l_0 is the unit cell length, ρ_0 is the initial density and $f_1(\rho)$ is a dimensionless function of the density of the film.

Equation 3.05 shows that as the stress is increased, the densification rate will decrease.

The expression given in equation 3.05 for calculating the stress in a thin constrained film should give the same result as the expression shown in equation 3.02a. One can only imagine that this additional calculation was determined due to the experimental difficulty of calculating the free strain rate, $\dot{\varepsilon_f}$, when the model was proposed. It will be shown later in this section that those calculations are now possible.

Scherer and Garino also commented on the assumptions of their model, in particular, that it neglected the possibility of anisotropy developing in the sintering film. To examine this complaint, the authors proposed a second model, in which the microstructure of the constrained film was represented by a series of tubes positioned with axes normal to the substrate. It was concluded that due to the similarities in results produced by these models, the effect of pore orientation (i.e. the anisotropy development) on the microstructure would not be a cause of serious error and therefore the original model, as described above, could still be used. The effect of anisotropy on the Scherer and Garino model will later prove an area for intense debate.

This model has been discussed in great length due to its importance in the field on sintering constrained electrolyte films. Much of the literature to be discussed in the remainder of this section is based on these original findings.

2.2 Additional Models – The Model of Bordia and Raj

R. K. Bordia and R. Raj[5] devised a model to describe both the densification behaviour and stresses in a constrained ceramic film in terms of a dimensionless parameter, denoted as β.

Using spring-dashpot analogies, both the shear and densification rates could be modelled. The ratio of these rates is the parameter defined above as β.

The authors suggest that a constraint will produce a tensile stress in the film, which in turn will produce a shear stress. The magnitude of this tensile stress will then be dependent on the relative aforementioned rates within the material. If the material can deform quickly to the response of the shear stress, the size of the tensile stress will be reduced.

In terms of the densification rate, if $\beta \gg 1$, then the sintering rate of the constrained film is predicted to be similar to that of the unconstrained film, whilst if $\beta \ll 1$, then the sintering rate of the constrained film will be hindered. In situations where $\beta < 0.1$, the tensile stress developed in the film will be capable of forming defects via the diffusional growth of pores.

According to the model, the upper-bound value of the tensile stress occurs when $\beta = 0$. Using approximations for the sintering potential, Bordia and Raj surmised that the maximum tensile stress developed in the film would be in the range of 1 – 10MPa.

However, to maintain a level of simplicity in this model, several geometric assumptions had to be made about the relative dimensions of the film and substrate. Rahaman[6] also notes that achieving values of $\beta \ll 1$ are physically unrealistic.

2.3 Validation of the Scherer and Garino Model

There have been several attempts to validate the model proposed by Scherer and Garino[1] as outlined in section 3.1. The first attempt was made post-publication by Garino and H. K. Bowen[7].

In this study the validity of the theorem was tested by examining three constrained particle films, each of which sinters by a different mechanism. Silica was chosen to represent densification via a viscous mechanism, whilst Bi_2O_3 doped zinc oxide was chosen to represent sintering in the presence of a liquid phase. Alumina was selected for sintering via a solid state diffusional process – the same mechanism as YSZ.

Each set of films was subsequently sintered over a range of various conditions. The alumina films, for example, were sintered at either 1250°C or 1500°C at 50°C/min, half of which had previously been held for an hour at an intermediate temperature of 1100°C. The resulting microstructures in the alumina films are shown in figure 1.

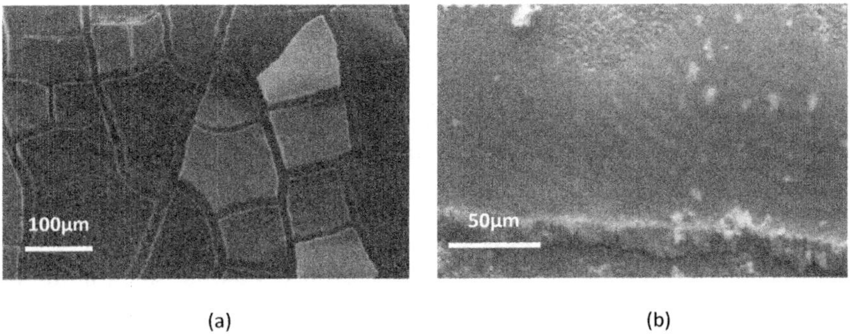

Fig. 1. SEM micrographs showing (a) an alumina film that cracked during sintering when heated directly to 1500°C and (b) a partially torn alumina film, where the unconstrained edge has sintered to theoretical density and the remaining constrained section shows residual porosity[7].

Garino and Bowen observed that cracking occurred in the films sintered directly to 1500°C (figure 1a), however noted that deformation was avoided when the intermediate temperature hold was introduced. Both samples sintered to 1250°C also experienced minor cracking.

The kinetics of the film densification were also examined (after sintering at 1450°C for 1 hour) in a rudimentary fashion using SEM. Before firing, the substrate was broken in two, revealing a section of unconstrained film at the interface; this section would be able to sinter without the conditions of the constraint. The resulting microstructure is shown in figure 1b, where the unconstrained section has sintered to theoretical density and the constrained region shows residual porosity.

Using these micrographs, the authors concluded that the qualitative results attained are in agreement with the model of Scherer and Garino for the case of a constrained film sintered by solid state diffusion mechanisms. The cracks produced when sintering to 1500°C are the result of rapid densification giving rise to large stress formation; which was suggested fits the prediction that the in-plane stress is proportional to the free densification rate. When the film was held at 1100°C prior to sintering at 1500°C, coarsening of the grains occurred, increasing the neck size and lowering the stress, again fitting the prediction of the model.

The kinetics investigation appears to show a decrease in densification rate caused by the developing stress, which is also in agreement with the model. The authors believed that the model was suitable for all three materials tested.

Despite these claims, the results available are solely qualitative, and the analysis of micrographs alone can be open to interpretation. To test the model of Scherer and Garino effectively, a more quantitative validation is required.

This was later provided by Garino and Bowen[2] in 'Kinetics of Constrained-Film Sintering'.

The paper experimentally applies the theoretical model proposed in section 3.1 to both constrained and unconstrained films in order to determine the validity of the constrained film model; samples were chosen on the basis that they sinter by either a viscous flow or solid state diffusional mechanism. Viscous flow sintering was measured using a typical soda-lime silica glass, whilst solid state diffusion sintering was measured using samples prepared from both zinc oxide and high purity alumina (to prevent an intergranular glassy phase forming).

The aim of the study was to measure both the constrained and unconstrained sintering rates experimentally, before using the densification measurements made for the unconstrained film, together with an expression derived for v_p, Poisson's ratio, to predict the constrained film sintering, as was previously shown in equation 3.02b. This theoretically derived model could then be compared to the experimental data gathered for constrained film sintering.

Shrinkage in the unconstrained samples was measured by determining the difference in area before and after isothermal sintering using an SEM. Each set of micrographs were developed from film before being cut-out and weighed. The weight difference between the micrographs (as measured using a digital balance) was used to calculate the areal shrinkage in the films.

The constrained film sintering was then determined using a more elaborate in-situ laser reflectance technique. Details of which will be given later in this review.

Figure 2 shows the sintering kinetics as determined for the glass samples. The results attained show that the model of Scherer and Garino is appropriate for both unconstrained and constrained films, although it is only valid up to ~25% volume shrinkage in the constrained case. It is believed that this anomaly could be caused

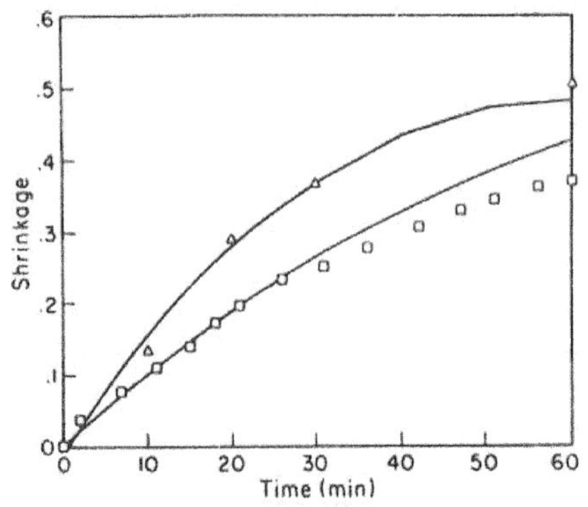

Fig. 2. "Isothermal shrinkage data for constrained (□) and unconstrained (Δ) films of the glass powder sintered at 650°C. The curve through the unconstrained-volume-shrinkage data is the best fit of the isotropic free-sintering model, and the curve through the constrained-shrinkage data is the curve predicted by the isotropic constrained-sintering model using the fitting constants determined from the unconstrained data"[2].

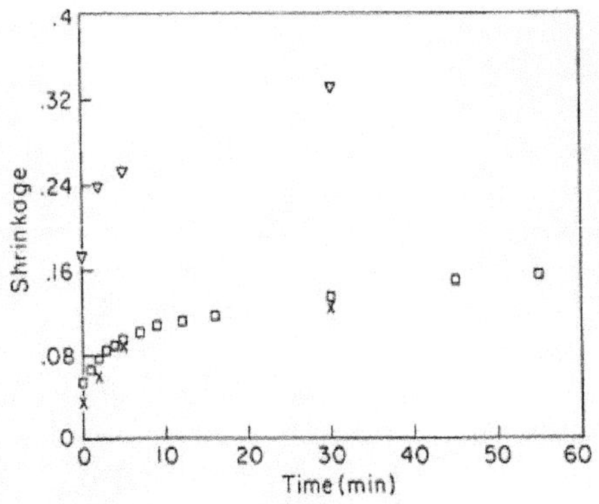

Fig. 3. "Linear-shrinkage data for the constrained and unconstrained films of alumina at 1285°C, after a 1 hour hold at 1060°C. Both the volume (∇) and the linear (×) shrinkage of the unconstrained films is shown, with the shrinkage of the constrained film (□)"[2].

by the value of v_p deviating due to differences in microstructure between the film and that assumed in the model. However, if the model was modified to take account of anisotropy, as was briefly touched upon in section 3.1, the model becomes appropriate for constrained sintering across the whole shrinkage range, although no anisotropy could be seen using SEM.

Figure 3 shows the sintering kinetics attained from the alumina sample. The results show a difference in the constrained densification rate relative to the unconstrained densification rate between the alumina and glass films. Whilst the shrinkage in the unconstrained alumina film rises to a final shrinkage value, that of the constrained film approaches a constant value below that attained in the former. Similar results were found for the zinc oxide shrinkage profiles.

These results are in contradiction to the model, which states that the constrained film should continue to sinter until it reaches the same density as that found in the unconstrained film, as shown with the silica glass. Several explanations were offered for these discrepancies, although they were dismissed by the authors. These included issues involving the assumptions made in the model, miscalculations of v_p in the viscous sintering experiment and the development of significant directional anisotropy in the constrained films, which was shown *not* to be present during microstructural analysis.

Garino and Bowen suggested that the retardation in the sintering of the constrained films could be due to the development of pores in the film. When two dense areas are separated by a region of lower density, the tendency of the constraint is to prevent particles from the denser areas diffusing to the lower density region. Through a rearrangement process, the particles in the low density region may also migrate to the higher density regions, forming pores. In an unconstrained film, this particle motion would contribute to densification as opposed to pore formation. The creation of pores will retard the densification of the constrained film, as pore size is inversely proportional to the sintering potential, as was shown in equation 2.02.

The second explanation offered by the authors, is that coarsening within the constrained film leads to a reduction in the driving force for densification. In the time taken for the unconstrained film to reach a limiting final density, the constrained shrinkage would appear to have reached a plateau. Since grain coarsening cannot occur in a glassy film, the reduction in driving force will not have occurred, and thus the sintering kinetics for the constrained and unconstrained glass films will be very similar.

Garino and Bowen tried to incorporate the coarsening phenomena into the original Scherer and Garino model. However even after compensating for grain growth, the model for constrained sintering failed to adequately match the experimental data[8].

2.4 The Isotropic Assumption

The issue of whether anisotropy exists in constrained thin films has previously been an intense area of debate. The earlier model of Scherer and Garino[1] made the assumption that the films were isotropic in microstructure. This was validated by a second model which had incorporated the effects of anisotropy in the films. The results of the two models showed enough similarity for the authors to conclude that anisotropy development was insignificant. Further experimental testing of these models[2,7] showed large discrepancies in the densification of constrained electrolyte films that sinter by a solid state diffusional mechanism. Rather than find fault in the model, the author's suggested other explanations for the differences. The effects of anisotropy were highlighted as a possible cause of the discrepancy, although it was dismissed as no evidence was palpable under SEM.

The isotropic formulation is nonetheless well developed, and has been used to gain a better understanding of issues involving the densification of thin films amongst other sintering problems[9,10]. For example, Jagota and Hui[11,12] provided an analysis for the mechanics of sintering thin films assuming that the material was viscoelastic, homogeneous and that the stresses in the film were uniformly distributed. The authors proposed a model which could predict the effects of densification and of damage initiation near a crack tip. It was shown that for a given friction (and film thickness), expressions derived for the stress intensity factor could be used to predict the behaviour of crack initiation and growth. These results were then confirmed using a finite element method. Zhao and Dharani[13] also used an isotropic finite element model to simulate the effect of sintering constrained thin films at a constant heating rate. They argued that many of the mechanical and physical properties that could be considered homogeneous in the green structure, e.g. viscosity and Poison's ratio, would alter after sintering due to changes in temperature and the associated shrinkage; thus renouncing an analytical model in favour of a finite element analysis. Using this computation, the authors were able to find both the relative density (figure 4a) and Young's modulus distribution within the thin film and show a shape distortion at the film ends. The principle stress distribution upon cooling was also calculated and is shown in figure 4b. This shall be discussed later in the review.

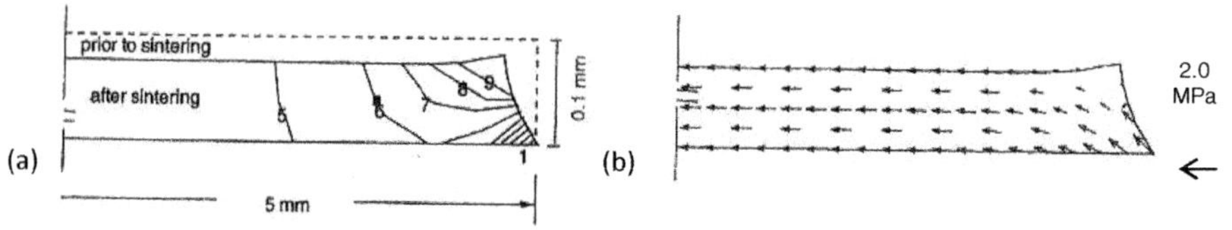

Fig. 4. Schematics showing (a) the density distribution within the constrained film, where the numbers 1-9 represent areas of relative increasing density and (b) the principle stress distribution within a constrained thin film at room temperature after a cooling rate of 50°C/min[13].

The aforementioned papers were theoretical models that assumed isotropy in the microstructure, however without an experimental parallel, their validity still can't be analysed.

In addition to the work of Garino *et al*, other novel experimental methods were used to test the sintering kinetics of constrained and unconstrained thin films[14,15], however none of the results gained could adequately match those of the initial isotropic models.

He and Ma[16] however studied the densification behaviour of micron-grain-sized alumina ceramics using an isotropic constitutive model formulated by Du and Cocks[17]. Using SEM images, they were able to attain measurements for the relative density in the constrained films, and stated that the constitutive model for grain boundary diffusion controlled sintering was consistent with the experimental results collected. This suggests that some isotropic models can be accurately used for measuring the sintering of constrained films. He and Ma also made an estimate, based on this model, for the magnitude of the induced tensile stress:

$$\sigma_c = \frac{4.5\gamma_s}{L_0}\rho^2\frac{2\rho-\rho_0}{1-\rho_0} \tag{3.06}$$

Where γ_s is the surface energy per unit area, L_0 is the mean grain size, and ρ and ρ_0 are the final and initial densities respectively. They calculated the induced tensile stress as having an average value of 0.3MPa at a relative density of 0.6.

Guillon[9] however exemplifies that all relative density measurements were limited to 60% and very few experimental data points were available, thus throwing possible doubt onto the conclusions of the model.

All of the literature mentioned previously has assumed isotropy in both the constrained and unconstrained thin films. However, since the number of studies in which the densification kinetics do not match the isotropic model is so large, some believe that the isotropic model should no longer be used and be replaced instead with one that accounts for anisotropy formation[18].

Anisotropy develops in thin films due to either the application of non-isotropic strain or stress fields during sintering. The directional diffusional fluxes which develop as a result can themselves lead to elongated grain growth and pore formation. The strain state will influence whether the anisotropy present is of an orthotropic or transverse symmetry. It should also be noted that anisotropy can be inherent to the green microstructure, having formed in the consolidation process[18].

The work of Bordia, Zuo and Guillon *et al*[18] proposed a modification to the isotropic model to include the formation of transverse anisotropy, however not before experimentally confirming its existence. This was achieved through dilatometry and sinter-forging studies of alumina thin films.

The anisotropy in their films had developed in such a way, that the microstructure in the plane of the film remained isotropic; thus presenting a transverse symmetry. In such a case the constitutive equations of formula 3.01 can now be re-written as[10,18]:

$$\dot{\varepsilon}_u = \dot{\varepsilon}_u^f + \frac{\sigma_u}{E_u^p} - \frac{\upsilon_{vu}^p\sigma_v}{E_v^p} - \frac{\upsilon_{wu}^p\sigma_w}{E_w^p} \tag{3.07a}$$

$$\dot{\varepsilon}_v = \dot{\varepsilon}_v^f + \frac{\sigma_v}{E_v^p} - \frac{\upsilon_{uv}^p\sigma_u}{E_u^p} - \frac{\upsilon_{wv}^p\sigma_w}{E_w^p} \tag{3.07b}$$

$$\dot{\varepsilon}_w = \dot{\varepsilon}_w^f + \frac{\sigma_w}{E_w^p} - \frac{\upsilon_{vw}^p\sigma_v}{E_v^p} - \frac{\upsilon_{uw}^p\sigma_u}{E_u^p} \tag{3.07c}$$

Since the film is isotropic in the *u*-v plane, the material properties within the plane will be the same, therefore: $\dot{\varepsilon}_u^f = \dot{\varepsilon}_v^f$, $E_u^p = E_v^p$, $\upsilon_{uv}^p = \upsilon_{vu}^p$, $\upsilon_{wu}^p = \upsilon_{wv}^p$ and $\upsilon_{vw}^p = \upsilon_{uw}^p$.

Thus equation 3.07 can be reduced to[10,18]:

$$\dot{\varepsilon}_u = \dot{\varepsilon}_v = \dot{\varepsilon}_u^f + \frac{1}{E_u^p}\left[\sigma_u - \upsilon_{uv}^p\sigma_v - \frac{E_u^p}{E_w^p}\upsilon_{wu}^p\sigma_w\right] \tag{3.08a}$$

$$\dot{\varepsilon}_w = \dot{\varepsilon}_w^f + \frac{1}{E_w^p}\left[\sigma_3 - \upsilon_{uw}^p\frac{E_w^p}{E_u^p}(\sigma_u + \sigma_v)\right] \tag{3.08b}$$

Using the assumptions made for the constrained film in the original model i.e. that $\varepsilon_u = \varepsilon_v = 0$, $\sigma_u = \sigma_v = \sigma$ and $\sigma_w = 0$, the expressions for the in-plane stress, σ, and the constrained densification rate, $\dot{\varepsilon}_w$, can be re-written as[10,18]:

$$\sigma = -\frac{E_u^p \dot{\varepsilon}_u^f}{1-\upsilon_{uv}^p} \quad \text{and} \quad \dot{\varepsilon}_w = \dot{\varepsilon}_w^f + \frac{2\upsilon_{uw}^p}{1-\upsilon_{uv}^p} \dot{\varepsilon}_u^f \qquad \begin{matrix}(3.09a)\\(3.09b)\end{matrix}$$

Equations 3.09(a-b) are very similar to those in the original isotropic model, although the anisotropic viscosities and viscous Poisson's ratio are required. Due to the anisotropic nature of the microstructure the two perpendicular free strain rates will also be required.

The main difference in the model, as shown in equation 3.10, is that the constrained densification rate is no longer directly related to the free densification rate[18]:

$$\left(\frac{\dot{\rho}}{\rho}\right)_c = \frac{\upsilon_{uw}^p}{1-\upsilon_{uv}^p}\left[\left(\frac{\dot{\rho}}{\rho}\right)_f + \dot{\varepsilon}_w^f\right] - \dot{\varepsilon}_w^f \qquad (3.10)$$

The development of this anisotropic model has been further developed in the literature[9,19].

3 Summary and Outlook

The concept of constrained sintering stress theoretical measurements are introduced. From the perspective of this review, the constrained sintering stresses generated became theoretically understandable and experimental approaches should be carried out to support these methods. Continuing improvements in both theoretical and experimental work are needed.

References

[1] Scherer, G.W., Garino, T.: Viscous Sintering on a Rigid Substrate. J. Am. Ceram. Soc. 68(4), 216–220 (1985)
[2] Garino, T.J., Bowen, H.K.: Kinetics of Constrained-Film Sintering. J. Am. Ceram. Soc. 73(2), 251–257 (1990)
[3] Scherer, G.W.: Sintering Inhomogeneous Glasses: Application to Optical Waveguides. J. Non-Cryst. Solids 56, 239–256 (1979)
[4] Rahaman, M.N.: Ceramic Processing and Sintering, pp. 626–628. Marcel Dekker, New York (1995)
[5] Bordia, R.K., Raj, R.: Sintering Behaviour of Ceramic Films Constrained by a Rigid Substrate. J. Am. Ceram, Soc. 68(6), 287–292 (1985)
[6] Rahaman, M.N.: Ceramic Processing and Sintering, pp. 628–629. Marcel Dekker, New York (1995)
[7] Garino, T.J., Bowen, H.K.: Deposition and Sintering of Particle Films on a Rigid Substrate. J. Am. Ceram. Soc. 70(11), C315–C317 (1987)

[8] Rahaman, M.N.: Ceramic Processing and Sintering, pp. 629–632. Marcel Dekker, New York (1995)
[9] Guillon, O., Aulbach, E., Rödel, J.: Constrained Sintering of Alumina Thin Films: Comparison Between Experiment and Modelling. J. Am. Ceram. Soc. 90(6), 1733–1737 (2007)
[10] Green, D.J., Guillon, O., Rödel, J.: Constrained sintering: A delicate balance of scales. J. Eur. Ceram. Soc. 28, 1451–1466 (2008)
[11] Jagota, A., Hui, C.Y.: Mechanics of sintering thin films – I. Formulation and Analytical Results. Mech. Mater. 9, 107–119 (1990)
[12] Jagota, A., Hui, C.Y.: Mechanics of sintering thin films – II. Cracking due to self-stress. Mech. Mater. 11, 221–234 (1991)
[13] Zhao, Y., Dharani, L.R.: Theoretical model for the analysis of a ceramic thin film sintering on a non-sintering substrate. Thin Solid Films 245, 109–114 (1994)
[14] Choe, J.W., Calata, J.N., Lu, G.Q.: Constrained-film sintering of a gold circuit paste. J. Mater. Res. 10(4), 986–994 (1995)
[15] Lin, Y.C., Jean, J.H.: Constrained Sintering of Silver Circuit Paste. J. Am. Ceram. Soc. 87(2), 187–191 (2004)
[16] He, Z., Ma, J.: Constitutive modelling of the densification of micron-grain-sized alumina ceramics. Philos. Mag. 83(16), 1889–1916 (2003)
[17] Du, Z.Z., Cocks, A.C.F.: Sintering of Fine-Grained Materials by Interface Reaction Controlled Grain Boundary Diffusion. Int. J. Solids. Struct. 31(10), 1429–1445 (1994)
[18] Bordia, R.K., Zuo, R., Guillon, O., Salamone, S.M., Rödel, J.: Anisotropic constitutive laws for sintering bodies. Acta Mater. 54, 111–118 (2006)
[19] Guillon, O., Weiler, L., Rödel, J.: Anisotropic Microstructural Development During the Constrained Sintering of Dip-Coated Alumina Thin Films. J. Am. Ceram. Soc. 90(5), 1394–1400 (2007)

Uniaxial Compressive Behaviour of Carbon Fibre-Epoxy Laminates – Part 2: Notched

Joung Hwan Lee[1]

Abstract. Compressive failure behavior of composite laminates containing open hole was investigated systematically through the mechanical tests and damage monitoring studies of the multidirectional composite laminates (T300/924C $[45/-45/0/90]_{3s}$ and T800/924C $[45/0/-45/90]_{3s}$). Fibre microbuckling in the 0° plies, surrounded by delamination, occurred in the vicinity of the hole prior to catastrophic fracture. The microbuckling, which leads to extensive in and out-of-plane fibre kinking, extends stably under increasing load before becoming unstable at a critical length of 2-3mm (depends on the specimen geometry) along the specimen width. The effect of a single 3mm diameter hole in a 30mm wide plate is to reduce the gross section strengths by about 40% and 38%, respectively when compared to the unnotched failure strength. The failure strengths were predicted successfully the open hole strength for both materials, less than 7% difference when compared to the measured experimental results.

1 Introduction

The design of fibre-polymer composite structures in aerospace applications is often controlled by the compressive strength of the material. The compressive strength of currently used composites is usually 40 ~ 50% of the tensile strength, which reduces their potential advantages. In addition, the design of composite structures frequently includes discontinuities introduced either intentionally as cutouts and fastener holes, or unintentionally due to manufacturing defects and in-service damage (impact damage). Experimental studies of compression loaded laminates indicate that much severer strength reduction can occur due to the presence of these stress raisers; they may reduce the compressive failure load of the composite laminate by more than 50% [1-3].

The aim of this study is to identify the uniaxial compressive failure behaviour of composite laminates containing an open hole. The compressive properties are presented from the mechanical tests of open-hole multidirectional (T300/924C $[45/-45/0/90]_{3s}$ and T800/924C $[45/0/-45/90]_{3s}$) laminates. Two different test fixtures which have a different loading system for an uniaxial compressive load (the modified ICSTM fixture, end-loading and compressive wedge grip fixture, shear-loading) to examine loading effects. Systematic investigations with the data

[1] University of Sheffield, The AMRC with Boeing, Sheffield, United Kingdom
j.h.lee@sheffield.ac.uk

J.H. Lee et al. (Eds.): EU-Korea Conference on Science and Technology, SPPHY 135, pp. 175–189.
springerlink.com

published and obtained from the current work were carried out. The progression of compressive damage was also monitored using optical microscopy, scanning electron microscopy (SEM) and penetrant enhanced X-ray radiography. Finally open hole compressive failure strength predictions are carried out for unidirectional and multidirectional composites with various commonly used models.

2 Compressive Test Results

The results obtained from the uniaxial compressive test of open hole multidirectional ($[45/\text{-}45/0/90]_{3s}$ for T300/924C and $[45/0/\text{-}45/90]_{3s}$ for T800/924C) specimens are presented. The modified ICSTM test fixture (Figure 1 (a)) and compressive wedge grip fixture (Figure 1 (b)) are used for the multidirectional specimens. The dimensions for the multidirectional specimens are 30mm × 30mm in the gauge section and specimen width with a circular hole of diameter 3mm drilled at the centre of the specimen.

Five specimens are used to determine mechanical strengths, each derived from a minimum of five satisfactory tests.

(a) Modified ICSTM Fixture

(b) Wedge Grip Fixture

Fig. 1. Compressive test fixtures (a) The modified ICSTM fixture and (b) The wedge grip fixture

2.1 Open Hole Specimens

2.1.1 Stress-Strain Behaviour

Compressive tests on specimens with a 3mm diameter hole for the multidirectional T300/924 ($[45/\text{-}45/0/90]_{3s}$) and T800/924C ($[45/0/\text{-}45/90]_{3s}$) laminates were carried out. A plot of the axial stress as a function of the remote axial strain for a selected specimen of the T300/924C material is shown in Figure 2. The gauges

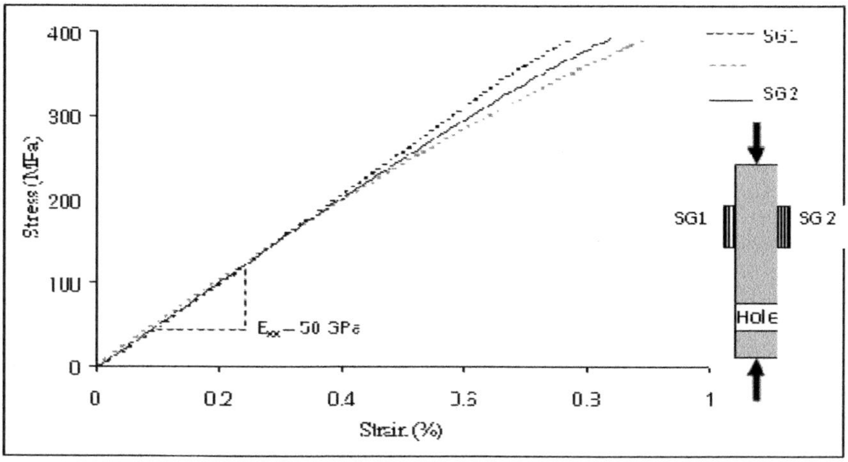

Fig. 2. Typical longitudinal stress-strain response of a multidirectional open hole specimen (Diameter = 3mm, $[45/-45/0/90]_{3s}$) for the T300/924C

were placed 12mm away from the hole centre on the mid-width line. Their positioning ensured that the registered strains were not affected by the presence of the hole or the end tabs.

The longitudinal strains on the two faces of the test piece are initially almost the same but, as the applied load is increased, the strains diverge like the stress-strain response of the unnotched specimens.

In order to investigate how the strains in different positions vary with the remote longitudinal stress, MacKinley [4] performed an experimental work with the specimen (hole diameter of 6.35mm), which has the same material and stacking sequence. These were also used in the current study. Strain gauges were placed in the through thickness direction inside the hole, a further two were placed on the compression plane, 1mm and 6.8mm from the hole edge (see Figure 3). The strain gauges were oriented along the loading direction. The gauges 1 and 2, shown in Figure 3, exhibited higher strain than that of the remote gauge 3 because they are located within the strain gradient surrounding the hole. This indicates that the strains are quite sensitive to the hole in the specimens and can be different according to hole sizes.

Figure 4 illustrates the measured mean compressive failure strengths of the open hole specimens of the T300/924C and T800/924C tested with the modified ICSTM fixture, end-loading (Figure 1 (a)) and compressive wedge grip fixture, shear-loading (Figure 1 (b). The mean failure strengths of the specimen with a 3mm hole for both T300/924C and T800/924C tested with the modified ICSTM fixture were 404 MPa and 403 MPa, respectively. When compared with the unnotched strength of 673 MPa for the T300/924C material and 647 MPa for the T800/924C material, the 3mm hole reduces the strength of the laminate by about 40 % for the T300/924C laminate and 38 % for the T800/924C laminate.

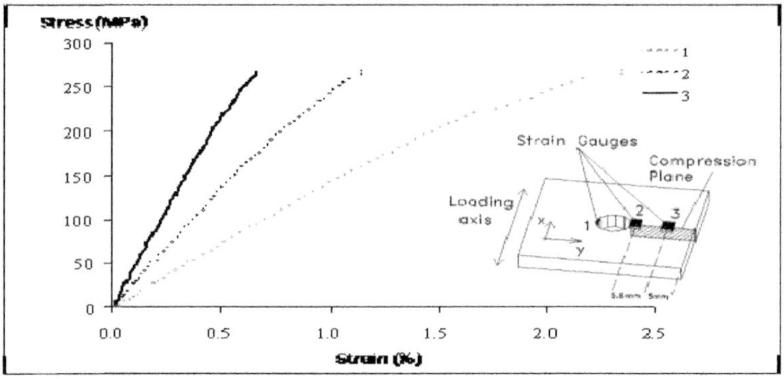

Fig. 3. Typical longitudinal stress-strain response of a multidirectional open hole specimen (Diameter = 6.35 mm, $[45/-45/0/90]_{3S}$) for the T300/924C[4]

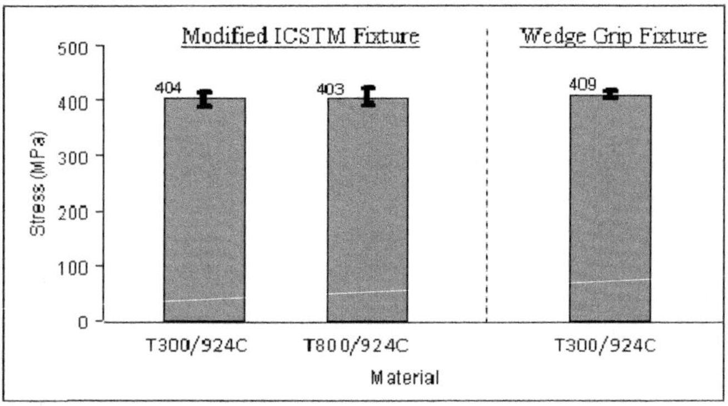

Fig. 4. Measured mean compressive strength for the multidirectional open hole T300/924C ($[45/-45/0/90]_{3S}$) and T800/924C ($[45/0/-45/90]_{3S}$) laminates (Diameter:3mm)

In addition, it is identified from Figure 4 that the open hole specimens are not influenced by the loading methods, i.e. test fixtures unlike unnotched compressive specimens when the open hole specimen failure strengths tested with the modified ICSTM fixture and wedge grip fixture are compared. These results imply that the hole directly controls the fracture strength rather than a defect remote from the hole.

2.1.2 Characteristics of Fractured Specimens

All specimens failed from the hole in a transverse direction to the loading axis by emitting a cracking type sound just prior to catastrophic failure. Post failure examination shows much delamination and fibre microbuckling along the plane of fracture but little damage away from the hole. Typical failed open hole specimens for the T300/924C and T800/924C material are shown in Figure 5.

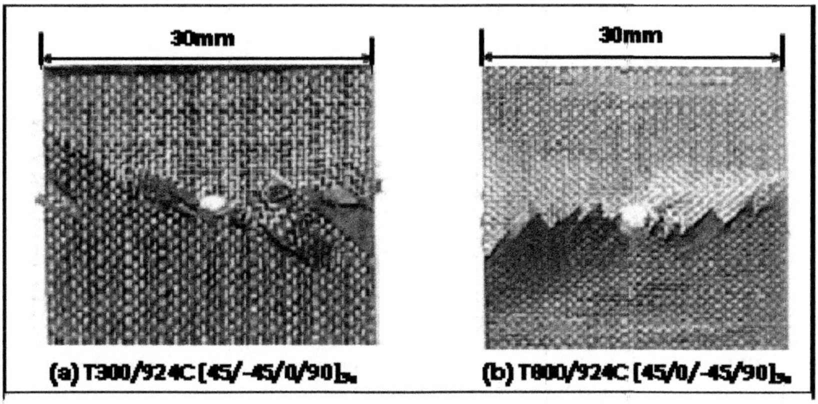

Fig. 5. Photographs of typical failed multidirectional specimens containing 3mm hole (a) T300/924C system and (b) T800/924C system

2.1.2.1 Scanning-Electron-Microscope (SEM) Examination

0° plies of the broken open hole T300/924C specimens after compression tests were produced by using the deply technique. The damaged surfaces of 0° plies were investigated in the SEM.

Figure 6 (a) and (b) illustrate the damage detected on a 0° ply at the hole boundary. The long vertical splits tangent to the hole in the 0° ply are observed at the hole edge, accompanied by fibre kinking as shown in Figure 6 (a). It is possible to see further fibre splitting which stemmed from the extensive damage at the kink band. The fibre kink inclination angle becomes almost perpendicular to the loading axis away from the hole. Figure 6 (b) presents a close-up view of the damage shown in Figure 6 (a). The kink band is inclined at $\beta = \sim 25°$ to the horizontal axis, which is similar to that of the unnotched specimens.

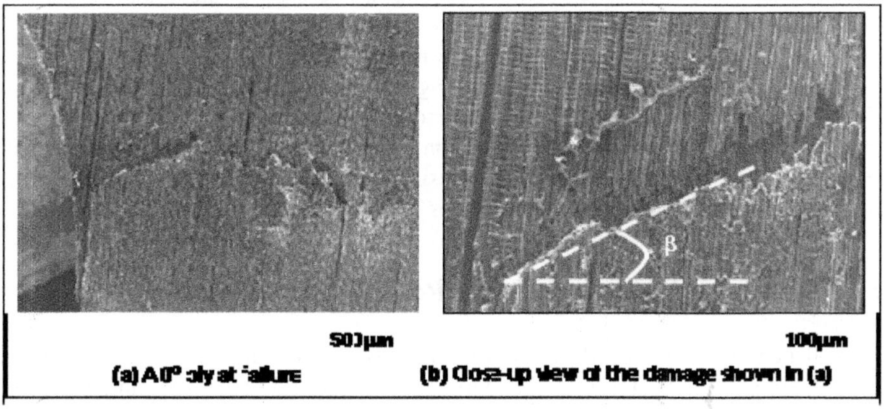

Fig. 6. SEM micrograph illustrating tensile and compressive surfaces on an individual failed fibre due to bending (Open hole specimen)

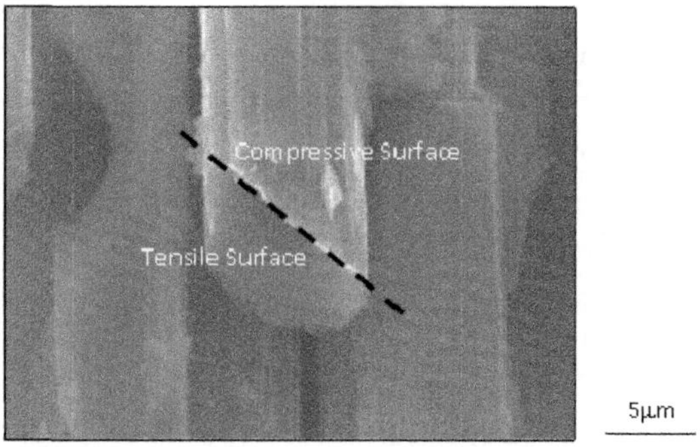

Fig. 7. SEM micrograph illustrating tensile and compressive surfaces on an individual failed fibre due to bending (Open hole specimen)

Figure 7 illustrates the areas of local tension and compression on an individual fibre that exist during fibre microbuckling. It is identified that all the fibres of the open hole specimens also failed due to a bending type deformation like the case of the unnotched specimens.

3 Damage Monitoring

The investigation of failure phenomena is very important since it provides insight into the type and extent of damage the composite experiences at extreme loads. The damage monitoring for unnotched specimens is almost impossible because of the extensive post-failure damage. The release of strain energy from both the specimen and test machine causes a substantial amount of delamination, making it difficult to study the relationships between failures in different plies at a micro-structural level.

To study the type and extent of damage in detail it is necessary to detect the initiation of failure and arrest or control its progress across the specimen. In general, specimens containing a hole or notch are used because the possible locations at which failure may be initiated are predetermined.

In this section, in order to monitor damage, two methods are employed: penetrant X-ray radiography method and optical microscopy using sectioning techniques. Figures presented in this section are based on those obtained by Mackinley's experimental work [4]. He gained the figures using the multidirectional ($[45/\text{-}45/0/90]_{3s}$) open hole (6.35mm diameter) specimens of T300/924C materials.

3.1 Penetrant Enhanced X-Ray Radiography Results

Typical penetrant enhanced X-ray photographs of the damage induced by static loading are shown in Figure 8. The photographs were taken for the ranges [85 % - 90 %], [90 % - 95 %] and [95 % - 100 %] of the average failure load. The damage consists of matrix cracks parallel to the fibres, delamination between the layers

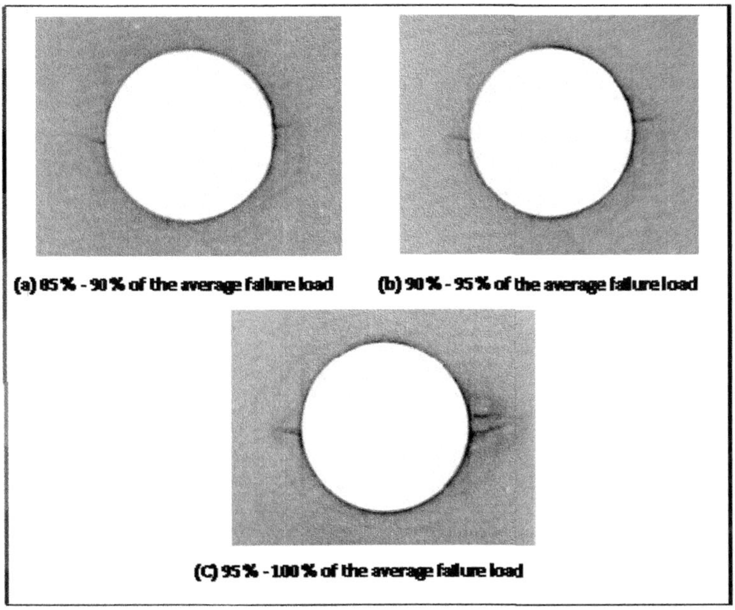

Fig. 8. X-ray radiographs showing compressive damage in the vicinity of an open hole (Diameter = 6.35 mm, T300/924C $[+45/-45/0/90]_{3s}$)

and fibre microbuckling. Delaminations appear as dark shaded regions and the matrix cracks are superimposed through the thickness in the flat radiographs, appearing as a single, thick crack.

At the first load level, kinkbands perpendicular to the load axis at the hole edge, delamination around the hole and the long vertical splits tangent to the hole on 0° plies are observed (Figure 8 (a)). Tensile cracking between the +45° fibres at the bottom of the hole is also shown. As the load was increased, the kinkbands became more prominent and extended slightly with a bit larger delaminated region around the hole (Figure 8 (b)). Finally just prior to failure, the damage is a lot more noticeable. The kinkbands, surrounded by delamination, are longer (2-3 mm) and it is also possible to see two running parallel to each other developing at the edge of the hole at areas of high in-plane compressive stress. Additional splits have developed in the ± 45° plies and 0° plies tangent to the hole (Figure 8 (c)).

3.2 Optical Microscopy Results

Mackinley [4] examined extensively the damage of open hole specimens through in-plane views and out-of-plane views using sectioning technique. For out-of-plane damage investigation, two different planes in depth were monitored starting from the hole edge: Plane A, where the fibre kinkbands are initiated; Plane B, separated by approximately 1mm from Plane A, grinding the specimens to the desired depth (see Figure 9 (a)).

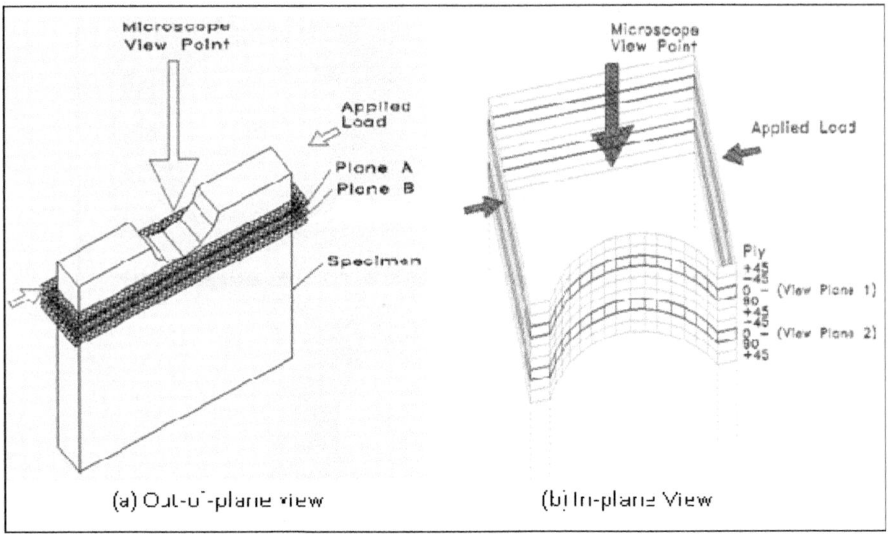

Fig. 9. Section planes for (a) out-of-plane and (b) in-plane damage investigation

For an investigation of in-plane damage, first 0° ply (plane 1) and second 0° ply (plane 2) from the surface of the laminate were selected since they were likely to show evidence of fibre microbuckling and kinking (see Figure 9 (b)).

3.2.1 Out-of-Plane Damage

Figure 10 and Figure 11 show the damages obtained from Plane A and Plane B of specimens, which were loaded up to 90 % - 95 % of the average failure load and then were sectioned, respectively.

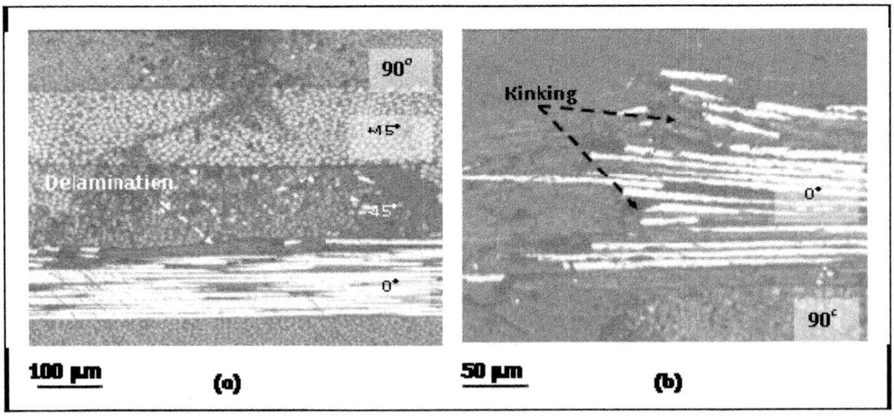

Fig. 10. Optical micrographs for compressive out-of-plane damage in the vicinity of an open hole - Plane A (Diameter = 6.35 mm, $[+45/-45/0/90]_{3s}$ - T300/924C). Applied load = 90 % - 95 % of the average failure load (287 MPa)

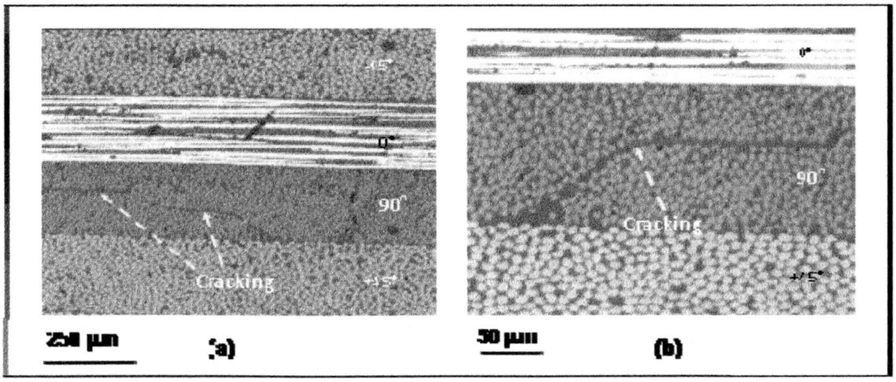

Fig. 11. Optical micrographs for compressive out-of-plane damage in the vicinity of an open hole - Plane B (Diameter = 6.35 mm, $[+45/-45/0/90]_{3s}$ - T300/924C). Applied load = 90 % - 95 % of the average failure load (287 MPa)

Local delamination, Figure 10 (a), is present at the –45°/0° interface to accommodate fibre displacements and rotations that occur locally and fibre kinking in 0° plies at the hole edge, Figure 10 (b), is visible. On Plane B some matrix cracking in the 90° plies is revealed in Figure 11.

In the higher load range of 95 % - 100 %, the damage was much more prominent. On Plane A, fibre kinking with the inclination angle, $\beta = \sim 30°$, delamination and matrix cracking are all present (Figure 12). The severity of damage increased in Plane B, where multiple kinkbands are visible (Figure 13).

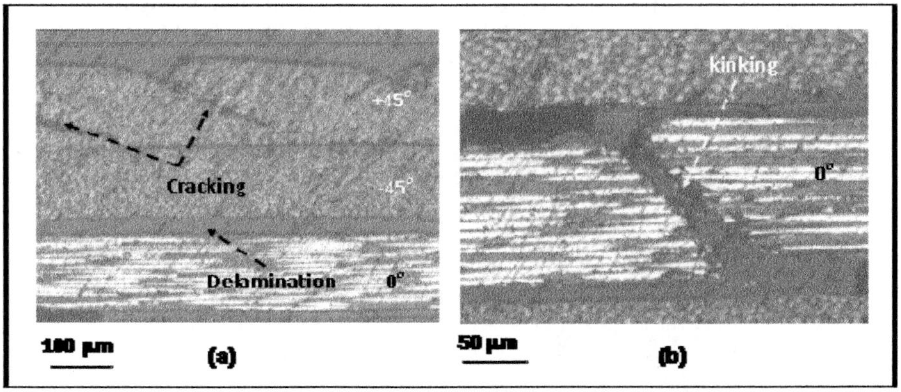

Fig. 12. Optical micrographs for compressive out-of-plane damage in the vicinity of an open hole - Plane A (Diameter = 6.35 mm, $[+45/-45/0/90]_{3s}$ - T300/924C). Applied load = 95 % - 100 % of the averge failure load (302 MPa)

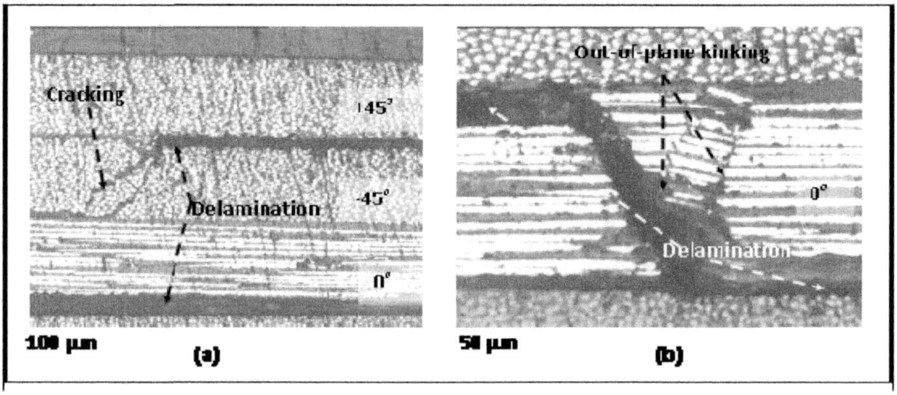

Fig. 13. Optical micrographs for compressive out-of-plane damage in the vicinity of an open hole - Plane B (Diameter = 6.35 mm, $[+45/-45/0/90]_{3s}$ - T300/924C). Applied load = 95 % - 100 % of the average failure load (302 MPa)

3.2.2 In-plane Damage

The in-plane damage on the outermost 0° ply (Plane 1) of open hole specimens, loaded up to 90 %- 95 % of the average failure strength, was first investigated. The initiation of a kinkband (individual fibres' microbuckling) at the hole boundary was observed in Figure 14. No damage was, however, detected at that load range away from the hole edge.

Figure 15 exhibits much more extensive damage for the same plane (Plane 1) in the next load range (95% - 100%). A clear kinkband propagating from the hole edge is visible with extensive fibre splitting (Figure 15 (a)). Closer examination of the kinkband zone reveals a crushing area. The angle of propagation of the kinkband, β varies from approximately 0° to 35°. From Figure 14 and Figure 15, it can be identified that the damage zone initiates with the 0° fibres buckling into the unsupported hole, allowing the kinkband to form.

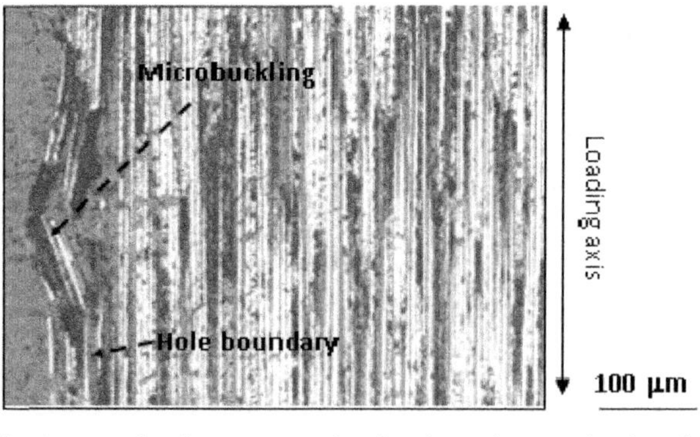

Fig. 14. Optical micrographs for compressive in-plane damage in the vicinity of an open hole-Plane 1 (Diameter = 6.35 mm, $[+45/-45/0/90]_{3s}$ - T300/924C). Applied load = 90 % - 95 % of the average failure load

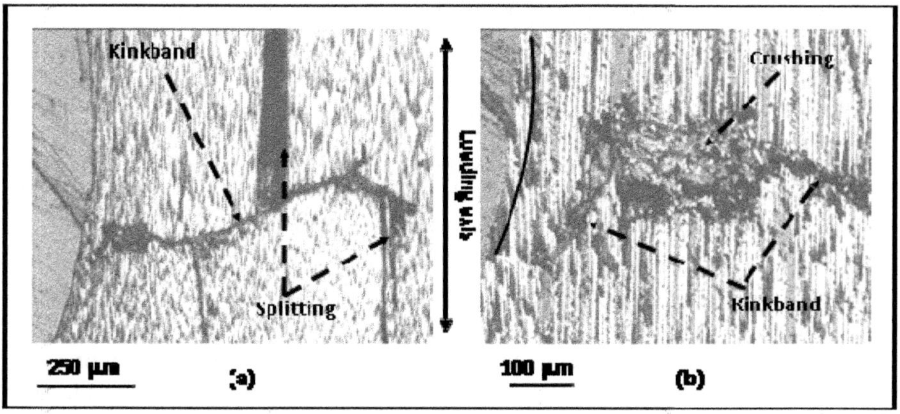

Fig. 15. Optical micrographs for compressive in-plane damage in the vicinity of an open hole-Plane 1 (Diameter = 6.35 mm, $[+45/-45/0/90]_{3s}$ - T300/924C). Applied load = 95% - 100% of the average failure load

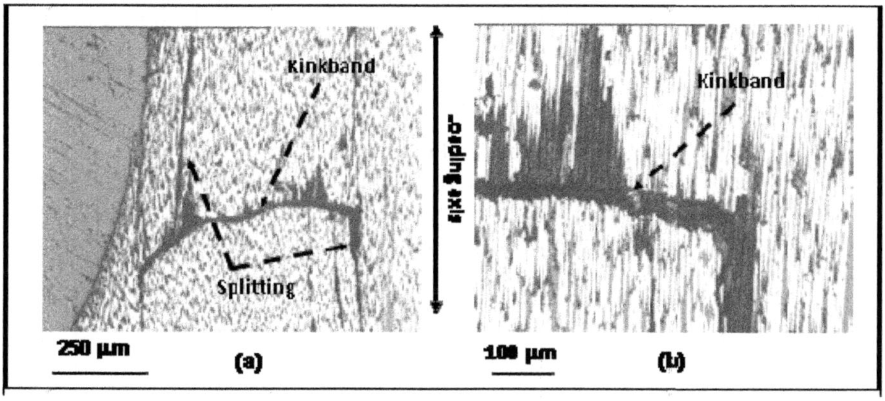

Fig. 16. Optical micrographs for compressive in-plane damage in the vicinity of an open hole-Plane 2 (Diameter = 6.35 mm, $[+45/-45/0/90]_{3s}$ - T300/924C). Applied load = 95% - 100% of the average failure load

Figure 16 shows the damage on Plane 2 (seventh ply from the surface) of open hole specimens at the same load range (95% - 100%). It can be observed that the damage on Plane 2 is quite similar in type and location, compared with the damage on Plane 1. The damage, however, does not appear to be as severe as in Plane 1 (third ply from surface, 0°). Finally the average kinkband inclination angle, β is lower than that observed on Plane 1.

4 Prediction of Composite Compressive Strength

The theoretical analysis used during the duration of the current project is explained in this section. Compressive strengths for the unnotched unidirectional laminates of T300/924C and T800/924C materials are predicted with various commonly used models. Compressive strength predictions are also performed for the unnotched multidirectional laminates of T300/924C and T800/924C materials using a fibre microbuckling model together with stiffness-ratio method. Finally, compressive analysis for the multidirectional laminates with an open hole is carried out to predict the strengths employing a linear softening cohesive zone model. All the predictions are compared to each other and to the experimental data. The models are discussed and their limitations are pointed out.

4.1 Open Hole Multidirectional Laminates

When a laminated composite plate with an open hole undergoes compression, damage in the form of fibre microbuckling and delamination initiates at the edge of the hole at approximately 80% of the failure load and extends stably under increasing load before becoming unstable at a critical length, which depends on specimen geometry. In the current study, this damage growth and failure are analysed by a linear cohesive zone model, developed by Soutis and coworkers [5,16].

In the cohesive zone model it is assumed that microbuckling initiates when the local compressive stress parallel to the 0° fibres at the edge of the hole equals the unnotched strength of the laminate σ_{un}, that is

$$k_T \sigma^\infty = \sigma_{un} \tag{1}$$

where k_T is the stress concentration factor and is the remote axial stress. Damage development is represented by replacing the damage zone by an equivalent crack, loaded on its faces by a normal traction, T, which decreases linearly with the crack closing displacement (CCD), 2v, as a result of material softening

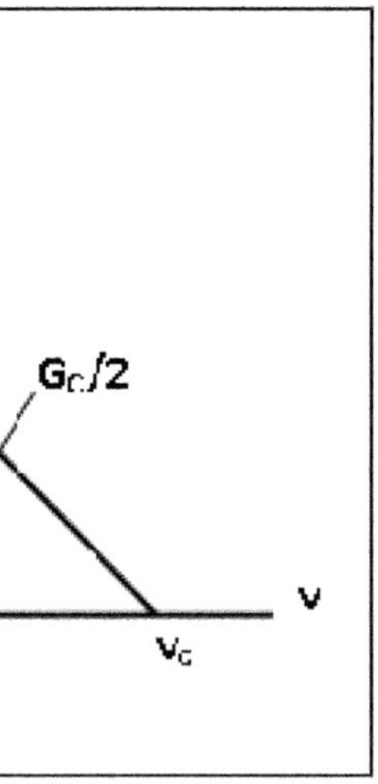

Fig. 17. Local stress carried by the buckled zone is linearly reduced with the vertical displacement

due to damage formation. The critical CCD can be estimated from a fibre micro-buckling analysis [5] or determined experimentally by independent fracture toughness test [6,7].

It is also assumed that the length of the equivalent crack, l, represents the length of the microbuckle. When the remote load is increased the equivalent crack grows in length, thus representing microbuckle growth. The crack bridging law is shown schematically in Figure 17. The area under the σ - v curve shown in Figure 17 corresponds to the fracture energy G_c

$$G_c = 2\int_0^{v_c} \sigma(v)dv = \sigma_{un} v_c \tag{2}$$

where σ_{un} is the unnotched compressive strength of the laminate and $2v_c$ is the critical crack closing displacement via fibre microbuckling. The toughness G_c represents the total energy dissipated by the microscopic failure processes: fibre microbuckling, matrix plasticity and cracking, and delamination.

The evolution of microbuckling is determined by requiring that the total stress intensity factor at the tip of the equivalent crack, k_{tot}, equals zero,

$$k_{tot} = k^{\infty} + k_t = 0 \tag{3}$$

where is the stress intensity factor due to the remote stress, and k_T is the stress intensity factor due to the local bridging traction, T, across the faces of the equivalent crack. The equivalent crack length from the circular hole is deduced as a function of using the following algorithm. For an assumed length of equivalent crack, l, it is solved for and T by matching the crack displacement profile from the crack bridging law (Figure 17) to the crack profile deduced from the elastic solution for a cracked body. The cracked body is subjected to a remote stress, and crack traction, T. The crack is discretised into a number of n elements and the loading T on the crack flanks is represented by piecewise, constant loads, each of magnitude T_i. As the crack advances, the number of elements increases. The linear stress-displacement relationship in the crush zone (Figure 17) allows direct calculation of the local tractions, T_i from the local crack surface displacement, v_i using the expression,

$$v_i = v_c\left(I_i - \frac{T_i}{\sigma_{un}}\right) \tag{4}$$

where I_i = 1 for i = 1, 2, 3, ----, n is the number of segments into which the cohesive zone is divided and v_c is the critical crack displacement. The normal displacement v_i of an element of the crack surface is calculated by adding the displacement due to remote and v_T due to local stress acting over the buckled zone,

$$v_i = v_i^{\infty} + v_i^{T} \tag{5}$$

Equation (5) is combined with Equation (4) and (3) to give an expression for the applied compressive stress as a function of microbuckling length, l, unnotched strength σ_{un}, critical crack closing displacement, v_c, laminate elastic properties, E and geometry (plate width, W and hole radius, R):

$$\sigma^{\infty} = \sum_{i=1}^{n} \beta_i T_i = f(l, \sigma_{un}, v_c, E, W, R) \qquad (6)$$

Detailed expression for the β_I and T_i are given in Soutis *et al.* [7]. At a critical length of equivalent crack, l_{cr}, the remote stress, , attains a maximum value, designated by σ_{cr} and catastrophic failure occurs.

In the present study the measured unnotched strengths of the multidirectional T300/924C ($[45/-45/0/90]_{3s}$) and T800/924C ($[45/0/-45/90]_{3s}$) laminate and an estimated fracture toughness [5] $K_c = 40$ MPa$\sqrt{m}$ were used to calculate the open hole specimen's failure strength of the T300/924C and T800/924C laminate (hole diameter: 3mm and specimen width: 30mm). The theoretical predictions are in good agreement with the experimental measurements, less than 10% difference, Figure 18.

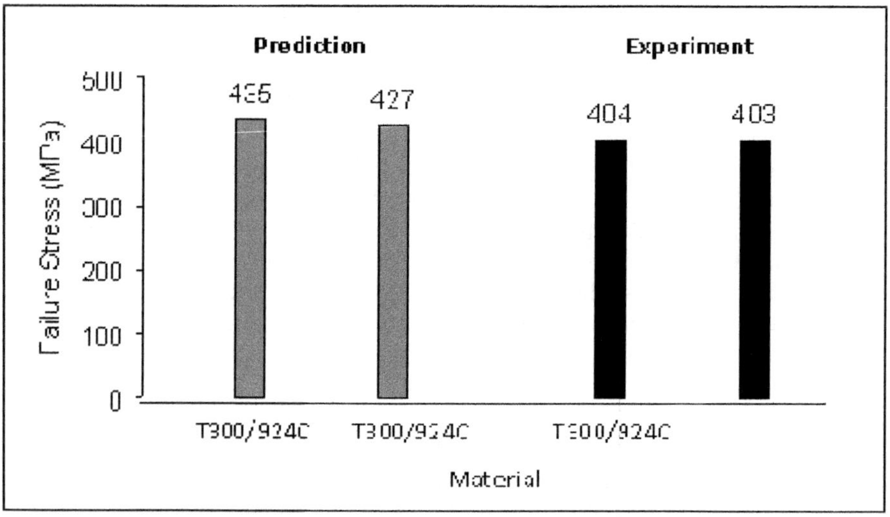

Fig. 18. Comparison of predicted and measured compressive failure strengths of the open hole multidirectional T300/924C ($[45/-45/0/90]_{3s}$) and T800/924C ($[45/0/-45/90]_{3s}$) laminates. (Hole diameter: 3mm and specimen width: 30mm)

5 Conclusion

A systematic experimental investigation with damage monitoring of the composite laminates was carried out to identify uniaxial compressive behaviour of the multidirectional laminates fabricated from the T300/924C and T800/924C material. Failure strengths of the open hole were predicted analytically using the commonly used models.

From the uniaxial compressive strength test, open hole (404 MPa for T300/924C and 403 MPa for T800/924C) multidirectional compressive failure strengths for both T300/924C and T800/924C laminate were also almost equal to each other. The failure was initiated from the hole edge and propagated in an almost transverse direction to the loading axis by emitting a cracking type sound just prior to catastrophic failure. The dominant failure mechanism was also 0° fibre microbuckling. The effect of a single 3mm diameter hole in a 30mm wide plate for both T300/924C and T800/924C materials is to reduce the gross section strengths by about 40% and 38%, respectively when compared to the unnotched failure strength.

In damage monitoring studies performed by X-ray radiography, SEM and optical microscopy method using open hole specimens, it is revealed that fibre microbuckling in the 0° plies, surrounded by delamination, occurred in the vicinity of the hole prior to catastrophic fracture. The microbuckling, which leads to extensive in and out-of-plane fibre kinking, extends stably under increasing load before becoming unstable at a critical length of 2-3mm (depends on the specimen geometry) along the specimen width.

Finally, the strengths for both T300/924C and T800/924C laminates were predicted analytically for open hole multidirectional specimens using the linear cohesive zone model developed by Soutis *et al.*[5,7]. The model predicted successfully the open hole strength for both materials, less than 7% difference when compared to the measured experimental results.

References

[1] Starnes, J.H., Rhodes, M.D., Williams, J.G.: Effect of Impact Damage and Holes on the Compressive Strength of a Graphite/Epoxy Laminate. In: Pipes, R.B. (ed.) Nondestructive Evaluation and Flaw Criticality for Composite Materials, ASTM STP. American Society for Testing and Materials, vol. 696, pp. 145–171 (1979)

[2] Potter, R.T.: The Experimental Degradation of Notched CFRP in Compression. Composites 14(3), 391–398 (1983)

[3] Soutis, C., Fleck, N.A.: Static Compression Failure of Carbon Fibre T800/924C Composite Plate with a Single Hole. Journal of Composite Materials 24, 241–256 (1990)

[4] Mackinley, C.P.: Compressive failure of CFRP laminates containing pin-loaded holes, PhD thesis, Imperial College of Science Technology and Medicine, UK (2000)

[5] Soutis, C., Curtis, P.T.: A Method for Predicting the Fracture Toughness of CFRP Laminates Failing by Fibre Microbuckling. Composites, Part A 31(7), 733–740 (2000)

[6] Soutis, C.: Compressive failure of notched carbon fibre-epoxy panels, PhD thesis, University of Cambridge, UK (1989)

[7] Soutis, C., Fleck, N.A., Smith, P.A.: Failure Prediction Technique for Compression Loaded in Carbon Fibre-Epoxy Laminate with Open Hole. Journal of Composite Materials 25, 1476–1498 (1991)

Comparative Study of Risk Assessment Approaches Based on Different Methods for Deriving DNEL and PNEC of Chemical Mixtures

Jongwoon Kim[1,2,*], Sanghun Kim[1], and Gabriele E. Schaumann[2]

Abstract. Most living organisms are actually exposed to chemical mixtures rather than individual substances. But current chemical risk assessment frequently focuses on the single chemical substances. The European Union presented some methods in the draft technical guidance notes for deriving DNEL (Derived No Effect Level) and PNEC (Predicted No Effect Concentration) for risk assessment of mixtures. This case study shows that the differences of DNELs and PNECs of mixtures (2 coating materials) based on two methods (KCC and CR methods) and explains why such differences are generated. The differences of DNELs and PNECs of the mixtures based on the two methods were estimated by 1.21 and 2.31 times respectively. Through the theoretical analysis on influence factors affecting DNEL and PNEC, it could be recognized that the difference between two methods can significantly increase in proportion to the number of substances having similar hazard and weight fraction.

1 Introduction

Generally most living organisms and environment are exposed to chemical mixtures rather than single substances in everyday life. However, chemical risk assessments frequently focus on the individual substances to estimate risk characterisation of products (e.g. substances, mixtures, and articles) [1, 2, 3]. The latest EU chemical regulation (Reg. 1907/2006), REACH (Registration, Evaluation, Authorisation and Restriction of Chemical Substances) entered into force in June 2007, puts less focus on the chemical mixtures [4, 5].

The EU presented some approaches in the draft technical guidance notes for deriving PNEC (Predicted No Effect Concentration) and DNEL (Derived No Effect Level) for risk assessment of mixtures [6, 7, 8]. The main advised methods are

[1] KIST Europe Forschungsgesellschaft mbH., Knowledge Research Group, Saarbrücken, Germany
jwkim@kist-europe.de
[2] University of Koblenz-Landau, Department of Organic and Environmental Chemistry, Institute of Environmental Sciences, Landau, Germany
[*] Corresponding author.

J.H. Lee et al. (Eds.): EU-Korea Conference on Science and Technology, SPPHY 135, pp. 191–202.
springerlink.com

'Key Critical Component (KCC)', 'Composite Reciprocal DNEL and PNEC (CR)', and 'Risk-based Emission Threshold (RET)' [6].

DNEL and PNEC are used as part of the environmental risk assessment and human health risk assessment. Both values are estimated on the bases of the available data on the intrinsic toxicological characteristics of substances. That is, DNEL and PNEC indicate the dose that is safe for environment and human individually. Any exposure at or below the DNEL and PNEC can be considered as a safe level for environment and human health. They are used for the risk characterization in the process of chemical risk assessment [6, 7].

Recently, a result of case study on testing EU REACH draft technical guidance notes for conducting chemical safety assessments was reported [5]. According to this study, a significant difference, 2 orders of magnitude, may be calculated by the risk assessment approaches for mixture based on the EU technical guidance notes. Nevertheless, reasons of the difference were not mentioned.

The conventional mixture effects are described by concentration addition (CA), independent action (IA), synergism, antagonism and potentiation [9, 10]. The current risk assessment is conducting with generic safety factors though the mixture toxicity is not generic [9, 11]. However, the knowledge of mixture toxicity is unsatisfying and still includes many unsolvable questions. The KCC and CR methods are assuming the IA and CA models respectively. These two mixture toxicity models are fundamentally mutual contradictory concepts because IA model is basically assuming dissimilar acting chemicals, on the other hand, CA model is assuming similar acting chemicals.

The objective of this study is to compare the risk assessment approaches for mixtures and investigate the reasons for discrepancy of results derived by the different approaches. This case study was carried out on risk assessment of two mixtures (coating products) using the KCC method and the Composite Reciprocal DNEL and PNEC method in the EU draft technical guidance notes.

The fundamental hypothesis is that the current KCC and CR methods for deriving DNEL and PNEC of mixtures result in significantly different results due to their contrary concepts. That is, KCC method can underestimate mixture toxicity rather than Composite Reciprocal DNEL and PNEC method. Therefore, it should be noticed that if the difference is significant, the results of current methods cannot be valid, without testing for the whole mixture.

2 Methodology

2.1 Data Collection

2.1.1 Coating Products Applied to Derive DNEL and PNEC

To derive the DNEL and PNEC of mixture, two coating products which contain all substances with officially published risk assessments were selected. And the dangerous substances in the product were confirmed by manufacturers' SDSs (Safety Data Sheets) [12, 13]. Table 1 shows the components of each coating product. The weight fractions (wt/wt) of each substance in the mixtures were optionally determined within the range of indicated weight fractions in the SDSs.

2.1.2 Data Sources for Chemical Risk Assessment

In order to estimate DNELs and PNECs of each substance in the mixtures, the following official data sources were utilized:

- International Uniform Chemical Information Database (IUCLID), European Chemical Bureau;
- European Substance Information System (ESIS), European Chemical Bureau;
- OECD Screening Information Data Set (SIDS) High Production Volume Chemicals, UNEP.

2.2 Exposure Scenarios for Mixtures as Coating Products

2.2.1 Environmental and Human Exposure Routes

The coating materials are applied for protection or decoration of surfaces. The coating products can be formulated with a wide variety of substances. Common components of coating materials are composed of pigments, additives, binders and the carrier fluid or solvent.

The critical components are those substances that determine the risk for one or more adverse effects via one or more exposure routes. To determine whether or not a substance is 'critical', a comparison of the substance's risk potential is necessary, taking into account its hazardous properties (DNELs/PNECs), mobility, vapour pressure, water solubility etc.) and its concentration in the mixture [6].

In this study, the Risk-phrases defined in the SDSs were used to determine the environmental and human exposure routes of substances.

2.2.2 Methods for Deriving DNEL and PNEC of Mixtures

The EU draft technical guidance note advised mainly the following approaches to reduce complexity in the chemical risk assessment for mixtures [6]:

- Key Critical Component (KCC) approach leads to the identification of a key substance of highest concern (i.e. the substance with the lowest weighted DNEL or PNEC) in the mixture;
- Composite Reciprocal (CR) approach treats a multi-component mixture as a single chemical entity by calculating a composite DNEL or PNEC for the mixture based on the DNELs or PNECs of substances, or a group of the substance (e.g. those which are classified as hazardous to environment or human health). These can be derived by using the following equation:

$$DNEL_{mixture}(or\ \ PNEC_{mixture}) = \frac{1}{({}^{f_a}/_{DNEL_a} + {}^{f_b}/_{DNEL_b} + {}^{f_c}/_{DNEL_c} + \cdots + {}^{f_n}/_{DNEL_n})} \tag{2.1}$$

where, f_a : weight fraction, substance (a)

- Risk-based Emission Threshold (RET) approach is to calculate maximum emission values without environmental risk for each substance in the mixture then uses the lowest emission value (lowest RET to derive an estimate of supportable use) for the mixture as a whole (e.g. mg/capita/day).

In this case study, the KCC and CR methods are used to derive the DNEL and PNEC of mixtures. RET method is excluded because this method is only relevant within an environmental risk assessment [6].

2.3 Theoretical Analysis of Influence Factors on Deriving DNEL and PNEC of Mixtures

In order to conduct the theoretical analysis of influence factors on deriving DNELs and PNECs based on the KCC and CR methods, the following four factors were considered:

- Ratio of toxicities between two substances;
- Ratio of weight fractions between two substances;
- Ratio of assessment factors between two substances;
- The number of substances in a mixture.

The ratio of toxicities means the ratio of DNELs or PNECs between two substances. And the weight fraction of substance is the weight of substance divided by the gross weight of the mixture. The assessment factor used in the assessment factor methods is a safety factor applied for derivation of the DNEL and PNEC based on the available toxicity data. DNELs and PNECs of substances are estimated by the lowest toxicity value, exposure conditions (e.g. operation conditions, risk management measures, etc.) and assessment factor (i.e. the sparser the available toxicity data, the higher is the assessment factor) [7, 8]. The differences between results of the KCC and CR methods, in terms of the different number of substances in a mixture, were estimated with change of the number of substances having a same toxicity and weight fraction.

3 Results

3.1 Exposure Scenarios for Mixtures

Environmental and human exposure routes of critical components in each mixture were determined by Risk phrases (R-phrases), defined in Annex III of European Union Directive 67/548/EEC: Nature of special risks to dangerous substances and preparations (mixtures), from Manufacturers' SDSs (Safety Data Sheets).

According to the mentioned approach above, DNEL inhalation and PNEC seawater were determined as foreseeable exposure routes for human (especially for workers) and environment respectively. The coating products in this case study are mainly applied to offshore or indoor. Other human exposure routes (e.g. dermal, oral, eyes) and environment exposure routes (e.g. soil, air, etc.) can be excluded because these coating products are only used by professionals [5].

The foreseeable exposure routes of the mixture products are shown in Table 2. The four substances (remarked 'Y' in Table 2) having the relevant exposure routes for inhalation in the mixture product A were used to derivate the DNEL inhalation (for workers) of mixture. In case of mixture A, it is inappropriate to compare the PNEC (for seawater) based on the KCC and CR methods because it has only one substance having relevant exposures for aquatic environment. To compare the KCC and CR methods for derivation of the PNEC of mixture, mixture B including four substances having relevant exposures for the aquatic environment was used in this study.

3.2 *Results of the Derivation of DNELs and PNECs of Mixtures*

DNELs and PNECs of substances were estimated based on the hazardous data in the hazardous/risk assessment reports shown in Table 3, in accordance with the EU REACH technical guidance [7, 8].

As mentioned above, each coating product was used for derivation of DNEL inhalation and PNEC seawater of mixture separately. Product A is not able to be applied for the CR method since it has only one substance having the aquatic environmental exposure routes.

Table 3 presents the DNELs of substances in mixture A. In the case of mixture A, xylene could be selected as the key critical component because it has the lowest DNEL value. Substance no. 5 and 6, shown in Table 2, without the relevant exposure routes for inhalation were excluded for deriving DNEL of mixture A. Table 4 presents the DNELs of mixture B based on the two methods and the ratio of the results between KCC and CR was 1.21.

The PNECs of substances in mixture B were shown in Table 3. In the case of mixture B, zinc was estimated as the key critical component. Substance no. 5, 6 and 7, shown in Table 2, were excluded to derivate the PNEC of mixture A since they don't have the relevant exposure routes for aquatic environment during their professional uses. The PNECs of mixture B based on the two methods are shown in Table 4. The ratio of the PNECs between KCC and CR methods was 2.31 (KCC/CR).

3.3 *Theoretical Analysis of Influence Factors on Deriving PNECs of Mixtures*

Theoretical analysis of influence factors affecting on the differences between the PNECs of KCC and CR methods was conducted with ratios of toxicities, weight fractions, assessment factors between two substances, and the number of substances having same toxicity and weight fraction in a mixture. Figure 1 illustrates the results of the theoretical analysis of influence factors.

Table 1. Substances in the mixture product A and B [12, 13]

No.	Product A		Product B	
	Substances	CAS No.	Substances	CAS No.
1	Xylene	1330-20-7	Zinc	7440-66-6
2	N-Butanol	71-36-3	Solvent Naphtha	64742-94-5
3	Iso-butanol	78-83-1	1,2,4-Trimethylbenzene	95-63-6
4	Toluene	108-88-3	Mesitylene	64742-88-7
5	Zinc phosphate	7779-90-0	Naphtha (petroleum), hydrotreated heavy	64742-48-9
6	Epoxy resin	25036-25-3	2-Ethyl hexanol	104-76-7
7	-	-	Stoddart solvent	8052-41-3

Table 2. Information of exposure routes of each substance in the mixture A and B based on the Risk phrases

No.	Product A	Exposure route	Product B	Exposure route
	Substances	Inhalation for worker	Substances	Aquatic environment
1	Xylene	Y	Zinc	Y
2	N-Butanol	Y	Solvent Naphtha	Y
3	Iso-butanol	Y	1,2,4-Trimethylbenzene	Y
4	Toluene	Y	Mesitylene	Y
5	Zinc phosphate	N	Naphtha (petroleum), hydrotreated heavy	N
6	Epoxy resin	N	2-Ethyl hexanol	N
7	-	-	Stoddart solvent	N

Y: substance with relevant risk phrases (R50/53; R51/53)
N: substance without relevant risk phrases (R20/21; R37; R67)

Through this analysis, it is recognised that the differences of toxicities, weight fractions and assessment factors between two substances are in proportion to the ratio of PNECs between KCC and CR methods. On the other hand, only the

Table 3. DNELs and PNECs of hazardous substances in the mixture A and B

No.	Mixture Product A		Mixture Product B	
	Substances	DNEL inhalation for worker (conc. weighted, mg/L)	Substances	PNEC seawater (conc. weighted, mg/L)
1	Xylene	1.74E+01	Zinc	8.40E-03
2	N-Butanol	1.02E+02	Solvent Naphtha	9.50E-03
3	Iso-butanol	9.74E+02	1,2,4-Trimethylbenzene	2.00E-02
4	Toluene	9.05E+02	Mesitylene	2.00E+00

Table 4. DNEL and PNEC of the mixture products and the ratios of results derived by KCC and CR methods

Methods	DNEL inhalation of Mixture A (mg/L)	PNEC seawater of Mixture B (mg/L)
KCC	1.74E+01 (Xylene)	8.40E-03 (Zinc)
CR	1.44E+01	3.64E-03
KCC/CR	1.21	2.31

KCC: Key Critical Component method
CR: Composite Reciprocal PNEC and DNEL method

Table 5. Data sources used for derivation of DNELs and PNECs of each substance in the mixture product A and B

No.	Mixture Product A		Mixture Product B	
	Substances	Data sources for DNEL	Substances	Data sources for PNEC
1	Xylene	• OECD Integrated HPV Database (OECD, 2003) [16]	Zinc	• EU Risk Assessment Report (ECB, 2008) [21]
2	N-Butanol	• OECD SIDS Initial Assessment Report (UNEP, 2001) [17]	Solvent Naphtha	• IUCLID – Dataset (ECB, 2000) [22]
3	Iso-butanol	• OECD SIDS Initial Assessment Report (UNEP, 2004) [18]	1,2,4-Trimethylbenzene	• IUCLID – Dataset (ECB, 2000) [23]
4	Toluene	• EU Risk Assessment report (ECB, 2003) [19] • OECD Integrated HPV Database (OECD, 2001) [20]	Mesitylene	• IUCLID – Dataset (ECB, 2000) [24]

number of substances in a mixture is in inverse proportion to the ratio of PNECs between KCC and CR methods.

In this analysis, it is determined that the main influence factor being able to increase the difference of the results based on the two methods is the number of substances having similar toxicity and weight fraction in a mixture.

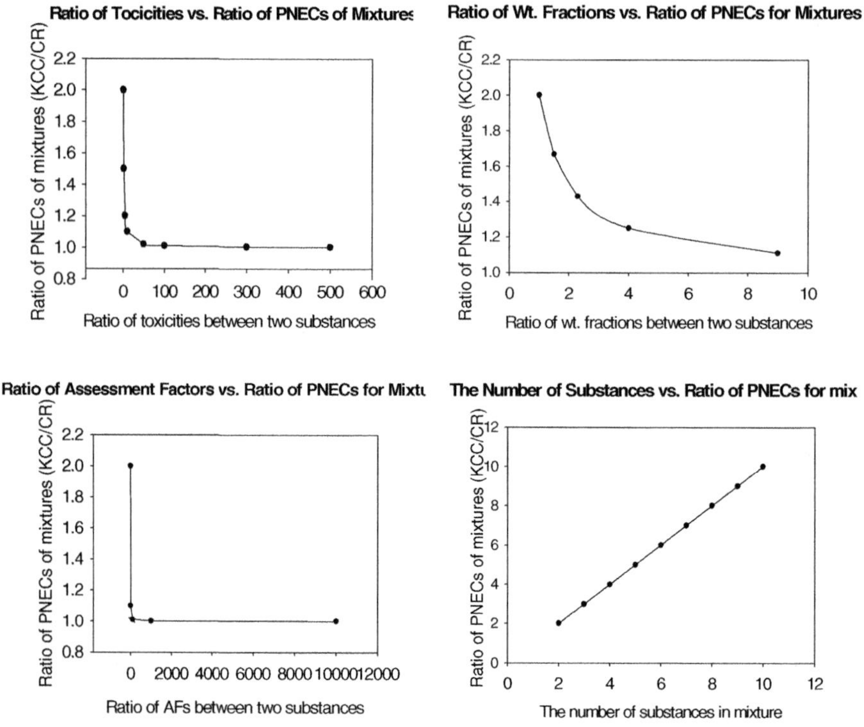

Fig. 1. Results of the theoretical analysis of influence factors on deriving PNECs of mixture based on KCC and CR methods.

4 Discussion

4.1 DNEL and PNEC of Mixture Based on KCC and CR Methods

To derive DNEL or PNEC of mixture without any toxicological testing for mixture, the DNELs or PNECs of substances in the mixture were required. However, the lack of hazard data for individual substances placing on the global market is still an obstacle to derive DNEL and PNEC of mixtures. According to the US EPA report, 93% of about 3,000 high production volume chemicals (≥1,000 ton/annum) are missing one or more of the six basic screening tests (i.e. acute toxicity, chronic toxicity, developmental and reproductive toxicity, mutagenicity, ecotoxicity, and environmental fate). 43% of these chemicals do not have these tests and only 7% of them have all six basic screening tests [14].

However, European Chemical Agency expects that toxicity database of around 30,000 substances, 1 ton or more per annum manufactured or imported into the EU, can be constructed until 2018 under the REACH Regulation. In the case of the substances manufactured or imported more than 1,000 and 100 ton annum, the registration deadlines are 2010 and 2013 respectively.

Nevertheless, this case study shows that the current risk assessment approaches, KCC and CR methods, for estimating mixture toxicity are not yet satisfied. Through this case study and theoretical analysis, it is expected that the derivation results of DNEL and PNEC between two methods are differed by in proportion to the number of substances having similar toxicities and weight fractions in a mixture, even if the hazard data of all substances in a mixture is established.

The differences between these methodologies can be an inevitable result since the KCC and CR methods are essentially based on the risk management concept rather than scientific backgrounds of mixture toxicity.

The KCC approach is a simplified methodology to select a substance of the highest concern to reduce the resources required to carry out a risk assessment of mixture. In order to use this approach, the independent action (IA), one of the mixture toxicity models, should be assumed. However, this model concept may or may not be applicable to the mixture (i.e. total toxic loading of substances may be under- or over-estimated). In this case study, the DNEL and PNEC of the mixtures based on KCC method were higher than CR method (i.e. the results of KCC method present lower overall toxic value than CR method). Contrary to the KCC method, CR method is assuming the concentration addition (CA) model (i.e. additivity of toxicity) in mixture toxicity.

According to the recent research paper, approximately 50% of 158 toxicity data sets for mixture toxicity could not be adequately predicted by the conventional mixture toxicity models, CA and IA models. The research showed that about 20% and 10% of the mixtures were predicted by IA and CA respectively, and another 20% of the experiment could be predicted by either CA or IA [15].

4.2 Results of the Theoretical Analysis of Influence Factors on Deriving PNECs of Mixtures

Theoretical analysis of influence factors on deriving the DNEL of mixture was not conducted additionally. In the case of the DNEL of mixture, the number of substances having similar toxicity and weight fraction is also considered as the main influence factor to cause the difference of result between two methods because a common formula and calculation mechanisms based on the DNEL and PNEC of substances are applied for deriving the DNEL and PNEC of mixture.

4.3 Data Sources

Table 5 presents the data sources were referred to estimate DNELs and PNECs of each substance in the mixtures.

4.4 What Does the Difference between Two Methods Imply for the Industry?

Under the EU REACH regulation, manufacturers or importers should register their substances to the ECHA to place his substances itself or substances in mixtures or in articles to the EU market.

For the registration, they have to conduct chemical safety assessment to prove that their substances are using without any reasonable risk during life cycle of the product, if necessary. During this process, DNELs and PNECs are used in the risk characterisation of product. Underestimated DNELs and PNECs may cause unnecessary RMMs (Risk Management Measures) to reduce the risk and finish the chemical safety assessment. However, additional RMMs are economical burden to industry. On the other hand, Overestimated DNELs and PNECs may cause a problem to overlook the possible risk of the product.

5 Conclusions and Outlook

5.1 Conclusions

- In the real 2 case studies, the ratios of DNELs and PNECs of KCC and CR methods (the differences of DNELs and PNECs derived) were 1.21 and 2.31 respectively.
- The number of substances having similar toxicity and wt. fraction is the main factor to make the difference between 2 methods.
- KCC and CR approaches may under- or overestimate the mixture toxicity (In this case, they cannot be validated without testing)
- KCC and CR approaches ultimately need to be substituted by more scientific concepts and methodologies.

5.2 Outlook

The study for understanding mixture toxicity using various computational algorithms will be carried out.

Acknowledgments

This study is funded by the Korean Ministry of Knowledge Economy and Korea Institute of Science and Technology.

References

[1] Altenburger, R., Greco, R.W.: Extrapolation concepts for dealing with multiple contamination in environmental risk assessment. Integrated Environment Assessment and Management 5(1), 62–68 (2008)

[2] Yang, R.S.H., Thomas, R.S., Gustafson, D.L., et al.: Approaches to developing alternative and predictive toxicology based on PBPK/PD and QSAR modeling. Environ Health Perspect 106(suppl.), 1385–1393 (1998)

[3] Heather, L., Martin, C.S., Lindsay, J.L., et al.: Measurement and modelling of the toxicity of binary mixtures in the Nematode Caenorhabditis Elegans - A test of independent action. Environmental Toxicology and Chemistry 28(1), 97–104 (2009)

[4] Regulations (EC) No 1907/2006 of the European Parliament and of the Council of 18 December 2006 concerning the Registration, Evaluation, Authorisation and Restriction of Chemicals (REACH), establishing a European Chemical Agency, amending Directive 1999/45/EC and repealing Council Regulation (EEC) No. 793/93 and Commission Regulation (EC) No. 1488/94 as well as Council Directive 76/769/EEC and Commission Directives 91/155/EEC, 93/67/EEC, 93/105/EC and 2000/21/EC, http://ecb.jrc.it/legislation/2006R1907EC.pdf (accessed June 19, 2009)

[5] Gade, A.L., Øvrebø, S., Hylland, K.: Testing REACH draft technical guidance notes for conducting chemical safety assessments -The experience of a downstream user of a preparation. Regulatory Toxicology and Pharmacology 51(2), 168–180 (2008)

[6] European Chemical Industry Council (co-ordinator), Considerations on Safety Data Sheets and Chemical Safety Assessments of Preparations - Final report. Service Contract Nos. 22551-2004-12 F1SC ISP BE and 22552-2004-12 F1SC ISP BE (2005b)

[7] European Chemical Agency, Guidance on information requirements and chemical safety assessment chapter R.8: Characterisation of dose (concentration)-response for human health. Guidance for the implementation of REACH, http://guidance.echa.europa.eu/docs/guidance_document/information_requirements_r8_en.pdf?vers=20_08_08 (Accessed June 19, 2009)

[8] European Chemical Agency, Guidance on information requirements and chemical safety assessment chapter R.10: Characterisation of dose (concentration)-response for environment. Guidance for the implementation of REACH, http://guidance.echa.europa.eu/docs/guidance_document/information_requirements_r10_en.pdf?vers=20_08_08 (Accessed June 19, 2009)

[9] Jonker, M.J., Svendsen, C., Bedaux, J.M., et al.: Significance testing of synergistic/antagonistic, dose level-response analysis. Environmental Toxicology and Chemistry 24(10), 2701–2713 (2005)

[10] Ahmad, M.: Potentiation/antagonism of deltamethrin and cypermethrins with organophosphate insecticides in the cotton bollworm, Helicoverpa armigera (Lepidoptera: Noctuidae). Pesticide Biochemistry and Physiology 80(1), 31–42 (2004)

[11] Silva, E., Rajapakse, N., Kortenkamp, A.: Something from "nothing" - Eight weak estrogenic chemicals combined at concentrations below NOECs produce significant mixture effects. Environ Sci. Technol. 36, 1751–1756 (2002)

[12] TEAL & MACKRILL (2007) Safety Data Sheet - EPIDOX 2 zinc phosphate primer (11/09/2007), http://www.coo-var.co.uk/pdfs/04410845.pdf (accessed June 19, 2009)

[13] COO-VAR (2005) Safety Data Sheet - zinc phosphate primer (26/08/2005), http://www.coo-var.co.uk/pdfs/04410594.pdf (accessed June 19, 2009)

[14] US Environmental Protection Agency, High Production Volume chemical hazard data availability study (2007), http://www.epa.gov/HPV/pubs/general/hazchem.htm (accessed June 21, 2009)

[15] Cedergreen, N., Christensen, A.M., Kamper, A., et al.: A review of independent action compared to concentration addition as reference models for mixtures of compounds with different molecular target sites. Environmental Toxicology and Chemistry 27(7), 1621–1632 (2008)

[16] OECD, OECD Integrated High Production Volume Database: OECD Agreed Conclusions and Recommendations - Xylene (2003), http://cs3-hq.oecd.org/scripts/hpv/Status/DownloadFile.ASP?CASNUM=1330207&StatusCode=SIARC&DataNo=1 (accessed June 21, 2009)

[17] UNEP, OECD Initial Assessment Report for High Production Volume Chemicals including Screening Information DataSet (SIDS) - N-Butyl alcohol (2001), http://www.chem.unep.ch/irptc/sids/OECDSIDS/71363.pdf (accessed June 21, 2009)

[18] UNEP, OECD Initial Assessment Report for High Production Volume Chemicals including Screening Information DataSet (SIDS) - Iso-butanol (2004), http://www.chem.unep.ch/irptc/sids/OECDSIDS/78831.pdf (accessed June 21, 2009)

[19] European Chemical Bureau, European Union risk assessment report – Toluene. Office for official publications of the European Communities (2003), http://ecb.jrc.ec.europa.eu/esis/index.php?GENRE=CASNO&ENTREE=108-88-3 (accessed June 21, 2009)

[20] OECD, OECD Integrated High Production Volume Database: OECD Agreed Conclusions and Recommendations - Toluene (2001), http://cs3-hq.oecd.org/scripts/hpv/Index2.asp?CASNUM=108883 (accessed June 21, 2009)

[21] European Chemical Bureau, European Union risk assessment report - Zinc metal. Office for official publications of the European Communities (2008), http://ecb.jrc.ec.europa.eu/DOCUMENTS/Existing-Chemicals/RISK_ASSESSMENT/REPORT/zincmetalreport072.pdf (accessed June 21, 2009)

[22] European Chemical Bureau, IUCLID Dataset Solvent naptha (petroleum), heavy arom (2000), http://ecb.jrc.ec.europa.eu/esis/index.php?GENRE=CASNO&ENTREE=108-88-3%20 (accessed June 21, 2009)

[23] European Chemical Bureau IUCLID Dataset 1,2,4-Trimethylbenzene (2000), http://ecb.jrc.ec.europa.eu/esis/index.php?GENRE=CASNO&ENTREE=108-88-3%20 (accessed June 21, 2009)

[24] European Chemical Bureau, IUCLID Dataset Solvent naptha (petroleum), medium aliph (Mesitylene) (2000), http://ecb.jrc.ec.europa.eu/esis/index.php?GENRE=CASNO&ENTREE=108-88-3%20 (accessed June 21, 2009)

Search for New Physics with AMS-02 Transition Radiation Detector

Chanhoon Chung[1]

Abstract. Today the universe consists of 4.6% of ordinary matter, 23.3% of dark matter and 72.1% of dark energy. The dark matter is generally assumed be stable, non-relativistic and only weakly interacting. But we do not know what the dark matter is made of and how it is distributed within our Galaxy. In general, the cosmic antiparticles are expected as secondary products of interactions of the primary cosmic-rays (CRs) with the interstellar medium during propagation. While the measurements of CR positrons have become more precise, the results still do not match with the pure secondary origins. AMS-02 is a large acceptance precision particle spectrometer approved for installation on the International Space Station (ISS). A key feature of AMS-02 is precise particle identification for measurements of primary cosmic ray anti-particle spectra with negligible background up to a momentum of 500 GeV/c to allow indirect searches for dark matter. To efficiently separate positrons/electrons from protons/anti-protons, AMS-02 will be equipped with a Transition Radiation Detector (TRD) with 5248 straw tube proportional counters filled with a Xe/CO_2 (80/20) mixture. The AMS-02 TRD was fully assembled and integrated into AMS-02 in 2007. In 2008 AMS-02 had recorded cosmic ray particles on ground to demonstrate full functionality of the device. For the AMS-02 TRD it will be shown that the detector response is as expected and the gas tightness will allow operation in space for 20 years with a gas supply of 25 kg.

1 Introduction

Recently published data from ATIC [4], PPB-BETS [5], Fermi-LAT [6] and H.E.S.S. [7] instruments have opened a new window in the study of Cosmic Ray (CR) electrons as shown in Fig. 1. These experiments report spectral deviations from the conventional CR propagation model and imply the presence of unknown primary sources such as nearby pulsars or exotic dark matter annihilation in the Galatic halo. However, a lack of accurate measurement of high energy CR positrons is a major disadvantage for understanding their origin and propagation in the local interstellar medium. Positron fraction data from PAMELA [3] and electrons data from other experiments are not consistent with any CR propagation model.

[1] RWTH Aachen University, I. Physikalisches Institut 1B, Aachen, Germany
chchung@physik.rwth-aachen.de

J.H. Lee et al. (Eds.): EU-Korea Conference on Science and Technology, SPPHY 135, pp. 203–212.
springerlink.com

The Alpha Magnetic Spectrometer (AMS-02) is a large acceptance precision particle spectrometer approved for installation on ISS to measure primary CR spectra in space. It will precisely determine the fluxes of elemental abundances with charge separation up to Z = 26 (Fe) in the energy range from 100 MeV/nucleon to 1 TeV/nucleon. As shown in Fig. 2 the spectrometer uses a Superconducting Magnet at its core and consists of a Silicon Micro-strip Tracker, a Transition Radiation detector (TRD), a Time of Flight (ToF) system with Anti-Coincidence Counter (ACC), a Ring Image Cherenkov Counter (RICH) and an Electromagnetic Calorimeter (ECAL). For details refer to Ref. [9].

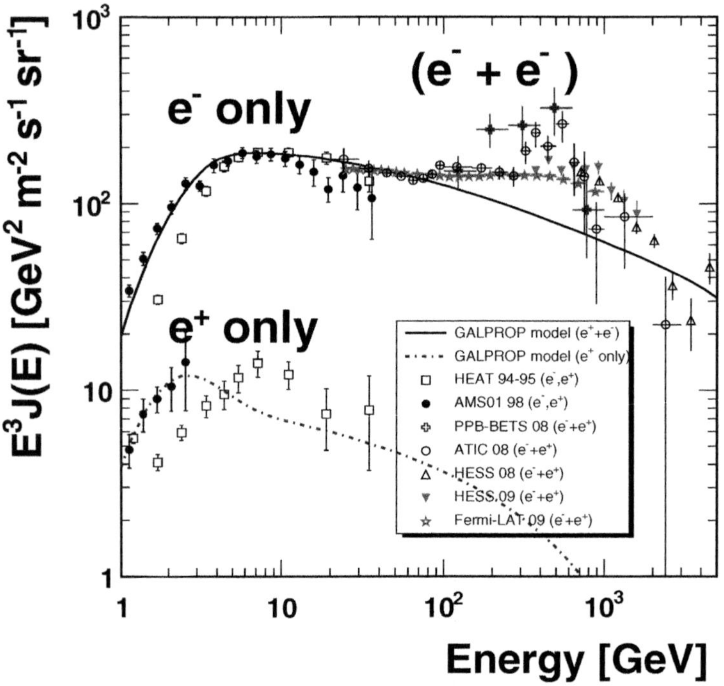

Fig. 1. The energy spectrum of CR electrons as measured by AMS [1], HEAT [2], ATIC [4], PPB-BETS [5], Fermi-LAT [6] and H.E.S.S. [7] experiments. A CR conventional model based on GALPROP codes [8] is included to compare the electron and positron spectra.

In AMS-02 the TRD is a key element to improve positron identification from the dominant proton background in a momentum range from 10 to 500 GeV/c by a proton rejection factor of $10^3 \sim 10^2$ at 90% positron efficiency [10, 11]. In conjunction with the ECAL and Tracker, overall rejection power can be improved to better than 10^6 by distinguishing lateral shower development between electromagnetic showers and wider hadronic showers and matching between shower energy and momentum.

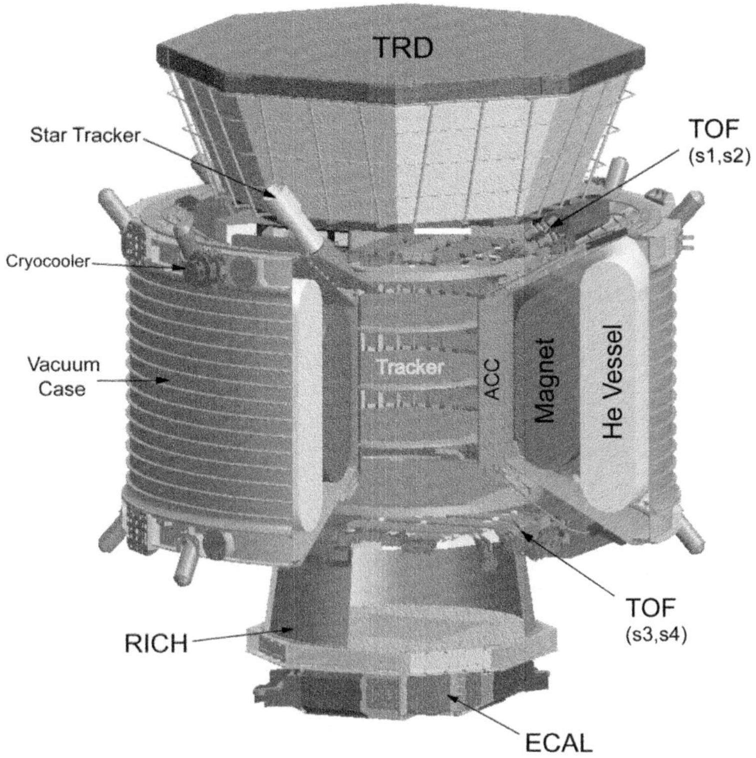

Fig. 2. AMS-02 Detector

2 AMS-02 TRD

Transition Radiation (TR) is observed as X-rays when a highly relativistic charged particle ($\gamma = E/mc^2 > 10^3$) passes through a material with varying index of refraction. In practice, the TRD is used to provide electron separation for 1 GeV/c $\leq p \leq$ 500 GeV/c because electrons emit TR whereas protons do not. To provide an efficient X-ray photon absorption a high-Z gas mixture of Xe/CO_2 (80/20) is used.

2.1 *Mechanical Structure*

The TRD consists of 20 layers of straw modules interleaved with 20 mm thick fleece radiators supported in a conical octagon support structure. The upper four and lower four layers (yz plane in the AMS-02 coordinates) are parallel to the magnetic field, and the middle 12 layers (xz plane) are perpendicular in order to provide 3D tracking of the CRs. This detector has 328 individual straw modules and 5248 readout channels. Each straw module is composed of 16 single straw tubes with a diameter of 6 mm and a straw wall thickness of 72 μm. A straw module is closed with two polycarbonate end-pieces for gas supply and signal readout. Its mechanical structure is re-enforced by longitudinal and transversal stiffeners. The radiator is made of 10 μm polypropylene/polyethylene fibre fleece with a density of 0.06 g/cm^3. The octagon wall made of carbon-fibre-aluminium honeycomb is the main support structure for radiators and straw modules. It is manufactured with a precision better than 100 μm to center the wires for a gas gain homogeneity better than 2 % for 5248 straw tubes in total.

2.2 *Quality Control of Straw Module Production*

For use in space, all flight straw modules have to comply with the following requirements: gas-tight, low dark current and high gas gain homogeneity performed with a ^{55}Fe radioactive source. A detailed description of the straw module production can be found in Ref. [12].

2.3 *DAQ and Slow Control Electronics*

The electronics for the TRD is divided into the 82 Front-End boards (UFE) mounted on the octagon walls, two U-Crates for data acquisition, and one UG-Crate to control the gas supply system.

Each UFE has two VA32 charge sensitive amplifiers to multiplex the analogue signals from four straw modules to serial ADCs (AD7476). It has a peaking time of 2.4 μs and the input is linear up to 2000 fC, corresponding to 50 MIPs signals at gas gain of 3000. The nominal gain of the UFE is 2 ADC/fC and its uniformity between channels of the same chip is better than 1%.

The U-Crate [13] includes three UFE power supply boards (UPSFEs), six data reduction boards (UDR2s), six high voltage supply boards (UHVGs) and a high rate interface board (JINF). The power for the UFE is controlled and regulated by the UPSFE. Each UPSFE consists of 14 linear regulators, two Actel FPGAs and slow control circuitry to switch on and off other U-Crate boards. Due to limited usable bandwidth of high level data link (HRDL) interface to the ISS each subdetector uses its own data reduction board based on a common digital part (CDP). The UDR2 comprises two CDPs for redundancy and an interface for 7 UFEs. Each CDP consists of a FPGA and a DSP where data is zero-subtracted and buffered. The UHVG can supply up to 1600 V using 16 stages of Cockroft-Walton voltage multiplier controlled by LeCroy MHV100 chips.

On request, the JINF sends commands, collects the data from all UDR2 boards and transfers it to the main AMS DAQ system by AMSWire[2] protocol for storage or direct downlink to earth. The time to readout 5248 channels is less than 80 μs and the data processing time is faster than 230 μs/event capable of handling a 10 kHz peak and 4 kHz average trigger rate. Also JINF is connected in parallel by LeCroy links to the UPSFE, UHVG boards and UPD control electronics to control and monitor the crate status.

---

[2] AMSWire: a serial point-to-point protocol based on IEEEE-1355.

The power for TRD electronics is generated by custom-made DC/DC converters housed inside the UPD which is connected to the AMS-02 main power distribution system (PDS). Power consumption of the complete TRD electronics is less than 100 W. To meet the NASA requirements, all of the flight front-end/crate electronics boards went EMI, vibration and thermal vacuum tests without any failure.

2.4 Gas Supply System

The gas system stores 20.0 kg Xe and 5.0 kg CO_2, filters, mixes, transfers and circulates a weekly resupply of Xe/CO_2 gas losses for at least three years. It can be divided into a supply system (Box S), a circulation system (Box C) and four gas distribution manifolds. Two mixing circuits from Xe and CO_2 vessels (Box S) convey the gases to the mixing vessel where the required mixture is made using partial pressures. A system of valves allows the transfer of gas from the mixing vessel to Box C. In Box C, two redundant pumps circulate the gas through the TRD volume in order to keep the gas mixed. The CO_2 fraction is monitored with a spirometer by measuring the speed of sound in Box C. Manifolds distribute the gas to the 5248 straw tubes segmented into 10 parallel loops. Each loop has two valves and one differential pressure sensor across a flow restrictor on inlet and outlet. The manifold pressure sensors are constantly monitored and relevant valves are closed to isolate a loop in case a leak is detected.

3 Performance of TRD during Pre-integration

The TRD was fully assembled at RWTH Aachen and successfully integrated into AMS-02 at CERN in 2007. The Tracker, ToF, ACC, RICH and ECAL were also integrated in the main DAQ system for complete detector/trigger operations and data quality monitoring as a whole. From December 2007 to June 2008, AMS-02 has recorded cosmic muons on ground to demonstrate full functionality of the instrument except of the magnet. The full software chains have been developed in order to reconstruct and analyse the recorded events as well as detector alignment and calibration. Within the framework of the AMS-02 data scheme, slow control data indicating the state of TRD such as measured voltages, currents, pressures, temperatures and performance statistics with operational information were also produced to asses the condition of the experiment. The excellent noise performance, position resolution, occupancy and signal homogeneity is acquired by analyzing the collected cosmic muons data.

3.1 Gas Tightness

The gas leak rate of the TRD is measured by increasing the ArCO_2 pressure to 1.8 bar and monitoring the pressure drop as a function of time. Figure 3 shows the pressure drop as a function of time. Precise pressure sensors with a precision of

0.4 mbar are used to monitor pressure drop at both inlet and outlet. The TRD has a volume of about 230 liters and its leak rate of $(0.30 \pm 0.01) \times 10^{-2}$ l·mbar/s is in agreement with individual He pressure drop measurements during module production. This result confirms that the TRD is still gastight to the diffusion level and allows TRD operation in space for 20 years with a gas supply of 25 kg.

3.2 Muon Signal, Position Resolution and Noise Performance

The muon signal is obtained by using a tracking algorithm, based on a statistical approach for all hit pairs found in the TRD, within the AMSRoot offline software framework. AMSRoot is a CVS standalone distribution for user analysis of AMS data and it provides an object oriented framework for event simulations and reconstructions in the AMS-02 detector. The event selection requires 4/4 layers in ToF, no hit in ACC and track events in Tracker. The implemented tracking algorithm detects single cosmic muon tracks with a tight constraint on number of hits in both xz and yz planes. Figure 4 shows a distribution of energy deposited on single straw tubes for selected cosmic muon events. The most probable value (MPV) of the Landau distribution is about 74 ADC counts and its monotonic tail extends to the limit of the Frontend dynamic range.

A position resolution is calculated from the width of the residual distribution of the reconstructed hits at tube center with respect to the track fit as shown in Figure 5. Regardless of the incident particle angles all reconstructed events are included. At signal to noise ratio of 45, the achieved single hit resolution of 2.0 mm is already close to the expected 1.8 mm for a pitch of 6.2 mm. In both xz and yz planes, the position resolution is the same value as designed.

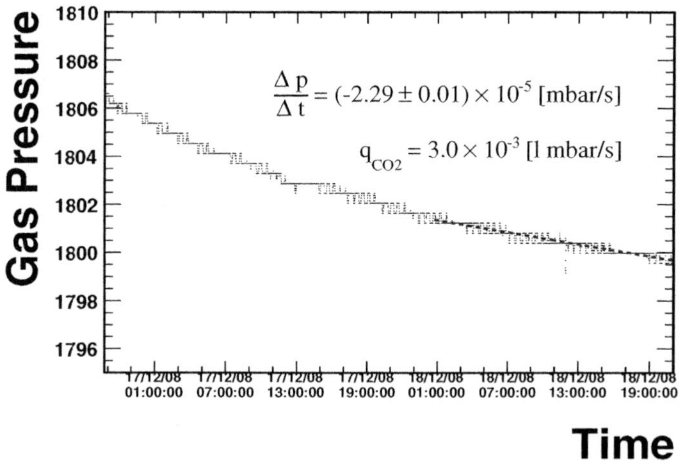

Fig. 3. Gas tightness testing for a TRD gas volume of 230 liters with ArCO2 (80/20) at 1800 mbar and constant temperature of 23.0 °C. Pressure sensors monitor on both gas inlet and outlet sides.

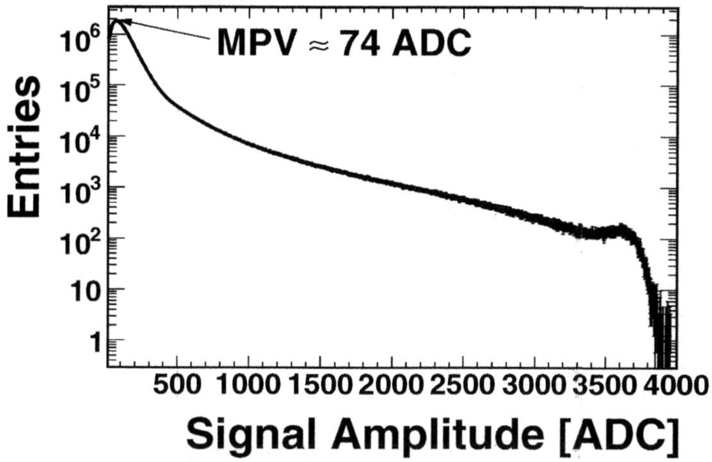

Fig. 4. Distribution of energy deposited on single straw tube by single muon track events. The ArCO2 gas is supplied at a constant flow rate of 1 l/h and the same high voltage of 1400 V is applied after calibration of each UHVG channel.

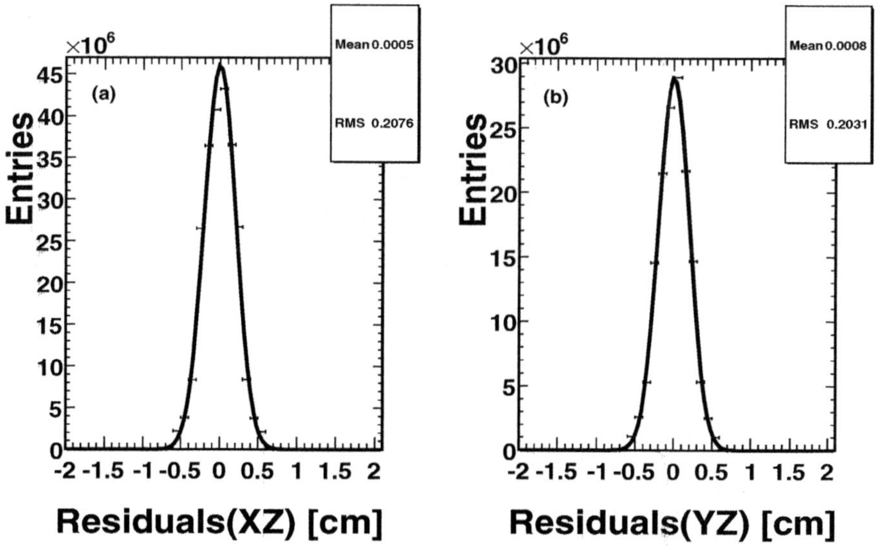

Fig. 5. Distribution of the residuals of the straw tubes from the reconstructed particle tracks in both xz (a) and yz (b) planes. Two residual plots are almost identical in the position resolution of 2.0 mm.

In the normal operation, a noise measurement gives an average value of less than 2 ADC counts corresponding to an equivalent noise charge of 4000 electrons. The spread of the noise with common mode noise of 1.6 ADC counts subtracted is in the order of 5 % as shown in Figure 6.

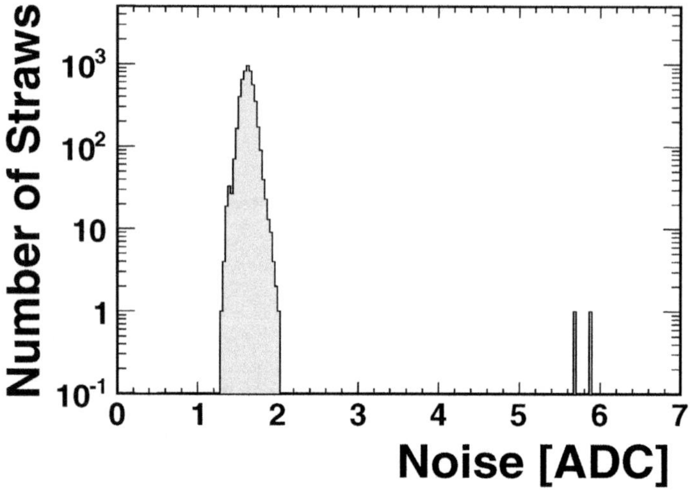

Fig. 6. The distribution of noise for all straw tubes. The average noise is 1.62 ± 0.08 ADC and spread of the noise is within 5 % except of two noisy channels out of 5248 channels.

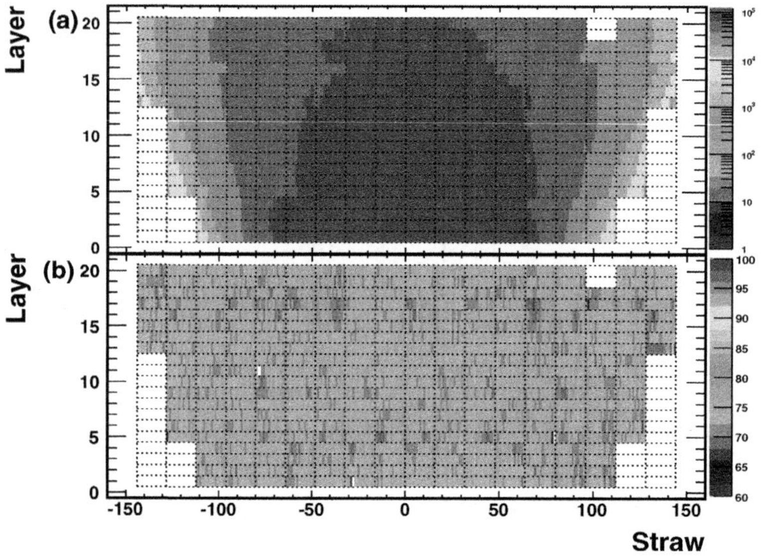

Fig. 7. Track occupancy (a) and gas gain homogeneity (b) with cosmic muons. The obtained average deposited energy is 73.56 ± 1.81 ADC counts and overall signal inhomogeneity is 2.5 % after corrections. The palette scales show the number of hit per straw tube (a) and the most probable value of the Landau fit on each straw tube (b). Two straw modules mounted 19th and 20th layers are excluded due to an unstable preamplifier chip which was replaced after pre-integration stage.

3.3 Occupancy and Homogeneity

In Figure 7-(a) an occupancy plot of reconstructed tracks from cosmic muons shows the homogeneous acceptance of the detector. To examine the signal homogeneity of the TRD, energy deposited on the single straw tube is calculated by considering only good single track events in a subset of good runs. The homogeneous response to cosmic muons is improved after corrections of HV and UFE calibrations as well as gas density variations in straw modules. As shown in Figure 7-(b) the signal in-homogeneity is 2.5 % which is negligible for the positron/proton separation.

4 Summary and Milestone

AMS-02 is approved for launch to the ISS in 2010 and its new data with increased statistics, excellent particle identification and wide energy coverage promise significant advances in the study of CR origin and propagation. The AMS-02 TRD is designed to achieve excellent positron identification in conjunction with ECAL and Tracker with a proton rejection of better than 10^6 at 90 % positron efficiency up to 500 GeV/c, and 3D particle tracking in 20 layers with a single hit resolution of about 2 mm as well as a signal homogeneity better than 3 %. During pre-integration of the detector from December 2007 to June 2008, AMS-02 has continuously recorded cosmic muons without any sign of performance deterioration. The results of a performance study with cosmic muons prove that the TRD fulfills the design requirements. In 2009 AMS-02 is scheduled to conduct a beam test at CERN and thermal vacuum tests at ESA-ESTEC in the Netherlands. Afterwards it will be delivered to NASA-KSC to prepare for the launch on flight (STS-134) scheduled for September 2010.

Acknowledgments. This project is funded by the German Space Agency (DLR) under contract Nr.50OO0501, the US Department of Energy DOE and NASA.

References

[1] Alcaraz, J., et al.: Physics Letter B 484, 10 (2000)
[2] DuVernois, M.A., et al.: Astrophysical Journal 559, 296 (2001)
[3] Adriani, O., et al.: arXiv:0810.4995 [astro-ph] (2008)
[4] Chang, J., et al.: Nature 456, 362 (2008)
[5] Torii, S., et al.: arXiv:0809.0760 [astro-ph] (2008)
[6] Abdo, A.A., et al.: arXiv:0905.0025 [astro-ph] (2009)
[7] Aharonian, F., et al.: arXiv:0905.0105 [astro-ph] (2009)
[8] Strong, A.W., Moskalenko, I.V., Reimer, O.: Astrophysical Journal 613, 962 (2004)
[9] Chung, C.H., et al.: Proceedings of the EPS-13, Bern, Switzerland (2005), arXiv:0710.2428, Proceedings of the 15th International Conference on Supersymmetry and the Unification of Fundamental Interactions, Karlsruhe, Germany (2007), Proceedings of the 31st International Cosmic Rays Conference, Lodz, Poland (2009)

[10] Doetinchem, P.v., et al.: Nuclear Instruments Methods A 558, 526–535 (2006)
[11] Siedenburg, T., et al.: Nuclear Physics B – Proceeding Supplements V150, 30–33 (2006), V113, 154–158 (2002)
[12] Kirn, T., et al.: Proceedings of TRDs for the 3rd millennium. Frascati Physics Series, vol. XXV, p. 161 (2002); Nuclear Instruments Methods A 522, 69–72 (2004)
[13] Hauler, F., et al.: IEEE Trans.Nucl.Sci. 51, 1365–1372 (2004)

The State of the Art of Visual Analytics

Dong-Han Ham[1]

Abstract. One of the critical issues challenging human decision makers in the information age is how to find out data relevant to their tasks and how to derive meaningful information from these data. As a new discipline for dealing with this issue, visual analytics has recently emerged. Visual analytics is defined as the science of analytical reasoning facilitated by interactive visual interfaces. It is a multidisciplinary subject that is related to data mining, information visualization, knowledge science, human factors, and so on. This paper reviews the state of the art of visual analytics and claims that the problems of visual analytics should be considered in the context of human-computer interaction and joint cognitive systems. Based on the review results, this paper proposes a conceptual framework for organizing research problems studied so far and identifying viable future research directions. The author hopes that this paper will provide well-organized information about visual analytics and be a good source for researchers who are interested in this new discipline.

1 Introduction

With the rapid growth of information and communication technologies, decision making and problem solving tasks can be conducted based on a large amount of data and information [17]. Although this seems to be beneficial to human decision makers, this has negative aspect at the same time, which is information overload problem [25]. It is highly likely that human decision makers have difficulty identifying data relevant to their tasks and obtaining meaningful information from the data, thereby getting lost in a mass of data [20]. Thus a lot of studies have been devoted to the problem of how to provide the right information in the right form at the right time to the right person under the umbrella of intelligent decision support [16, 25].

Regarding the problem above, it is worth considering two points in relation to human decision making context and performance. Firstly, in many times, decision makers cannot expect all kinds of task situations and don't know always what information should be and can be obtained from data available to them [4]. Accordingly they want to be supported in three aspects: (1) getting information relevant to expected task situations, (2) exploring meaningful information from data sets, which can be useful to unexpected task situations, and (3) discovering unknown

[1] School of Engineering and Information Sciences, Middlesex University,
London, United Kingdom
d.ham@mdx.ac.uk

J.H. Lee et al. (Eds.): EU-Korea Conference on Science and Technology, SPPHY 135, pp. 213–222.
springerlink.com

information from data sets, which can be used to create new knowledge [5, 18]. Secondly, considering the forms of information presented to decision makers, it is important to visualize the semantics of information in a way that they easily perceive and utilize it without much cognitive workload [21].

In order to address these two points under a unified framework, visual analytics (VA) has recently emerged as a new discipline [25]. VA is defined as the science of analytical reasoning facilitated by interactive visual interfaces [33]. VA aims to develop techniques and tools to support people in synthesizing information and deriving insight from massive, dynamic, unclear, and often conflicting data. In terms of decision making process, VA is an interaction between humans and computers that involves information gathering, data pre-processing, knowledge acquisition, knowledge modelling, knowledge representation, perception of knowledge representation, and decision making [33]. As such, VA emphasizes the cooperation between humans and computers [20, 25].

As a multidisciplinary subject, VA has been influenced by several disciplines. Two academic areas, however, are most influential to VA: data mining and visualization [20]. Data mining or knowledge discovery aims to find out new kind of information from a set of data by usually employing methods of artificial intelligence (AI) and statistics. The general process of getting new knowledge contains four operations: classification, clustering, association, and prediction of pattern and trend [18].

Visualization is generally composed of two distinct areas: scientific visualization and information visualization [30]. Scientific visualization studies how to visualize large amounts of scientific data obtained from sensors or laboratory tests. In contrast, information visualization aims to explore large amounts of abstract data, which cannot be directly mapped onto the geographic coordinates or into virtual 3D environments, to derive new insights. In recent years, to emphasize the roles of visualization in knowledge transfer among people, new visualization area named knowledge visualization has emerged [1, 10, 12].

VA as a discipline is in its infancy and thus needs a range of research activities [33]. Now is a right time to review the state of the art of VA and establish a framework for directing future research directions [11]. The purpose of this paper is to give a concise review of VA and then propose a conceptual framework that can be used a basis for identifying promising research areas and organizing research results of VA. Particularly, the author emphasizes that VA should be studied in the context of human-computer interaction (HCI) and joint cognitive systems (JCS).

2 Review of Studies on Visual Analytics

As space is limited, it is difficult to give very comprehensive reviews here. Thus the author aims to present essential information about VA research concisely, dividing it into four parts, and inspire readers to explore VA in more detail.

2.1 Scope and Process of VA

National Visualization and Analytics Centre (NVAC) established by the department of homeland security in USA is a world-leading organization for VA research [25]. NVAC recently published a book titled "*Illuminating the Path: The Research and Development Agenda for Visual Analytics*", which has become a bible in VA. According to this book, research activities of VA can be categorized into four parts: (1) analytical reasoning techniques enabling humans to obtain deep insights that support situation assessment, planning, and decision making, (2) visual representation and interaction techniques that facilitate visual perception and information exploration of decision makers, (3) data representations and transformations that convert several types of conflicting data to the formats with which coherent visualization is possible, and (4) production, presentation, and dissemination of analytical results to deliver information to a variety of people.

Based on these four major parts, VA experts of NVAC recently proposed a taxonomy for VA research [26]. The taxonomy has four high-level dimensions: domain/applications, analytic methods/goals, science/technology, and data types/structures. The science/technology dimension is composed of five sub areas: analytical reasoning and human processes, data representations and theory of knowledge, systems and evaluations, interactive visualization, and theory of communication. These dimensions and areas can be used to describe the purposes and characteristics of a research within VA.

Although there are several process models of VA, it is worth noting the process proposed by Keim et al. [19]. As shown in Fig. 1, VA process can be regarded as the transformation from data sets (S) to insight (I), which is a concatenation of several sub processes, such as visualizing data sets and generating hypothetical models from S.

The process emphasizes the harmonious collaboration between human decision makers and computing devices and visualization tools in performing cognitive tasks for discovering insight and managing data and information [15]. This is the point making VA different from other related disciplines [20]. Another point to note is that visualization for VA does not mean just only visualization of data sets of interest and information derived from them. Several types of reasoning artefacts (e.g., hypotheses, arguments, and causality) that are produced and used to carry out cognitive tasks also need to be visualized.

2.2 Visualization Techniques

As VA is a multidisciplinary subject, researchers always need to attempt to examine the state of the art of each discipline as well as VA itself. Particularly, because visualization is a fundamental process that significantly influences the method and process of VA, a systematic understanding of visualization techniques is absolutely needed [4, 22].

It is useful to develop a classification scheme of visualization techniques in consideration of their design implications to help designers select visualization

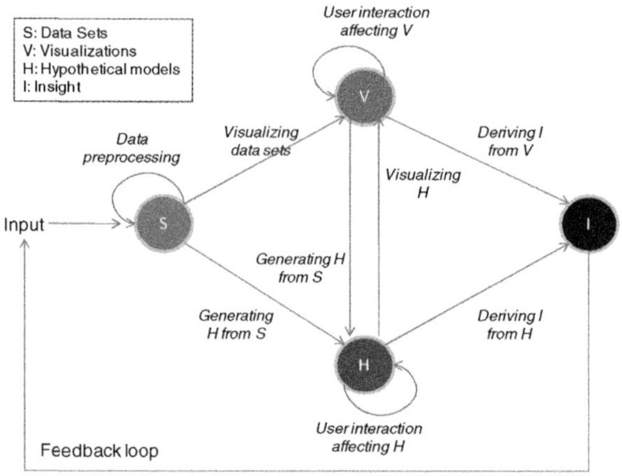

Fig. 1. General model of VA process [19]

techniques suitable to tasks of interest [17]. There are several good studies that listed and classified the currently available visualization techniques [4, 7-8, 24, 31]. Here three classification schemes are presented. Card [2] gives a comprehensive list of information visualization techniques and categorized them into four groups by visual structures. Visual structures mean the structures that express variable values through a vocabulary of visual elements like spatial substrates and graphical properties. The four groups are: simple visual structures composed visual structures, interactive visual structures, and focus + context attention-reactive visual abstraction. For example, the group of interactive visual structures contains six techniques: dynamic queries, magic lens, overview + detail, linking and brushing, extraction and comparison, and attribute explorer.

Tory and Möller [27-28] claims that the taxonomy of visualization techniques needs to be based on the type of data model they use, rather than the attributes of data itself. Thus visualization techniques can be divided into two groups: continuous model visualization and discrete model visualization. Continuous model visualization is broken down according to the number of independent and dependent variables, and the type of the dependent variables. Discrete model visualization is classified according to whether data points are considered connected or unconnected. Unconnected models can be further classified by the number of dimensions that the visualization supports (2D, 3D, or nD).

The third classification scheme suggested by Keim and Ward [18] uses three criteria: the data to be visualized, the display types, and interaction techniques. The data type to be visualized may be: one-dimensional data, two-dimensional data, multidimensional data, text and hypertext, hierarchies and graphs, and algorithms and software. There may be five types of information display: standard 2D/3D display, geometrically transformed display, iconic display, dense pixel display, and stacked display. Typical examples of interaction techniques include: standard, projection, filtering, zoom, distortion, and link and brush. These three criteria are supposed to be orthogonal, and any visualization can be characterized by the combination of elements within the three criteria.

Considering that two main components of information visualization systems are representation and interaction, most of classification studies are more focused on the part of representation [9]. Thus more studies on the classification of interaction techniques are needed. Regarding this problem, Yi et al. [33] propose an interesting classification of interaction techniques. They categorized interaction techniques into seven types according to the intent of tasks conducted with visualization systems, which include: select, explore, reconfigure, encode, abstract/elaborate, filter, and connect. They argue that the classification based on human tasks rather than low-level techniques offered by visualization systems can give more useful foundation toward a deeper understanding of visualization techniques.

2.3 Human Factors in Visualization

One mainstream in VA research is to examine why and how visualization influences the performance of human cognitive tasks, most of which are related to human factors [13, 29]. Many studies have reported the benefits of information visualization in relation to human cognitive performance [6, 23]. Card et al. [3] listed six ways that visualization can amplify cognition

- Increasing memory and processing cognitive resources available
  (high-bandwidth hierarchical interaction, parallel perceptual processing, offload work from cognitive to perceptual system, expanded working memory, and expanded storage of information)
- Reducing search for information
  (Locality of processing, high data density, and spatially indexed addressing)
- Enhancing the recognition of patterns
  (recognition instead of recall, abstraction and aggregation, visual schemata for organization, and value-relationship-trend)
- Enabling perceptual inference operations
- Using perceptual attention mechanism for monitoring
- Encoding information in a manipulable medium

Based on empirical results about the relation between visualization and human performance, several design principles for visualization become available to visualization designers [29]. Wickens et al. [32] present thirteen design principles that are classified into four groups, as follows.

<Perceptual principles>
- Avoid absolute judgment limits
- Top-down processing
- Redundancy gain
- Discriminability-similarity causes confusion

<Mental model principles>
- Principle of pictorial realism and principle of configural displays
- Principle of the moving part
- Ecological interface design

<Attention principles>
- Minimizing information access cost
- Proximity compatibility principle
- Principle of multiple resources

<Memory principles>
- Principle of predictive aiding
- Principle of knowledge in the world
- Principle of consistency

As various techniques and tools are developed in VA, research issues about evaluation and development methodology will be more important [5]. User-centred design methods and a range of usability evaluation methods developed in the field of human factors and HCI will give a good background for addressing them [14]. For example, human-centred visualization methodology proposed by Zhang et al. [34] gives a good answer to the question of how to incorporate user and task analysis results into the design process of visualization. But as Chen [5] pointed out, most of currently available evaluation and development methodologies are those that predated information visualization and VA era. Thus they might not address the critical details specific to VA. For this reason, there is a strong need to develop new kinds of evaluation and development methodologies for developing VA systems.

2.4 Application Areas and Software Tools of VA

Keim et al. [19] gives a good summary of the 10 most promising application areas of VA. They include: physics and astronomy, business, environmental monitoring, disaster and emergency management, security, software analytics, biology, medicine and health, engineering analytics, personal information management, and mobile graphics and traffic. For example, VA can provide a useful means to find out new patterns of terroristic activities, based on a set of related data, such as communication records between potential terrorists, political news in local areas, travelling paths of the terrorists, and so on.

The software tools listed below are the results collected from the author's survey and VA digital library (http://vadl.cc.gatech.edu/). It should be noted that this is never a complete list, but a very partial list of the tools. Detailed description of the following tools is not given due to the limitation of space.

- Thinkmap (www.thinkmap.com)
- Spotfire (www.spotfire.com)
- InfoZoom (www.infozoom.com)
- InfoScope (www.macrofocus.com)
- Table Lens/Eureka (www.inxight.com)
- Ezchooser (www.brisa.merl.com:8080/myezchooser)

- Advizor (www.advizorsolutions.com)
- ILOG Discovery (www2.ilog.com/preview/Discovery/)
- Polaris (www.tableausoftware.com)
- InfoVis Toolkit (ivtk.sourceforge.net)
- Prefuse (prefuse.sourceforge.net/)
- Piccolo (www.cs.umd.edu/hcil/piccolo/)
- Panopticon (www.panopticon.com)
- Jigsaw (www.cc.gatech.edu/gvu/ii/jigsaw/)

3 Research Framework for Visual Analytics

For the last several years, there has been the rapid growth of VA research in terms of the range of topics and practical applications to real world problems. However, there is a lack of conceptual frameworks for classifying and organizing various research results and suggesting viable research topics. This paper proposes a research framework emphasizing that VA research needs to be understood and conducted in the context of HCI and JCS (Fig. 2).

It is interesting to compare VA research activities with those of cognitive systems engineering (CSE). CSE is a new type of systems engineering that is concerned with analysis, design, and evaluation of complex socio-technical systems.

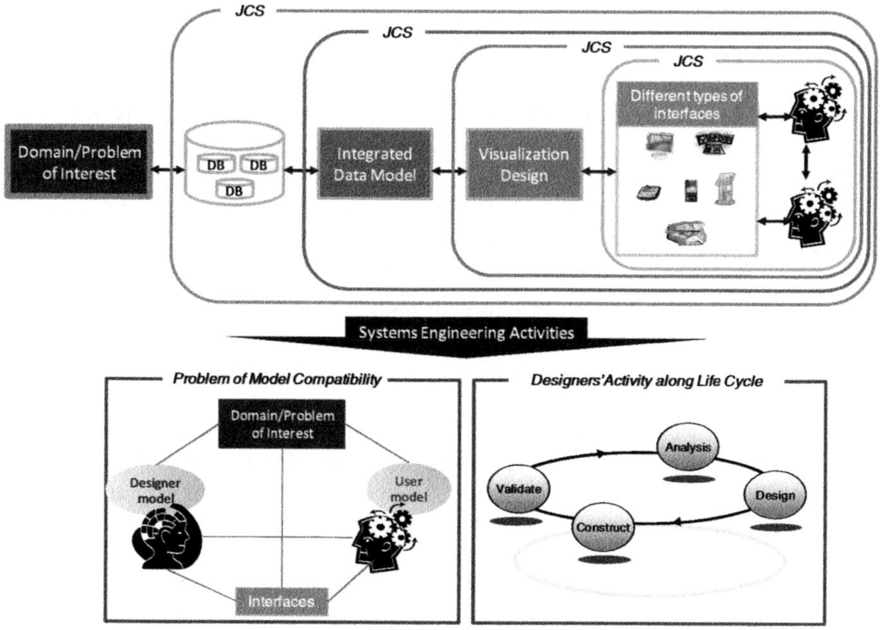

Fig. 2. A Framework for VA Research

The ultimate purpose of CSE is to realize a joint cognitive system that makes the most of cognitive strengths of humans and computers. One of critical means to achieve this purpose is to design intelligent decision support with advanced visualizations to facilitate human perception and cognition, which is very similar to VA. The success of VA systems is highly dependent on the effective interaction between humans and computers; thus HCI concepts and principles should be the basis for designing VA systems.

The framework suggests that researchers should give their more attention to some of research topics, which have not been much studied so far, to advance the state of the art of VA. The topics include the compatibility between designers' model and users' model on domain and interfaces, collaborative VA focusing on knowledge visualization, effects of technical constraints by the types of devices on design process, and the difference between exploration and presentation of information that visualization supports, and so on.

Whatever the scope of JCS under consideration is, three types of research activities should always be conducted in parallel: scientific study concerned with modelling, engineering activities for actual design implementation, and methodological development bridging science and engineering. Particularly, the scientific and methodological aspects of VA research can be much benefited by earlier studies in the field of CSE and HCI.

4 Conclusion

Due to a large amount of data generated from multiple sources on a daily basis, two research problems are increasingly significant to achieve effective decision support: (1) how to process a lot of data generated from multiple sources to secure useful information and (2) how to visualize the information to be easily perceived by humans. VA is a newly developed discipline to deal with these problems in an integrated manner. It is expected that human decision makers can enhance their situation awareness and sense-making with the use of VA technologies and tools. This paper reviews the earlier studies on VA from the four perspectives: scope and process, visualization techniques, human factors in visualization, and application problems and software tools. To help researchers organize the studies of VA and derive viable research issues, this paper proposes a conceptual framework for VA research.

VA and CSE share many things in common in terms of research objectives and process. They all aim to enhance the performance of decision making and other cognition-based tasks by seamlessly integrating the strengths of humans and computers-realizing joint cognitive systems. Future research activities within VA need to be planned and pursued with the aim of building human-computer jointly intelligent visualization systems.

Acknowledgments. The authors wish to acknowledge the assistance and support of all those who contribute to EKC 2009.

References

[1] Burkhard, R.A.: Towards a framework and a model for knowledge representation: Synergies between information and knowledge representation. In: Tergan, S.-O., Keller, T. (eds.) Knowledge and Information Visualization. LNCS, vol. 3426, pp. 238–255. Springer, Heidelberg (2005)
[2] Card, S.: Information visualization. In: Sears, A., Jacko, J. (eds.) The human-computer interaction handbook: fundamentals, evolving technologies and emerging applications. Lawrence Erlbaum Associates, NJ (2000)
[3] Card, S., Mackinlay, J., Shneiderman, B.: Information visualization: Using vision to think. Morgan Kaufmann, San Francisco (1999)
[4] Chen, C.: Information visualization: Beyond the horizon. Springer, London (2006)
[5] Chen, C.: Top 10 unsolved information visualization problems. IEEE Computer Graphics and Applications 25(4), 12–16 (2005)
[6] Chen, C., Yu, Y.: Empirical studies of information visualization: A meta-analysis. International Journal of Human-Computer Studies 53(5), 851–866 (2000)
[7] Chengzhi, Q., Chenghu, Z., Tao, P.: Taxonomy of visualization techniques and systems: Concerns between users and developers are different. In: Proceedings of Asia GIS Conference 2003, Wuhan, China (2003)
[8] Chi, E.H.: A taxonomy of visualization techniques using the data state reference model. In: Proceedings of the IEEE Symposium on Visualization 2000, Salt Lake City, USA, pp. 69–76 (2000)
[9] Chi, E., Riedl, J.T.: An operator interaction framework for visualization systems. In: Proceedings of IEEE Symposium on Information Visualization, North Carolina, USA, pp. 63–70 (1998)
[10] Dillon, K.M., Talbot, P., Hillis, W.D.: Knowledge visualization: Redesigning the human-computer interface. Technology Review Journal 13(1), 37–55 (2005)
[11] Duke, D., Brodlie, K., Duce, D., et al.: Do you see what I mean? IEEE Computer Graphics and Applications 25(3), 6–9 (2005)
[12] Eppler, M.J., Burkard, R.A.: Knowledge visualization: Towards a new discipline and its fields of application. ICA Working Paper #2/2004, University of Lugano (2004)
[13] Fekete, J.-D., van Wijk, J., Stasko, J., et al.: The value of information visualization. In: Kerren, A., Stasko, J.T., Fekete, J.-D., North, C. (eds.) Information Visualization. LNCS, vol. 4950, pp. 1–18. Springer, Heidelberg (2008)
[14] Graham, M., Kennedy, J., Benyon, D.: Towards a methodology for developing visualizations. International Journal of Human-Computer Studies 53(5), 789–807 (2000)
[15] Green, T.M., Ribarsky, W., Fisher, B.: Building and applying a human cognition model for visual analytics. Information Visualization 8(1), 1–13 (2009)
[16] Hollnagel, J., Woods, D.: Joint cognitive systems: Foundations of cognitive systems engineering. CRC Press, Florida (2005)
[17] Keim, D.: Information visualization and visual data mining. IEEE Transactions on Visualization and Computer Graphics 7(1), 100–107 (2002)
[18] Keim, D., Ward, M.: Visualization. In: Michael, B., Hand, D. (eds.) Intelligent data analysis: An introduction. Springer, Berlin (2002)
[19] Keim, D., Andrienko, G., Fekete, J.-D., et al.: Visual analytics: Definition, process, and challenges. In: Kerren, A., Stasko, J.T., Fekete, J.-D., North, C. (eds.) Information Visualization. LNCS, vol. 4950, pp. 154–175. Springer, Heidelberg (2008)

[20] Keim, D., Mansmann, F., Schneidewind, J.: Visual analytics: Scope and challenges. In: Simoff, S.J., Böhlen, M.H., Mazeika, A. (eds.) Visual Data Mining. LNCS, vol. 4404, pp. 76–90. Springer, Heidelberg (2008)
[21] Keller, T., Tergan, S.-O.: Visualizing knowledge and information: An introduction. In: Tergan, S.-O., Keller, T. (eds.) Knowledge and Information Visualization. LNCS, vol. 3426, pp. 1–23. Springer, Heidelberg (2005)
[22] Mazza, R.: Introduction to information visualization. Springer, London (2009)
[23] Scaife, M., Rogers, Y.: External cognition: how do graphical representations work? International Journal of Human-Computer Studies 45(2), 185–213 (1996)
[24] Spence, R.: Information visualization: Design for interaction. Prentice Hall, Essex (2007)
[25] Thomas, J., Cook, K. (eds.): Illuminating the path: The research and development agenda for visual analytics. IEEE Computer Society, Los Alamitos (2005)
[26] Thomas, J.: Taxonomy for visual analytics. VAC Views, May 6-7 (2009)
[27] Tory, M., Möller, T.: A Model-based visualization taxonomy. Technical Report SFU-CMPT-TR2002-06, Simon Fraser University (2002)
[28] Tory, M., Möller, T.: Rethinking visualization: A high-level taxonomy. In: Proceedings of the IEEE Symposium on Information Visualization, Austin, USA, pp. 151–158 (2004)
[29] Tory, M., Möller, T.: Human factors in visualization research. IEEE Transactions on Visualization and Computer Graphics 10(1), 1–13 (2004)
[30] van Wijk, J.: Views on visualization. IEEE Transactions on Visualization and Computer Graphics 12(4), 421–432 (2006)
[31] Ware, C.: Information visualization: Perception for design. Morgan Kaufmann, San Francisco (2004)
[32] Wickens, C., Lee, J., Liu, Y., Gordon-Becker, S.: Introduction to Human Factors Engineering. Addison-Wesley, New York (2003)
[33] Wong, P.C., Thomas, J.: Visual analytics. IEEE Computer Graphics and Applications 24(5), 20–21 (2004)
[34] Yi, J.S., Kang, Y.A., Stasko, J.: Toward a deeper understanding of the role of interaction in information visualization. IEEE Transactions on Visualization and Computer Graphics 13(6), 1224–1230 (2007)
[35] Zhang, J., Johnson, K.A., Malin, J.T., et al.: Human-centred information visualization. In: Proceedings of the International Workshop on Dynamic Visualization and Learning, Tubingen, Germany Proceedings of the IEEE Symposium on Visualization 2000, Salt Lake City, USA, pp. 69–76 (2002)

Development of Analytical Algorithm for the Performance Analysis of Power Train System of an Electric Vehicle

Chul-Ho Kim[1,*], Kee-Man Lee[2], and Sang-Heon Lee[3]

Abstract. Power train system design is one of the key R&D areas on the development process of new automobile because an optimum size of engine with adaptable power transmission which can accomplish the design requirement of new vehicle can be obtained through the system design. Especially, for the electric vehicle design, very reliable design algorithm of a power train system is required for the energy efficiency. In this study, an analytical simulation algorithm is developed to estimate driving performance of a designed power train system of an electric. The principal theory of the simulation algorithm is conservation of energy with several analytical and experimental data such as rolling resistance, aerodynamic drag, mechanical efficiency of power transmission etc. From the analytical calculation results, running resistance of a designed vehicle is obtained with the change of operating condition of the vehicle such as inclined angle of road and vehicle speed. Tractive performance of the model vehicle with a given power train system is also calculated at each gear ratio of transmission. Through analysis of these two calculation results: running resistance and tractive performance, the driving performance of a designed electric vehicle is estimated and it will be used to evaluate the adaptability of the designed power train system on the vehicle.

1 Introduction

Air pollution especially in the urban area is a serious issue to be solved all around the world. It is well known that one of the main causes of the air pollution in the

[1] Seoul National University of Technology, Department of Automotive Engineering, Seoul, South Korea
profchkim@snut.ac.kr

[2] Sunchon National University, School of Mechanical & Aerospace Engineering, Sunchon, Jeonnam, South Korea
kmlee@sunchon.ac.kr

[3] University of South Australia, School of Advanced Mechanical & Manufacturing Engineering, Adelaide, South Australia, Sang-Heon
Lee@unisa.edu.au

* Corresponding author.

J.H. Lee et al. (Eds.): EU-Korea Conference on Science and Technology, SPPHY 135, pp. 223–233.
springerlink.com

urban area is from the transportations such as public buses, taxis and private vehicles that use conventional gasoline or diesel engines. Because of not only the air pollution problem but the shortage of the crude oil, scientists have been studied on the alternative energy sources for the transportation system. One of the promising alternative energy sources that have been studied for the last two decades is electric energy that is known as an energy source of zero emission. Before the development of pure electric vehicle, several major automobile manufacturers have attempted to develop lower emission power train system. As a result, they have been released in the market known as hybrid engines which are operated with a conventional gasoline or diesel engine in half and with an electric motor in the other half. However, the hybrid engine still consumes hydro-carbon (HC) fuel and produces the harmful emissions. Global warming problem mainly caused by carbon dioxide became another serious issue forcing engineers to develop zero-emission vehicle that uses only electricity for an energy source since it is a clean and renewable energy obtainable from solar, wind, hydro-power and geothermal energy.

In this study, an analytical algorithm is developed to find out the optimum size of the power train of an electric vehicle designed. The principal theory of the algorithm developed is based on the conservation of energy in a control volume. All of the sink or source terms of the energy in the governing equation were evaluated numerically and experimentally.

2 Theoretical Background of an Analytical Algorithm

When a vehicle runs on the road, it experiences resistance forces called running resistance of the vehicle on a road. If a driver wants to keep its speed constant, he needs to have an equivalent engine power to overcome the running resistances. During the acceleration, it needs more power to overcome the acceleration resistance while during the deceleration the inertia energy of the vehicle can be restored by the inertia energy recovery system of the vehicle. The engine power required is always fluctuated due to the variation of driving condition of the motor vehicle on a road.

In here, the mathematical equations of the resistance terms are introduced to make the governing equation of the power train system of a vehicle. In general, the running resistance terms on a moving vehicle consist of:

1) Rolling resistance (R_R)
2) Aerodynamic resistance (R_A)
3) Gradient resistance (R_G)
4) Acceleration/Deceleration resistance (R_I)
5) Tractive resistance (R_{trac})

A general form of total running resistance force of a vehicle can be expressed as the summation of all of these terms given above:

$$R_{tot} = R_R + R_A + R_G + R_I + R_{trac} \tag{1}$$

With the changes in driving conditions, Eq. (1) needs to be modified. For example, if a vehicle moves at a constant speed on a leveled road, the terms; R_I and R_G, are ignored. For the deceleration of the vehicle, R_I should be negative singed.

2.1 Rolling Resistance

The friction resistance between road and tire surface is defined as rolling resistance of a vehicle. It is clearly affected by the surface roughness of road but not by the vehicle speed. From the principle of physics, the rolling resistance of a running vehicle can be obtained from Eq. (2):

$$R_R = \mu_R \times (W_{car} + W_x) \,, \tag{2}$$

where W_{car} is gross weight of a vehicle and W_x is the induced lift or down force on a running vehicle.

Table 1. Rolling resistance coefficient for various road surface conditions [2]

Conditions of surface	μ_R
Asphalt paved road	approx. 0.010
Concrete paved road	approx. 0.011
Stone block paved road	approx. 0.020
Well-maintained unpaved road	approx. 0.04
Unpaved road	approx. 0.08
Pebble-stone road	approx. 0.12
Sand and pebble-stone road	approx. 0.16
Sand road	approx. 0.2 -0.3

The rolling resistance coefficient (μ_R) depends on surface condition of road, material and tread pattern of tires and its charged air pressure and vehicle speed etc. Thus the multiple factors affecting rolling resistance cannot be taken into account at a time. In here, the most commonly used coefficient varied with road surface condition is incorporated in this study. Table 1 shows some of typical example of the rolling resistance coefficients.

2.2 Aerodynamic Resistance

As a vehicle runs on the road, the relative air movement occurs against to the driving direction of the vehicle even with no wind in air. Because of this air flow, the vehicle experiences aerodynamic forces such as drag and lift on the body. The aerodynamic drag forces generated on the frontal and rear side of the car body are acting like a driving resistance on the vehicle. This aerodynamic drag force can be obtained from the analytical equation, Eq. (3) [3].

$$R_A = C_D \frac{1}{2} \rho (V_{car} - V_{wind})^2 A \cos\theta \,, \tag{3}$$

where C_D is drag coefficient of a model vehicle, A is a projected frontal area of the vehicle and θ is the angle between wind and driving direction.

The projected frontal area (*A*) of a vehicle is normally obtained with Eq. (4) [7]:

$$A = 0.82 \times b \times h \,, \tag{4}$$

where *b* and *h* is the maximum width and height of a model vehicle. The drag coefficient (C_D)of a vehicle is calculated from the wind-tunnel test results and in general it is in the range of 0.35 ~ 0.55 for small sedans. Air density (ρ) is very sensitive to atmospheric pressure and temperature and can be obtained from Eq. 5.

$$\rho(P,t) = \frac{348.7 \times P\ (bar)}{273.15 + t(°C)} \ (\mathrm{kg/m^3}) \tag{5}$$

2.3 Gradient Resistance (R_G)

As a vehicle goes up or down the hill, it experiences gravitational resistance due to its weight. This is called the gradient resistance of the vehicle. Fig.1 shows a schematic diagram of gradient resistant.

The gradient resistant is calculated by Eq. (6):

$$R_G = \pm W_{car} \times \sin\theta_G \,, \tag{6}$$

where θ_G is gradient angle of road.

When a car goes down a hill, the gradient resistance is a negative value, that is, the total resistance power required is reduced and instead the inertia energy can be restored with an energy recovery system.

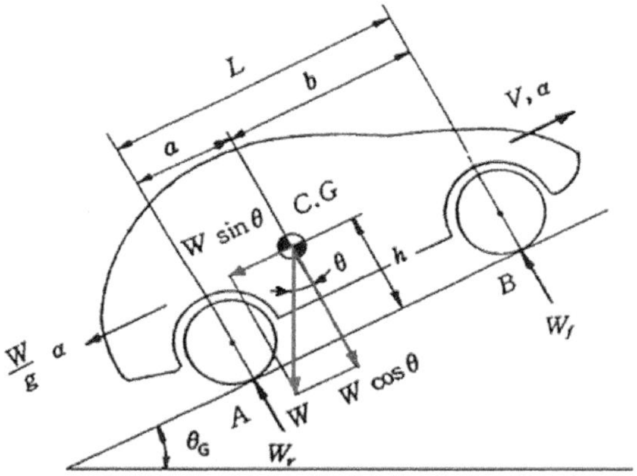

Fig. 1. Gradient resistance of a model vehicle

2.4 *Acceleration Resistance (R_I)*

When a vehicle accelerates or decelerates on the road, it will be under positive or negative inertia resistance. The inertia resistance of a running vehicle consists of two different types: an angular inertia of the rotating parts of power train system and the linear inertia of the running vehicle.

This inertia resistance can be determined by Eq. (7) [1]:

$$R_I = \frac{W_r}{g} \times \alpha = w\left(\frac{W_r}{W}\right)\frac{\alpha}{g}, \tag{7}$$

where W is the gross weight of a vehicle and W_r is the equivalent mass of all rotating components of the power train and α is acceleration speed of a vehicle. Table 2 shows the variations of (W/W_r) with the speed of power transmission.

Table 2. Variation of (W/W_r) with the speed of power transmission [6]

	Small Car	Large Car
Low gear	1.4	-
1-gear	0.5	0.3
2-gear	0.2	0.14
3-gear	0.11	0.09

For the decision of acceleration resistance of a designed car, the detailed acceleration performance schedule of the designed vehicle is needed. In this case study, steady state driving condition has only been considered.

2.5 *Towing Resistance (R_T)*

When a vehicle tows a trailer at the rear, the vehicle experiences the pulling resistance and it is called towing resistance of the vehicle. The size of the coupled resistance can be estimated with the analytical equations given in Eq. 2, 3, 6; rolling resistance, aerodynamic resistance, gradient resistance. This term is not considered in this study.

2.6 *Tractive Resistance (R_{trac})*

Engine power is transmitted to the tires through power transmission unit, driveshaft and axle in a vehicle. The force required to overcome the total running resistance at a given driving condition on tires is called the traction force of a vehicle.

The size of traction force on a tire is proportional to the torque generated at the engine crank shaft. The crank shaft torque is:

$$\tau_{engine} = F \times l, \tag{8}$$

where F is the torsion force (N) and l is the radius (m) of the engine crank shaft.

The tractive resistance of a vehicle is obtained from the Newton's 2nd Law (NSL):

$$R_{trac} = F_{engine} - (F_{IE} + F_{IT} + F_{ID} + F_{IWA}) \, , \tag{9}$$

where F_{IE}, F_{IT}, F_{ID} and F_{IWA} are the resistance of inertia force of engine, transmission, driveshaft and wheel & axles shaft, respectively.

The total resistance of angular inertia force of rotating parts of power train system can simply be expressed with the concept of the equivalent mass as given in Eq. (10).

$$F_{IE} + F_{IT} + F_{ID} + F_{IWA} = w(\frac{W_r}{W})\frac{\alpha}{g} \tag{10}$$

For the steady state driving condition of a vehicle, the acceleration resistance of the rotating parts of the power train system is ignored then the tractive resistance of the power train is calculated by Eq. (11) [1].

$$R_{trac} = \frac{\tau_{engine} \times \eta_{tm} \times i_{tm}}{R_{tr}} \, , \tag{11}$$

where τ_{engine} is the brake torque of engine, η_{tm} is the combined mechanical efficiency of transmission and final drive, i_{tm} is the combined gear ratio of transmission and final drive and R_{tr} is radius of the tire.

3 Performance Analysis of the Designed Power Train System of a Vehicle

An analytical algorithm to estimate the general driving performance of a designed power train system can be made with the resistance terms given in Section 2. The tractive force versus the vehicle speed curve can be obtained with the developed algorithm. Subsequently the maximum speed of the vehicle and the maximum climbing capacity can be estimated with the obtained performance curve. The maximum cruise distance of a vehicle with a given battery size can also be determined.

3.1 Initial Conditions of the Analytical Algorithm

To determine the driving performance of a designed power train system of a vehicle, the following initial conditions of the vehicle should be given:

(1) Specification of a vehicle
- Dry weight of a vehicle (kg)
- Vehicle size in width (m) and height (m)
- No. of passenger and payload capacity (kg)

(2) Specification of power transmission
- Mechanical efficiency of transmission (η_{tm})
- Gear ratio of total speed change of trans-mission

(3) Driving condition
- Acceleration performance (α)
- Aerodynamic drag coefficient (C_D)
- Rolling resistance coefficient (μ_R)
- ($\varepsilon = W/\Delta W$) (W/W_r) of vehicle

(4) Atmospheric condition
- Atmospheric pressure and temperature

(5) Performance curve of electric power motor
- (*torque-rpm*) curve on an operating range

3.2 Analytical Algorithm of Calculation

The total resistance force of a vehicle is varied with the road condition. The main parameters affecting the amount of the total resistance force are the weight of the vehicle and its speed, the aerodynamics drag coefficient and the gradient angle of road. Conceptually the total resistance force of a vehicle is energy sink term that is needed by the vehicle on its driving condition.

The power generated from a designed power train system is determined by the technical information of a proposed power train system, such as a performance curve (rpm-torque curve) of engine and the specification of power transmission. This is energy source term that is generated by the proposed power train system.

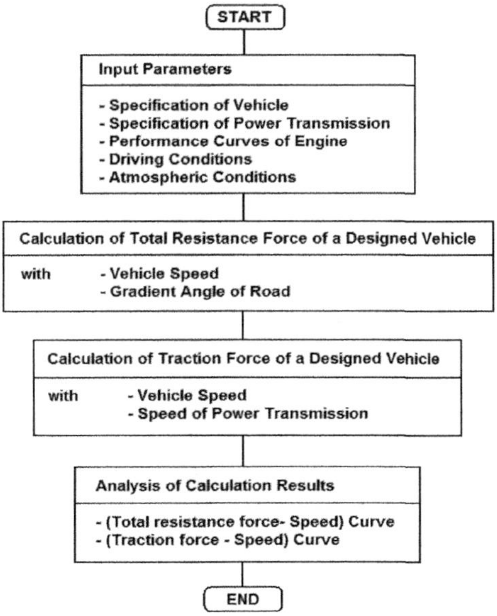

Fig. 2. Flow of analytical calculation algorithm

Fundamentally, the energy source generated by the power train system should keep its balance to the energy required by a designed vehicle to have its stable driving condition on the road.

Fig.2 shows the flow chart of the analytical calculation algorithm to get the driving performance of the power train system of a designed vehicle.

From the calculation results, the driving resistance force required of the designed vehicle and the traction force of the designed power train system are obtained. The driving performance of a designed vehicle with the power train system is estimated by the comparison of these two calculation results.

4 Analytical Simulation of an Example Model Vehicle

A case study is conducted to estimate the driving performance of a vehicle installed with the designed power train system with the analytical simulation algorithm developed in this study. One of the small-size hatchback type vehicles was selected for this case study and its specification is given in Fig. 3. The performance curve of the electric motor and the gear-ratio of the designed power transmission are given in Fig. 4 and Table 3 respectively.

(1) Specification of model vehicle [4]

(2) Driving conditions:
- Steady state driving ($\alpha = 0$)
- Aerodynamic drag coefficient (C_D=0.4)
- Rolling resistance coefficient (μ_R=0.02)

(3) Atmospheric conditions
- 25°C at 1atm

(4) Performance curve of electric power motor [5]

(5) Specification of a designed power transmission [4]

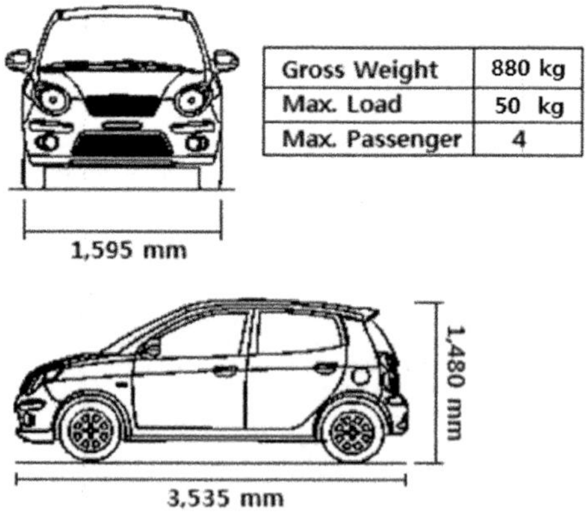

Fig. 3. Configuration of a model vehicle

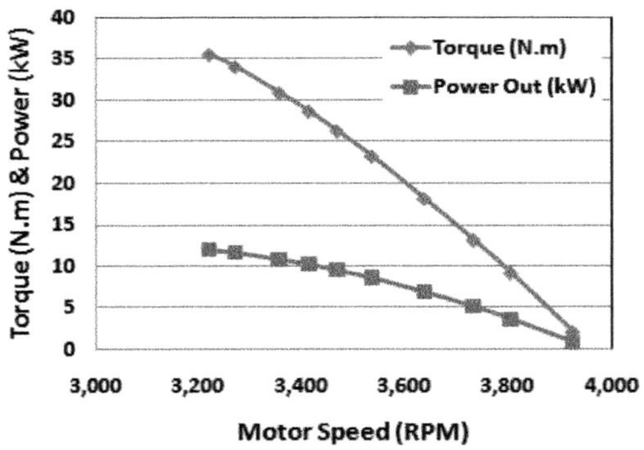

Fig. 4. Performance curve of an electric power motor chosen for this study

Table 3. Gear Ratio of a Designed Transmission System

	1-gear	2-gear	3-gear	4-gear
Gear Ratio	2.3	1.68	1.06	0.44
Reduction Gear Ratio	8	8	8	8
Final Speed Change Ratio	18.4	13.4	8.48	3.52
Mechanical Efficiency(η_{tm})	96.6	96.7	97.2	97.3

5 Results and Discussion

The driving performance of the designed power train system of a model vehicle was determined by the analytical simulation algorithm developed in this study. This driving performance curve obtained as a result will provide very important information to vehicle designers to get an optimum size of the power train system of a designed vehicle.

In this study, two cases: the first case with the original rotational speed of electric power motor and the second case with double of the original rotational speed of the power motor, are considered and compared the results to understand the effect of the rotational speed of the electric power motor to the driving performance of a designed vehicle.

Fig. 5 shows the driving performance of a model vehicle with the designed power train system over its operating range. As shown in Fig. 5, the maximum vehicle speed is about 94km/h at the level road. The vehicle can go up the hills with 41.4% of gradient at 1-speed gear ratio. Maximum slope that the vehicle can go up is about 26.8% on the constant speed of 37 km/h at 3-speed gear ratio.

Fig. 6 shows the driving performance of the model vehicle at the double of the original rotational speed of the electric power motor with the designed gear ratio of transmission over its operating range. To increase the maximum driving speed of the vehicle, the rotational speed of the power motor has been increased as twice as. In this case, the highest vehicle speed reached at a leveled road is about 125 km/h and increased about 33% in speed.

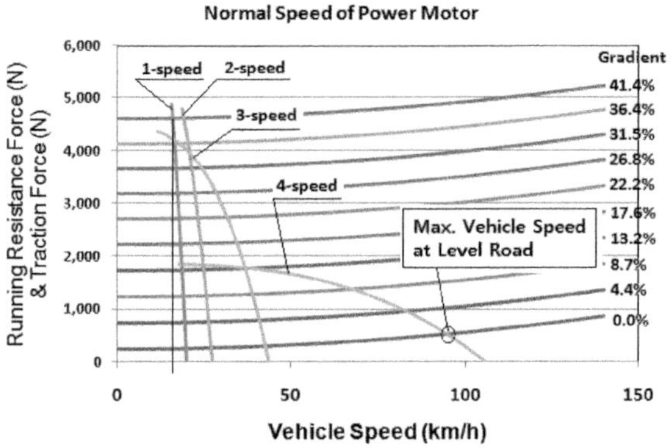

Fig. 5. Performance curve of the designed power train system of a model vehicle at its original rotational speed of the electric power motor with the designed gear ratio of transmission

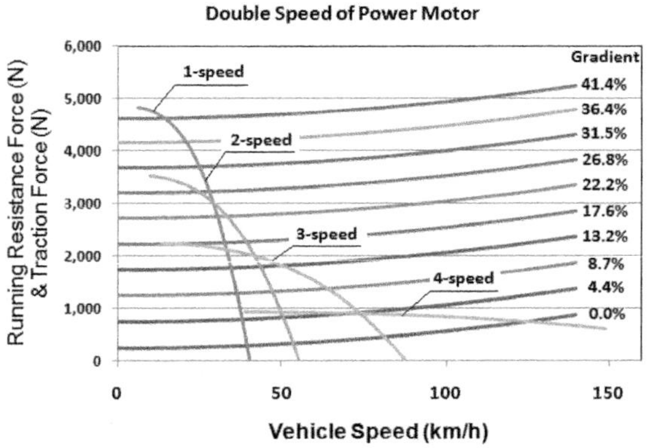

Fig. 6. Performance curve of the designed power train system of a model vehicle at the double of original rotational speed of the electric power motor with the designed gear ratio of transmission

6 Conclusion

In this study, an analytical simulation algorithm was developed to estimate the driving performance of a designed power train system of a model vehicle. With the simulation outputs of the analytical algorithm, the following results are estimated for an optimum design of a new power train system of a vehicle.

- Maximum driving speed of a designed vehicle at each gradient angle of a road
- Climbing ability of a designed vehicle
- Maximum driving distance with one charge of a designed electric battery
- Optimum gear ratio of a designed transmission to meet the required driving performance

The simulation result obtained from the analytical algorithm is significantly important information to the vehicle design engineers at the planning stage of power train design of a vehicle because it could save a reasonable amount of time for an optimum design of the new power train system.

Acknowledgments. This work was supported by the Industry-academic Cooperation Foundation of Seoul National University of Technology, Korea in 2008.

References

[1] Gillespia, T.D.: Fundamentals of Vehicle Dynamics. Society of Automotive Engineers, Warrendale (1992)
[2] The Traffic Institute, Traffic Accident Reconstruction. Northwestern University, 62–140 (1990)
[3] Munson, B.R., Young, D.F., Okiishi, T.H.: Fundamentals of Fluid Mechanics, 5th edn., pp. 518–521. John Wiley, Iowa (2006)
[4] KIA Motors Co., http://www.kiamotors.com/ (accessed on August 6, 2008)
[5] TMC(The Lynch Motor Company), Technical Date Sheet of Model 20D127, UK (2008)
[6] Cole, D.: Elementary Vehicle Dynamics course notes in Mechanical Engineering. The Univ. of Michigan, Michigan (1972)
[7] Hucho, W.H.: Aerodynamics of Road Vehicles, pp. 142–145. Warrendale PA, Society of Automotive Engineers (1998)

Stochastic Gene Expression Model Base Gene Regulatory Networks

Haseong Kim[1] and Erol Gelenbe[2,*]

Abstract. Gene regulatory networks consist of a number of genes and their interactions which regulate expressions of the genes. Along with the development of gene regulatory network studies, computer simulations have become a valuable tool to evaluate complex relationships between genes. Due to the stochastic nature of gene expressions, various stochastic approaches have attracted increasing interest. In this study, we build gene regulatory networks based on a stochastic gene expression model with delicate assumptions such as transcription, translation, DNA-protein, protein-protein associations and time delay for protein activation. Two simple in-silico gene regulatory network models are constructed and monitored their expression profiles reflecting the inhibition and activation of the gene regulations.

1 Introduction

Gene regulatory networks (GRNs) mainly consist of genes and their interactions such as inhibitions and activations. These interactions lead to dynamical changes of gene expression profiles which are crucial evidences to infer relations of genes in various GRN methods [1-4]. There are many attempts to explain the gene expressions by using mathematical and statistical models but there are still many arguments that try to define decisive expression processes. Recently, there are studies that gene expressions are described by discrete and probabilistic character of processes instead of assumptions of continuous and deterministic concentration changes [5-10]. Paulsson reviewed a general gene expression processes that is applicable to both prokaryotes and eukaryotes [7]. These discrete and probabilistic characters of processes are taken into account in the stochastic modeling of gene regulation.

The gene expressions can be described by two or three stage models (transcription, translation, and degradation) [10]. In order to describe complex gene expressions, we should concern not only the three stages but also DNA-protein

[1] Intelligent Systems & Networks Group, Imperial College London, United Kingdom
haseong.kim08@imperial.ac.uk
[2] Intelligent Systems & Networks Group, Imperial College London, United Kingdom
e.gelenbe@imperial.ac.uk
* Corresponding author.

J.H. Lee et al. (Eds.): EU-Korea Conference on Science and Technology, SPPHY 135, pp. 235–244.
springerlink.com

protein-protein associations/disassociations and time delays for protein activation. For example, McAdams and Arkin examined a required time for protein concentration to grow to effective signalling levels after a promoter is activated [5]. D. Bratsun *et. al.* proposed generalized Gillespie algorithm that can describe delay induced oscillations of gene expression by taking non-Markovian properties in stochastic gene expression processes [11]. Also there is a study that explains differences of degradation rates between monomers and dimers [12].

In this study, we model GRNs by assembling the delicate features of gene expressions and gene regulations. Figure 1 depicts the assumptions of our stochastic gene expression model. In order to evaluate our approaches, Toggle switch model [9, 13] and coupled link model [5] based Four-gene linked network are constructed.

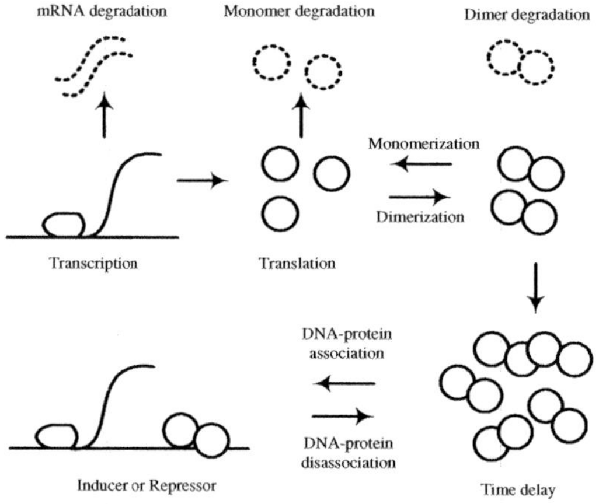

Fig. 1. Processes for the stochastic gene expressions

2 Stochastic Gene Expression Modeling

2.1 Gene Activation

Depending on cell growth conditions, there are several copies of partially replicated chromosomes [7]. Let n_1 and n_1^{max} are the number of active genes and a constant that maximum copies of genes (chromosomes), respectively. Then each copy of genes spontaneously switches ON and OFF with rate λ_1 and γ_1, respectively. The stationary distribution for the number of active genes is then Binomial where the probability of being ON is $P_{ON}= \lambda_1/(\lambda_1+ \gamma_1)$. However, for the simplicity, we assume the changing rate of the number of active genes is zero in this study. That is, $d<n_1>/dt=0$.

2.2 Transcription

Transcription is typically assumed to follow Poisson distribution [5]. Let *RP* be RNA polymerase and *P* is corresponding protein. The initiation of transcription is only activated by the RNA polymerase open complex which is represented by *RPo*. Then $RP + P \leftrightarrow RPc \leftrightarrow RPo$ where *RPc* is a RNA polymerase closed complex. The time interval between successive transcripts is considered to have an exponential distribution with $1/T_{avg}$ where T_{avg} is the average transcript initiation interval. Let λ_2 be the activation rate of *RPo* then $\lambda_2=1/T_{avg}$ and the production rate per unit time is $\lambda_2 n_1$. The probability of the transcription initiation reaction in a small time interval Δt is $\lambda_2 n_1 \exp(-t\lambda_2 n_1)\Delta t$ where t is time. The transcription initiation rate, λ_2, of mRNA is 0.0025sec^{-1} [8, 14]. The value of degradation rate will be described later.

2.3 Translation

In prokaryotic cells, ribosomes bind successively to an mRNA as soon as it is accessible behind the transcribing RNA polymerase. Multiple ribosomes space about 80 nucleotides [5]. Corresponding proteins are produced successively from the attached ribosomes. These translation processes are continued until the mRNA is degraded by an RNAse-E. An RNAse-E is known that it directly competing with rebosomes to bind corresponding promoters because these two types of molecules share their binding site of a promoter. So the competition leads to successful translation or degradation of mRNAs. Then the probability for the number of proteins, n_3, is $P(n_3 = n) = p^n(1-p)$ where p and $(1\text{-}p)$ are the probability that a ribosome and an RNase-E bind to the mRNA, respectively. Let T_D is an average time interval between successive competitions. Then translation initiation rate λ_3 is $\lambda_3=1/T_D$. T_D can be obtained in terms of p which will be described in next Section.

Many studies have interpreted variations of proteins in terms of burst which takes place in brief periods of high expression intensity followed by long periods of low intensity [14, 15]. Let b is the burst size and the size is equal to an average number of proteins per one mRNA. Then the probability that an ribosome binds an mRNA is computed by solving $b=p/(1\text{-}p)$. All these parameter values (λ_3, γ_3, b and p) will be explained in the next section.

2.4 Degradation of mRNA and Protein

Generally, an nth order reaction equation is defined $d[A]/dt = k[A]^n$ where k is reaction constant and $[A]$ is the concentration of substance A (mRNA or protein). When the order is one (n=1), the decrease in the concentration of the A over time is $log([A]/[A]_0)=-kt$ where $[A]_0$ is the initial concentration of A at time t=0. So the half-life of A is $T_{half}=log(2)/k$. Then, the surviving mRNAs, n_2, in the population after transcription is blocked would be $n_2 = n_{2,0}p^{t/T_D}$. This is equal to T_{half} = -($log(2)/log(p))T_D$ [5]. Therefore, the translation initiation rate $\lambda_3=1/T_D$ can be computed. The average half-life of mRNAs is assumed by 120sec [14]. The burst size of proteins, *b*, is varied; 100 for an average E. coli gene [7], 40 for lacZ, and 5 for lacA [14]. This burst size determine the translation initiation rate. In this study, *b* is 10 so the probability that a ribosome binds the mRNA, *p*, is 0.91. Then, the average time interval between successive competitions of ribosome and RNase-E, T_D, is 16.327sec. Therefore, the translation initiation rate λ_3 is 0.0612sec^{-1}. The degradation rate of mRNA is 0.00578sec^{-1}. The average of half-life of proteins is known as 15min~120min [14]. In our study, the half-life of monomer proteins and dimer proteins are assumed as 10min and 12min, respectively. It is known that dimers or oligomers are more stable than their monomeric components [12]. So the degradation rates of monomers and dimers are 0.001155sec^{-1} and 0.00096sec^{-1}.

2.5 Time Delay for Gene Expression

It is known that time delays in gene expression cause stochastic oscillations [11]. Bratsun et al. simulated stochastic models with delayed time of degradation and negative interactions processes [11] while McAdams and Arkin investigated a delayed time for protein activations after their promoters are activated [5]. The expression of a gene is directly affected by the other genes' products (proteins) which are needed to be accumulated for their activation. That is, a protein product (transcription factor) of a gene binds a specific region of the other gene's DNA string (promoter) and activates its expression with time delays. We use the generalized Gillespie algorithm [11, 16] which allow us to introduce the time delay in our model. In our study, the time delay, Δt, is assumed 1.

2.6 DNA-Protein, Protein-Protein Association and Disassociation

Proteins act in a form of a dimer or an oligomer rather in isolation. It is known that the protein multimerization is an important feature in regulations of biomolecules such as enzyme and transcription factors [17]. The dimer dissociation constant is $K_D=k_{d1}/k_{a1}$=10nM [12] where k_{2a} is dimer association rate and k_{2d} is dimer dissociation rate. So, in this study, k_{2a} =0.001 and k_{2d} =0.0001.

Along with the protein-protein associations, we consider the DNA-protein (operator-repressor) associations. A GRN consists of numerous coupled links where a protein product (e.g. transcription factor) encoded by one gene regulates expression of other genes. Two kinetic constants, association and dissociation, are required for the DNA-protein interaction mechanism. In this study, the kinetics of association (k_{1a}) and dissociation (k_{1d}) are 0.189 and 0.0157, respectively [18]. The protein concentration growth is affected by additional parameters: initial cell volume, and cell growth rate. We assume the cell growth rate is 1, *i.e.* the cell volume is fixed. The parameters and initial values of our models are shown in Table 1.

3 Modeling Gene Regulatory Networks

3.1 Toggle Switch Gene Expression Model

Our model is applied to construct gene toggle switch system [13]. The toggle switch system consists of two repressor molecules and two promoter sites. Repressors are DNA bind proteins regulating (inhibition) gene transcriptions. Figure 2 shows this toggle switch system. Each promoter is inhibited by the repressor that is transcribed by the opposing promoter. Also it has two inducers which have the ability to reactivate the inhibited promoter.

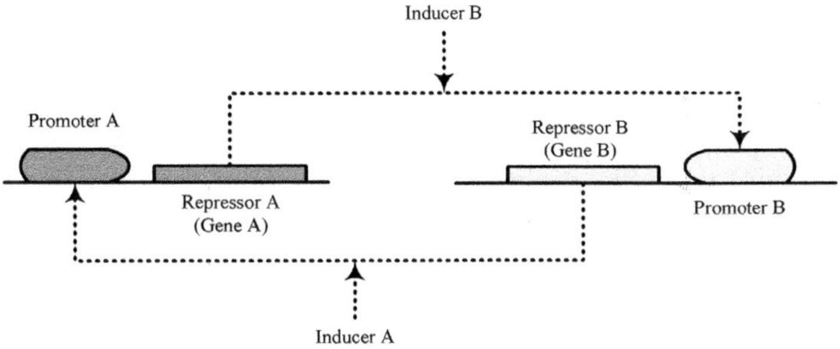

Fig. 2. The toggle switch system. Repressor *A* inhibits transcription from promoter *B* and is induced by inducer *B*. Repressor *B* inhibits transcription from promoter *A* and is induced by inducer *A*.

A switch is achieved by controlling the levels of inducers. The following chemical reaction equations describe processes of the toggle switch model.

$$\begin{aligned} Pro_A(t) &\xrightarrow{\lambda_2} Pro_A(t) + R_A(t) \\ Pro_B(t) &\xrightarrow{\lambda_2} Pro_B(t) + R_B(t) \end{aligned} \tag{1}$$

$$\begin{aligned} R_A(t) &\xrightarrow{\lambda_3} R_A(t) + P_A(t) \\ R_B(t) &\xrightarrow{\lambda_3} R_B(t) + P_B(t) \end{aligned} \tag{2}$$

$$\begin{aligned} P_A(t) + P_A(t) &\xrightarrow{k_{a1}} PP_A(t) \\ P_B(t) + P_B(t) &\xrightarrow{k_{a1}} PP_B(t) \end{aligned} \tag{3}$$

$$\begin{aligned} PP_A(t) &\xrightarrow{k_{d1}} P_A(t) + P_A(t) \\ PP_B(t) &\xrightarrow{k_{d1}} P_B(t) + P_B(t) \end{aligned} \tag{4}$$

$$\begin{aligned} \mathrm{Pro}_A(t) + PP_B(t) &\xrightarrow{k_{a2}} \mathrm{Pro}_A PP_B(t+\Delta t) \\ \mathrm{Pro}_B(t) + PP_A(t) &\xrightarrow{k_{a2}} \mathrm{Pro}_B PP_A(t+\Delta t) \end{aligned} \tag{5}$$

$$\begin{aligned} \mathrm{Pro}_A PP_B(t) &\xrightarrow{k_{d2}} \mathrm{Pro}_A(t) + PP_B(t) \\ \mathrm{Pro}_B PP_A(t) &\xrightarrow{k_{d2}} \mathrm{Pro}_B(t) + PP_A(t) \end{aligned} \quad (6)$$

$$\begin{aligned} PP_B(t) + Ind_A(t) &\xrightarrow{k_{d2}} PP_B Ind_A(t+\Delta t) \\ PP_A(t) + Ind_B(t) &\xrightarrow{k_{d2}} PP_A Ind_B(t+\Delta t) \end{aligned} \quad (7)$$

$$\begin{aligned} R_A(t) &\xrightarrow{\gamma_2} 0(t) \\ R_B(t) &\xrightarrow{\gamma_2} 0(t) \\ P_A(t) &\xrightarrow{\gamma_3} 0(t) \\ P_B(t) &\xrightarrow{\gamma_3} 0(t) \\ PP_A(t) &\xrightarrow{\gamma_4} 0(t) \\ PP_B(t) &\xrightarrow{\gamma_4} 0(t) \end{aligned} \quad (8)$$

where Pro_A is the promoter site of gene A. R_A and P_A are the $mRNA$ and the protein molecule, respectively. PP_A is a dimer form of P_A. Ind_A is inducer. Equations (1) and (2) represent the transcription and the translation processes, respectively. Equation (3) is the protein dimerization process and (4) is the protein dissociation process. Equation (5) shows the repression process of the promoters with time delay Δt. In Equation (7), inducer reactivates the promoters' expression ability with

(a)

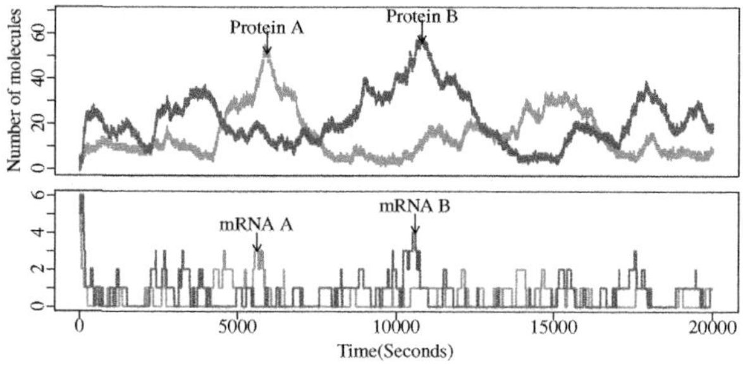

(b)

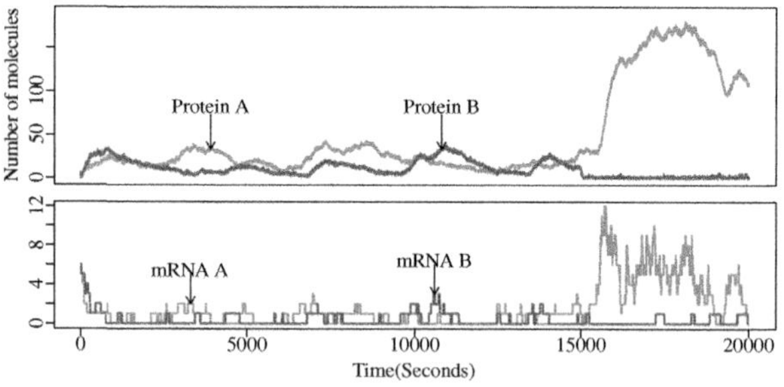

Fig. 3. mRNA and protein expressions in toggle switch model when both inducers are present (a) and inducer B is removed after at 15000sec (b).

time delay Δt. Equation (8) represents the degradation processes. This approach is different with the previous toggle switch study which performed by Ribeiro *et. al.* [9]. Their models illustrate protein productions without mRNA productions. Also, in our model, the inducers inactivate the repressors by binding to them while Ribeiro *et al* modeled an inducer disassociate a promoter-repressor complex.

Figure 3 (a) shows the simulated mRNA and protein expressions of the toggle switch model when both inducers are present. The system state switches depending on protein concentrations of the two genes. In Figure 3 (b), we remove inducer B from at 15000sec. So repressor A inhibits the expression of gene B and the expression of gene A increases. These results show our model properly work as the previous models [9, 11, 13].

3.2 *Four-Gene Linked Network with a Multiple Activator*

Along with the negative regulations mediated by repressors, positive regulations controlled by activators such as transcription factors also play an important role in gene regulatory networks. Figure 4 presents the Four-gene linked model. The protein of gene A activates the expression of gene B by binding the promoter of the corresponding DNA strand. Gene B inhibits expressions of gene D. Gene C is activated when both gene A and D are expressed.

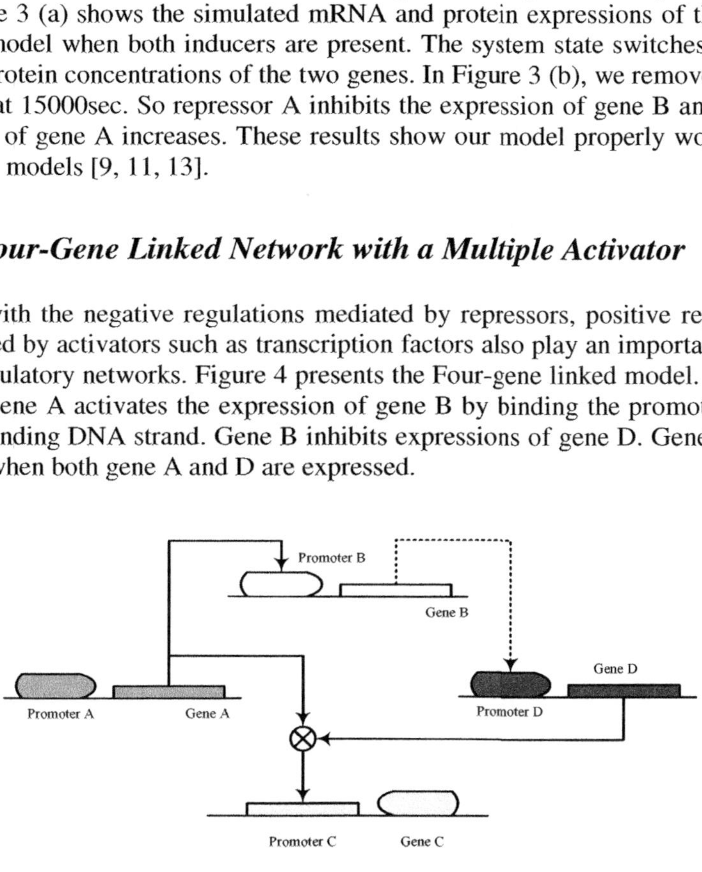

Fig. 4. Coupled link model.

$$\mathrm{Pro}_i(t) \xrightarrow{\lambda_2} \mathrm{Pro}_i(t) + R_i(t) \tag{9}$$

$$R_i(t) \xrightarrow{\lambda_3} R_i(t) + P_i(t) \tag{10}$$

$$P_i(t) + P_i(t) \xrightarrow{k_{a1}} PP_i(t) \tag{11}$$

$$PP_i(t) \xrightarrow{k_{d1}} P_i(t) + P_i(t) \tag{12}$$

$$\mathrm{Pro}_i(t) + PP_j(t) \xrightarrow{k_{a2}} \mathrm{Pro}_i PP_j(t + \Delta t) \tag{13}$$

$$\text{Pro}_i PP_j(t) + PP_l(t) \xrightarrow{k_{d2}} \text{Pro}_i(t+\Delta t) + PP_j(t+\Delta t) + PP_l(t+\Delta t) \tag{14}$$

$$\begin{aligned} R_i(t) &\xrightarrow{\gamma_2} 0(t) \\ P_i(t) &\xrightarrow{\gamma_3} 0(t) \\ PP_i(t) &\xrightarrow{\gamma_4} 0(t) \end{aligned} \tag{15}$$

where $i, j, l \in \{A, B, C, D\}$. Equations (9) to (15) are the same as that of the toggle switch model except Equation (14) which represents the processes of activators. The activators disassociate the promoter-repressor complexes while inducers bind to the repressor proteins (Equation (7)).

Figure 5 shows the protein expression profiles of the coupled link model. Note that through the profiles in Figure 5 represent protein expressions, we use the term "Gene" instead of the "Protein". In the top of Figure 5, Gene D (red) has low concentration through all the time points because gene B (purple) inhibits gene D. Gene C (blue) is only expressed when both gene A and D are all expressed. Even though the expression level of Gene D is not high enough, Gene C still have

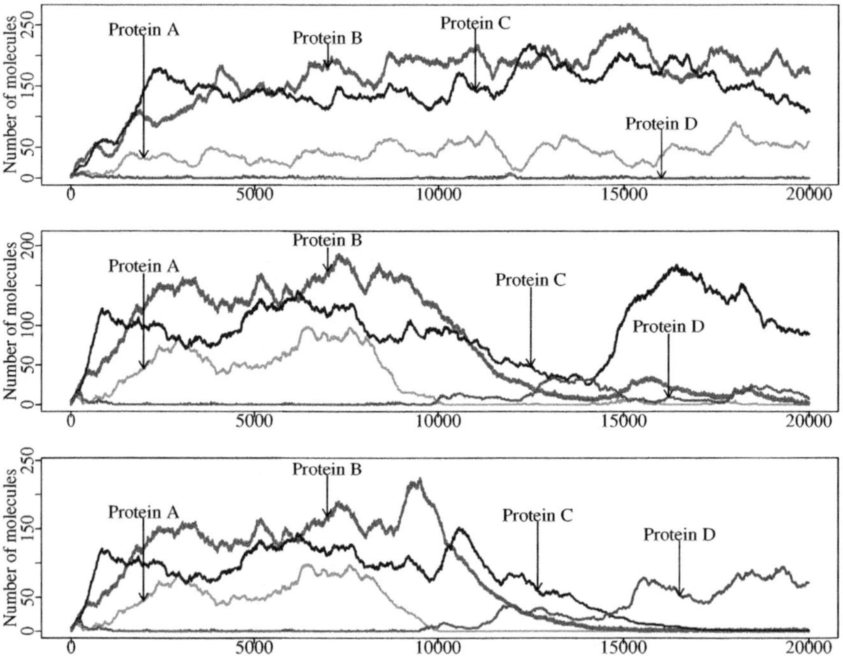

Fig. 5. Protein expression profiles of the Four-gene linked model. Gene A, B, C, and D are represented by green, purple, blue, and red colors, respectively. The profiles of the top figure are obtained under normal condition. Gene A is knocked down (middle) and knock out (bottom) from 10000sec.

Table 1. Parameters and initial values of the simulation models

Toggle switch and Four gene linked model			
Parameters		Initial values	
λ_2	0.0025	RPo_X	5
λ_3	0.0612	Pro_X	5
γ_2	0.00578	R_X	5
γ_3	0.00115	P_X	5
γ_4	0.00090	PP_X	5
k_{a1}	0.001	$ProPP_X$	0
k_{d1}	0.0001	Ind_X	5
k_{a2}	0.189	Rep_X	5
k_{d2}	0.0157		
Δt	1		

possibility of the activation because none of the two expressions (A and D) is zero. We knock down the gene A (Middle of Figure 5) from time 12000sec to the end of the simulation time by setting its transcription initiation parameter to λ_2=0.0002 which is 10 times lower than the normal transcription parameter. As consequence, the expressions of Gene B (purple) and C (blue) decrease but Gene C is expressed again at time 15000sec where Gene A and D are slightly activated. In the bottom of the Figure 5, Gene A is knocked out by setting λ_2=0. It makes the concentration of Gene B and C to be decayed while the expression of Gene D increases. These results show our model with the gene expression assumptions work properly.

4 Discussions

In this study, we build gene regulatory networks based on stochastic gene expression model. Our stochastic approach has included important biological assumptions such as time delays, protein dimerizations, and second order interactions. All the parameters of our models are obtained in the base of published experimental and theoretical gene expression studies. Two types of gene regulatory networks are examined. The Toggle switch model have evaluated by showing the oscillatory patterns of the gene expressions [9, 11, 13]. In Four-gene linked model, our method produces appropriate expression profiles reflecting the relations (interactions) between genes. We expect that our study enables us to evaluate not only the modeling studies of gene expression processes but also GRN inferring methods such as Bayesian and Boolean Networks [1, 2]. However, the kinetic parameters used in this study are needed to be optimized because the values are taken from independent experiment studies. Also the gene regulatory mechanisms with post-transcriptional interactions such as phosphorylations should be considered to generate more realistic GRN model. The successful study of this in-silico networks can be used to construct a in-vitro synthetic network some of which are already experimentally and numerically performed [19].

References

[1] Akutsu, T., Kuhara, S., Maruyama, O., et al.: Identification of gene regulatory networks by strategic gene disruptions and gene overexpressions. In: Proceedings of the ninth annual ACM-SIAM symposium on Discrete algorithms, pp. 695–702 (1998)

[2] Friedman, N., Linial, M., Nachman, I., et al.: Using Bayesian Networks to Analyze Expression Data. Journal of Computational Biology 7, 601–620 (2000)

[3] Shmulevich, I., Lahdesmaki, H., Dougherty, E.R., et al.: The role of certain Post classes in Boolean network models of genetic networks. Proceedings of the National Academy of Sciences 100, 10734–10739 (2003)

[4] Opgen-Rhein, R., Strimmer, K.: Learning causal networks from systems biology time course data: an effective model selection procedure for the vector autoregressive process. BMC Bioinformatics 8, S3 (2007)

[5] McAdams, H.H., Arkin, A.: Stochastic mechanisms in gene expression 94, 814–819 (1997)

[6] Arkin, A., Ross, J., McAdams, H.: Stochastic kinetic analysis of developmental pathway bifurcation in phage λ-infected Escherichia coli cells. Genetics 149, 1633–1648 (1998)

[7] Paulsson, J.: Models of stochastic gene expression. Physics of Life Reviews 2, 157–175 (2005)

[8] Golding, I., Paulsson, J., Zawilski, S.M., et al.: Real-Time Kinetics of Gene Activity in Individual Bacteria. Cell 123, 1025–1036 (2005)

[9] Ribeiro, A., Zhu, R., Kauffman, S.A.: A General Modeling Strategy for Gene Regulatory Networks with Stochastic Dynamics. Journal of Computational Biology 13, 1630–1639 (2006)

[10] Shahrezaei, V., Swain, P.S.: Analytical distributions for stochastic gene expression: Supporting information. Proceedings of the National Academy of Sciences 105, 17256–17261 (2008)

[11] Bratsun, D., Volfson, D., Tsimring, L.S., et al.: Delay-induced stochastic oscillations in gene regulation. Proceedings of the National Academy of Sciences 102, 14593–14598 (2005)

[12] Buchler, N.E., Gerland, U., Hwa, T.: Nonlinear protein degradation and the function of genetic circuits. Proceedings of the National Academy of Sciences 102, 9559–9564 (2005)

[13] Gardner, T.S., Cantor, C.R., Collins, J.J.: Construction of a genetic toggle switch in Escherichia coli. Nature 403, 339–342 (2000)

[14] Thattai, M., van Oudenaarden, A.: Intrinsic noise in gene regulatory networks. Proceedings of the National Academy of Sciences 98, 8614–8619 (2001)

[15] Cai, L., Friedman, N., Xie, X.S.: Stochastic protein expression in individual cells at the single molecule level. Nature 440, 358–362 (2006)

[16] Gillespie, D.T.: Exact stochastic simulation of coupled chemical reactions. The Journal of Physical Chemistry 81, 2340–2361 (1977)

[17] Marianayagam, N., Sunde, M., Matthews, J.: The power of two: protein dimerization in biology. Trends in Biochemical Sciences 29, 618–625 (2004)

[18] Goeddel, D.V., Yansura, D.G., Caruthers, M.H.: Binding of Synthetic Lactose Operator DNAs to Lactose Repressors. Proceedings of the National Academy of Sciences 74, 3292–3296 (1977)

[19] Cantone, I., Marucci, L., Iorio, F., et al.: A Yeast Synthetic Network for In Vivo Assessment of Reverse-Engineering and Modeling Approaches Cell (2009)

The Concentrations of Circulating Plasma Oxytocin and the Pattern of Oxytocin Release in Mare during Oestrus and after Ovulation

Sung Eun Bae[1]

Abstract. Mares susceptible to persistent mating-induced endometritis (PMIE) accumulate intrauterine fluid after mating. One of the factors causing delayed uterine clearance is thought to be impaired uterine contractility. Oxytocin is central in controlling myometrial contractility. The objective of the present study was to describe peripheral oxytocin release during estrus and in the early postovulatory period in reproductively-normal mares and to compare the baseline circulating oxytocin concentrations in reproductively-normal mares and mares with PMIE. Blood samples were collected from reproductively-normal mares (n=5) from day -5 of estrus to day 2 postovulation and every 5 min for 30 min from reproductively-normal mares (n=5) and mares with PMIE (n=5) on day 3 of estrus. Pulsatile secretion of oxytocin was observed in all mares. Mean plasma oxytocin concentrations were significantly higher ($P<0.05$) in estrus (day –5 to day –2) than on the day of ovulation (day 0). After ovulation, plasma oxytocin concentrations tended to increase. On day 3 of estrus, plasma oxytocin concentrations were significantly higher ($P<0.01$) in reproductively-normal mares than in mares with PMIE. The results showed there is a significant difference in plasma oxytocin concentrations between mares to PMIE. The low plasma oxytocin concentrations in mares with PMIE may contribute to predisposing factors in their poor uterine clearance in these mares.

1 Introduction

Oxytocin is known to be produced mainly by the magnocellular neurons in the posterior pituitary [1, 2] and is transported via the blood to the uterus, stimulating the release of prostaglandin $F_{2\alpha}$ ($PGF_{2\alpha}$) from the endometrium [3]. Oxytocin is central in controlling myometrial contractility and forms a positive feedback loop with $PGF_{2\alpha}$. In the previous chapter, the possibility of uterine oxytocin acting in a paracrine manner to elicit endometrial $PGF_{2\alpha}$ is also suggested.

As with all hypothalamic hormones, oxytocin is released discontinuously, and serial measurements reveal that plasma oxytocin levels fluctuate in the circulation,

[1] Department of Veterinary Clinical Studies, Royal (Dick) School of Veterinary Studies, University of Edinburgh, Easter Bush, Roslin, Midlothian, EH25 9RG, United Kingdom
dvmsbae@kseauk.org

J.H. Lee et al. (Eds.): EU-Korea Conference on Science and Technology, SPPHY 135, pp. 245–255.
springerlink.com

known as 'spurt' release [4, 5, 6]. Previous studies on the plasma concentrations of oxytocin in mares gave contradictory results because of its pulsatile manner of secretion and differences in assay methodology. Peripheral concentrations of oxytocin have been measured during the oestrous cycle. Burns *et al.* [7] collected samples from four different stages of the oestrous cycle and showed that concentrations of oxytocin were highest on day 2 of oestrus, and were greater on day 5 post-ovulation than on day 10 and day 15 post-ovulation. In contrast, Tetzke *et al* [8] showed that plasma concentrations of oxytocin were greater on day 15 post-ovulation than on day of ovulation or days 3 and 7 post-ovulation. By frequent sampling, these workers also showed a very irregular pattern of pulsatile secretion. Stevenson *et al.* [9] measured plasma oxytocin concentrations throughout the oestrous cycle but mean oxytocin concentrations remained at very low concentrations with no evidence of cyclic variation. Recently, Vanderwall *et al.* [10] collected samples from the intercavernous sinus of the mares during luteolysis and showed a clear pulsatile mode of oxytocin secretion.

Mares with delayed uterine clearance (e.g. persistent mating-induced endometritis; PMIE) are subfertile. Previous research showed that mares susceptible to PMIE accumulated six times more intrauterine fluid after bacterial challenge than mares resistant to PMIE [11]. The main factor causing delay in drainage of intrauterine fluid is thought to be impaired myometrial contractility in these susceptible mares [12]. As oxytocin has a major influence on uterine contractility, these results suggest that there may be differences in susceptible and resistant mares in the pattern of oxytocin release, plasma oxytocin concentrations, the concentrations of the uterine oxytocin receptors, or in post-receptor mechanisms [13].

The objective of the present study was to measure the plasma oxytocin concentrations in genitally-normal mares by collecting frequent blood samples and to compare the baseline circulating oxytocin concentrations in genitally normal mares and in mares with delayed uterine clearance. The experiment was designed to determine: 1) plasma oxytocin concentrations during oestrus and after ovulation and 2) the differences in the concentrations of plasma oxytocin between two groups, genitally-normal mares and mares with delayed uterine clearance.

2 Material and Methods

2.1 *Experiment 1. Concentrations of Plasma Oxytocin during Oestrus and Three Days after Ovulation in Genitally-Normal Mares (n=5)*

Animals

Five genitally-normal mares, aged 6 to 13 years, weighing 410-520 kg were used. Oestrus was defined by transrectal ultrasonographic examination of the genital tract as the presence of an ovarian follicle of >30 mm in diameter and uterine oedema, with the mare responding positively to teasing with a stallion. Ovulation was detected ultrasonographically by the disappearance of the follicle and the presence of a corpus luteum on the ovary. Day of ovulation was designated as Day 0.

Blood sampling

On the first day that the mares were considered to be in oestrus, a catheter (either 13 gauge, Presidio Medico, Ecouen, France, or Central Venous Catheter Set with peel-away sheath introduction technique, V7854, Cook®, Australia) was placed in the jugular vein after asceptic preparation of the area and local anaesthesia. The catheter was kept patent by flushing every 30 min with 20 ml heparin-saline (5,000 iu/l). Two blood samples of approximately 7 ml each were collected into evacuated heparinised tubes every 5 min for four hours between 0900 h and 1400 h on three to five days during oestrus and for three consecutive days starting from day 0 (ovulation).

2.2 Experiment 2. Concentration of Plasma Oxytocin in Genitally-Normal Mares (n=5) and in Mares with Delayed Uterine Clearance (n=5) during Oestrus

Animals

Five mares of mixed breeding, aged 5-15 years, weighing 370-520 kg and classified as genitally-normal mares, and five mares, aged 7 to 18 years, weighing 370 to 620 kg and classified as mares with delayed uterine clearance from their previous clinical history were used. Oestrus was defined by transrectal ultrasonographic examination of the genital tract as the presence of a follicle of >35 mm in diameter and uterine oedema, with the mare responding positively to teasing with a stallion.

Blood sampling

Jugular blood samples (15ml) were collected by venipuncture as described in Experiment 1 from genitally-normal mares (n=5) and mares with delayed uterine clearance (n=5), every 5 min for 30 min.

2.3 Sample Handling

Blood samples were immediately placed on ice until separation. The samples were centrifuged at 2000 × g for 15 min at 4°C. An aliquot of plasma (4ml) was acidified with 10M acetic acid (10 μl per 1ml plasma) for oxytocin assay to improve oxytocin recovery rates as already shown in the sow [14] and goat [15]. All plasma samples were frozen at -70°C and subsequently stored at -20°C until assayed.

2.4 Oxytocin Assay

Oxytocin was extracted from plasma using C18 SepPak cartridges (Waters Chromatography, Milford, MA, USA) and the radioimmunoassay carried out in duplicate samples as described by Thornton et al [16] using an antiserum (kindly donated from Prof. Flint, Nottingham University, UK) described previously [17]. The extraction recovery rate was 75 %. The detection limit for the assay was 0.7 pg/ml. The intra- and inter-assay coefficients of variation were 3.7 % and 7.9 %, respectively.

2.5 Statistical Analysis

All plasma oxytocin concentrations were calculated individually in both Experiment 1 and Experiment 2. Concentrations in samples below the detection limit of the assay were designated as equivalent to the detection limit of the respective assay. Data expressed as mean ± standard errors of the means (SEM). In Experiment 1, the statistical significance of the differences between days and mares was examined by analysis of variance (ANOVA). The source of variation that was used in SEM in each day of mean plasma oxytocin concentrations was the interaction term between horses and days. In this example, the variation was 1.419 (degrees of freedom = 21). The calculation for ANOVA was performed by the statistical computer package Genstat 5. Overall mean concentrations of plasma oxytocin on each day during oestrus and after ovulation were compared with mean plasma oxytocin concentrations on day of ovulation (Day 0) using t-test. Mean plasma oxytocin concentrations in oestrus and plasma oxytocin concentrations on the day of ovulation (Day 0) were also compared using t-test. In Experiment 2, difference in plasma oxytocin concentrations between genitally-normal mares and mares with delayed uterine clearance was compared using t-test. In all cases, a P value of <0.05 was considered significant.

3 Results

3.1 Experiment 1

Mean plasma oxytocin concentrations in oestrus were significantly higher ($P<0.05$) than on day of ovulation (Figure 1). Pulsatile secretion of oxytocin was observed in all mares (Figure 2). Plasma oxytocin concentrations were significantly higher ($P<0.05$) in oestrus (Day –5 to Day –2) than the day of ovulation (Day 0) and decreased towards the day of ovulation (Figure 3). Rapid fluctuations in oxytocin concentrations were observed during oestrous days (Figure 6.3). On day of ovulation (Day 0) and Day 1 postovulation, plasma oxytocin concentrations were lower than any other day in oestrus and the release pattern of oxytocin was relatively constant (between 5 pg/ml and below 10 pg/ml) compared to other days. After ovulation, plasma oxytocin concentrations gradually increased.

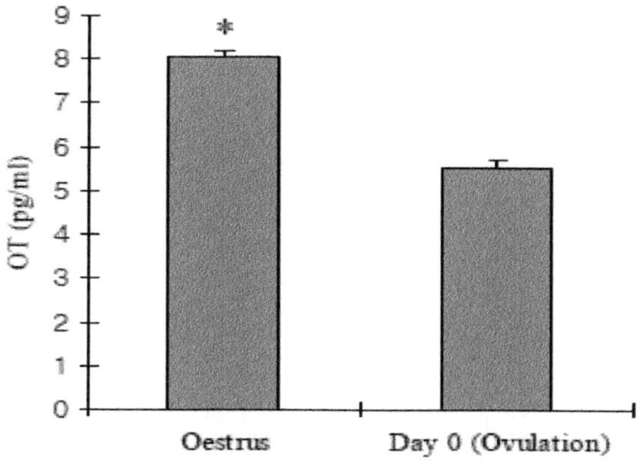

Fig. 1. Mean (± SEM) plasma oxytocin concentrations in oestrus and day 0 (ovulation) in genitally normal mares (n=5). The asterisk (*) indicates significant difference (P<0.05).

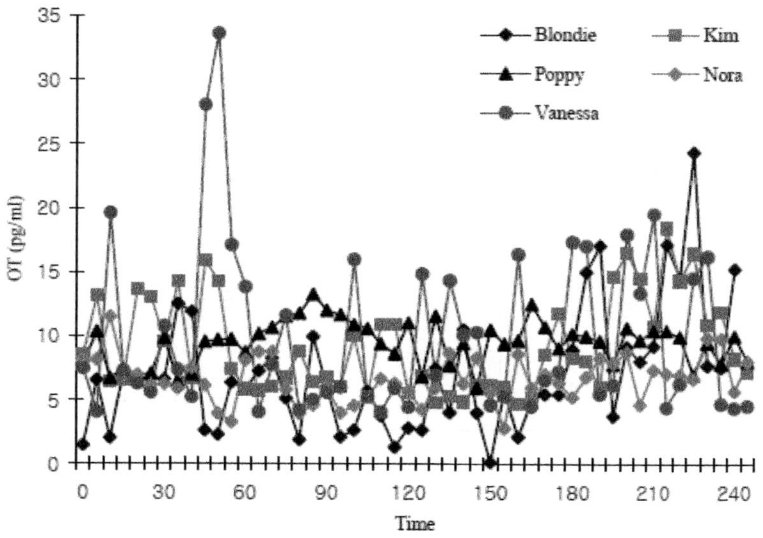

Fig. 2. Plasma oxytocin concentration on Day –2 (oestrus) sampled at 5-min intervals. Graphs show a pulsatile secretion of oxytocin from 5 individual mares.

3.2 *Experiment 2*

Plasma oxytocin concentrations were significantly higher (P<0.001) in genitally-normal mares (n=5) than in mares with delayed uterine clearance (n=5) (Figure 4).

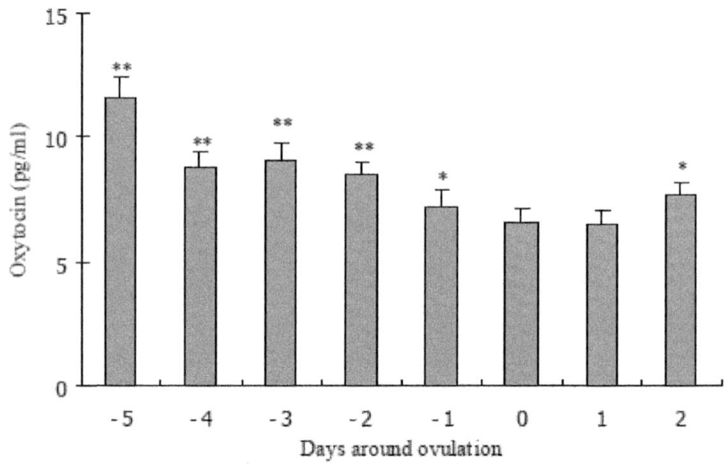

Fig. 3. Mean (± SEM) plasma oxytocin concentrations during oestrus and three days after ovulation in genitally normal mares. Asterisks indicate significant differences (*: $P<0.05$ and **: $P<0.01$) between each day during oestrus and after ovulation and Day 0 (ovulation).

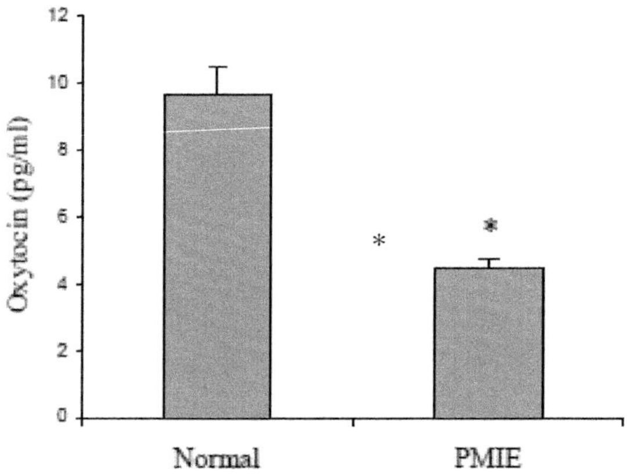

Fig. 4. On day 3 of estrus, mean (± SEM) plasma concentrations of oxytocin in reproductively-normal mares (n=5) and mares with PMIE (n=5). The asterisk (*) indicates a significant difference ($P<0.01$).

4 Discussion

Uterine contractility is mediated by hormonal factors in association with neurogenic and myogenic factors. In this chapter, we have studied changes in circulating oxytocin, one of the hormonal factors controlling uterine contractility. Plasma oxytocin concentrations were significantly higher during oestrus, a period chracterised by low progesterone and high oestrogen levels [18, 19] than on Day 0 (ovulation) and Day 1 after ovulation. Plasma oxytocin concentrations increased on day 2 after ovulation. A strong correlation between oxytocin release and uterine contractility has been reported in the mare [20]. This correlation between oxytocin and uterine contractility has been studied using electromyography (EMG) [20, 21] and intrauterine pressure (IUP) [22, 23]. Ko et al [22] reported that IUP was higher 3 days before ovulation, decreased towards ovulation (Day 0) and increased after ovulation in normal mares. These IUP results are in agreement with our plasma oxytocin concentrations. Increased uterine contractility during oestrus is involved in sperm transport and facilitates uterine clearance mechanisms. Since oestrogen has an excitatory effect on paraventricular neurons, oestrogen may increase oxytocin secretion. Behrendt-Adam et al [24] also showed effects of steroid hormones on endometrial oxytocin in the mare and demonstrated that oestrogen upregulated expression of the endometrial oxytocin gene.

Previous studies on plasma concentrations of oxytocin in mares have produced contradictory results. These differences probably result from methodology including different radioimmunoassay procedure, oxytocin extraction from plasma, the limit of detection of the assay as well as the frequency of blood sampling. Because of its pulsatile manner of release and short half-life, 6.78 min [25], the frequency of blood sampling could affect plasma oxytocin levels. In our study, we used measurements during oestrus and after ovulation by frequent blood sampling (5-min interval), longer periods of sampling (4 hours), larger numbers of mares (n=5), and a well-validated and sensitive radioimmunoassay. Therefore, our data could be used as a reliable baseline of plasma oxytocin concentrations for further studies.

The fact that plasma oxytocin concentrations are high during oestrus when uterine contractility at its most active [26] suggests a correlation between oxytocin/oxytocin receptors and sex steroid hormones. Oestrogen and progesterone control oxytocin receptor concentration/density and their sensitivity/affinity to oxytocin in the uterus in most species [27, 28, 29]. Our results suggest that the increased uterine contractility during oestrus may be related to the elevated concentrations of oxytocin. It is also reported that the levels of uterine oxytocin are high in oestrus [30]. Jones et al [31] measured uterine motility by electromyography and found that mares responded to oxytocin in all cycle stages, but they found a less pronounced response during dioestrus. Troedsson et al [21] also reported that an increased number of uterine myoelectrical activity bursts after oxytocin treatment was observed in oestrous mares compared to dioestrous mares. They also suggested that the different characteristics of myometrial activity between oestrus and dioestrus are the result of hormone-dependent changes in myometrial cell communication such as the formation of gap junctions.

At the time of mating (oestrus) in mares, semen and bacteria are introduced into the uterus, resulting in a uterine inflammatory response [21]. Most mares clear this inflammatory reaction from the uterus within 24-36 h of mating. However, inability to clear this inflammatory product from the uterus is a major cause of infertility. Between 11% and 39% of mares accumulate fluid at oestrus [32, 33]. Previous studies have shown that accumulation of intrauterine fluid in dioestrus is associated with a significant reduction in pregnancy rates and an increase in embryonic loss rate [34]. It is also reported that oestrous intrauterine fluid reduces spermatozoal motility in vitro, reduces embryo recovery rates [35] and pregnancy rates [32]. In this study, we have shown that there is a significant difference in the circulating plasma oxytocin concentrations between genitally-normal mares and mares with delayed uterine clearance. Previous studies showed that normal oestrous mares have significantly higher uterine contractile activity ($P<0.001$) than normal dioestrous mares and mares with delayed uterine clearance in both oestrus and dioestrus [26]. A recent study showed that there is no difference in endometrial oxytocin receptor concentrations between normal mares and mares with delayed uterine clearance [36], however myometrial receptors were not measured in that study. Our results suggest therefore that low uterine contractility in the mares with delayed uterine clearance is because of low plasma oxytocin concentrations, rather than oxytocin receptor concentrations. However, there may also be differences in these two groups of mares in post-receptor mechanisms as suggested in the study by Nikolakopoulos et al [13] where mares with delayed uterine clearance had significant lower release of PGF2α after oxytocin administration. Vanderwall et al [10] showed there is an eightfold difference between concentrations of oxytocin observed in blood samples from the intercavernous sinus versus the jugular vein. Therefore, the differences in pituitary oxytocin levels between genitally-normal mares and mares with delayed uterine clearance are larger than plasma oxytocin levels. The low oxytocin levels in the mares with delayed uterine clearance may lead to defective myometrial contractility which causes the accumulation of uterine fluid after mating, in contrast to genitally normal mares that evacuate the intrauterine fluid in oestrus.

Uterine contractility is an intricate interaction comprising myogenic [37], neurogenic [37, 38, 39] and hormonal factors [40, 41]. In this chapter, we have reported a reliable basal oxytocin levels in genitally-normal mares. Our data showed there is a significant difference in plasma oxytocin concentrations between genitally-normal mares and mares with delayed uterine clearance. These results suggest that there may be differences in these two groups in the pattern of basal oxytocin release that might predispose to differences in uterine contraction after challenge, although no difference was found in oxytocin response to artificial insemination [13] in these two groups. Further study is required to determine the exact nature of the relationship between pituitary oxytocin secretion, uterine oxytocin secretion, oxytocin receptors and uterine contractility in the mares with delayed uterine clearance.

References

[1] Bargmann, W., Sharrer, E.: The site of origin of the hormones of the posterior pituitary. Am. Scientist 39, 255–259 (1951)

[2] Melrose, P.A., Knigge, K.M.: Topography of oxytocin and vasopressin neurons in the forebrain of Equus caballus: further support of proposed evolutionary relationships for proopiomelanocortin, oxytocin and vasopressin neurons. Brain Behav. Evol. 33, 193–204 (1989)

[3] Roberts, J.S., McCracken, J.A., Garagan, J.E., Soloff, M.S.: Oxytocin-stimulated release of prostaglandin F2a from ovine endometrium in vitro: Correlation with estrous cycle and oxytocin-receptor binding. Endocrinology 99, 1107–1114 (1976)
[4] Gibbens, G.L.D., Chard, T.: Observation on maternal oxytocin release during human labour and the effect of intracenous alcohol administration. Am. J. Obstet. Gynecol. 126, 243–246 (1976)
[5] Dawood, M.Y.: Neurophpophysial hormones. In: Fuchs, F., Klopper, A. (eds.) Endocrinology of pregnancy, pp. 204–228. Harper and Row, Philadelphia (1983)
[6] Fuchs, A.R., Romero, R., Keefe, D., Parra, M., Oyarzun, E., Behnke, E.: Oxytocin secretion and human parturition: pulse frequency and duration increase during spontaneous labor in women. Am. J. Obstet. Gynecol. 165, 1515–1523 (1991)
[7] Burns, P.J., Kumaresan, P., Douglas, R.: Plasma oxytocin concentration in cyclic mares and sexually aroused stallions. Theriogenology 16, 531–539 (1981)
[8] Tetzke, T.A., Ismail, S., Mikuckis, G., Evans, J.W.: Patterns of oxytocin secretion during the oestrous cycle of the mare. J. Reprod Fert. Suppl. 35, 245–252 (1987)
[9] Stevenson, K.R., Parkinson, T.J., Wathes, D.C.: Measurement of oxytocin concentrations in plasma and ovarian extracts during the oestrous cycle of mares. J. Reprod Fert. 93, 437–441 (1991)
[10] Vanderwall, D.K., Silvia, W.J., Fitzgerald, B.P.: Concentrations of oxytocin in the intercavernous sinus of mares during luteolysis: temporal relationship with concentrations of 13, 14-dihydro-15-keto-prostaglandin F_2□. J. Reprod Fert. 112, 337–346 (1998)
[11] Troedsson, M.H.T., Liu, I.K.M.: Measurement of total volume and protein concentration of intrauterine secretion after intrauterine inoculation of bacteria in mares that were either resistant or susceptible to chronic infection. Am. J. Vet. Res. 54, 570–572 (1992)
[12] Troedsson, M.H.T., Liu, I.K.M., Ing, M., Pascoe, J., Thurmond, M.: Multiple site electromyography recordings of uterine activity following an intrauterine bacterial challenge in mares susceptible and resistant to chronic uterine infection. J. Reprod Fert. 99, 307–313 (1993)
[13] Nikolakopoulos, E., Kindahl, H., Watson, E.E.: Oxytocin and PGF_2□ release in mares resistant and susceptible to persistent mating-induced endometritis. J Reprod Fert 56(suppl.), 363–372 (2000)
[14] de Winter, P., Verdonck, M., de Kruif, A., Devriese, L., Haesebrouck, F.: Bacterial endometritis and vaginal discharge in the sow: precalence of different bacterial species and experimental reproduction of the syndrome. Ani. Reprod Sci. 37, 325–335 (1995)
[15] Homeida, A.M., Cooke, R.G.: Loss of biological and immunological activity of oxytocin from goat plasma on storage. In: Proceedings of the 3rd Meeting of British Endocrine Societies, vol. 134 (1984)
[16] Thornton, S.N., Leng, G., Bicknell, R.J., Chapman, C., Purdew, T.: Vasopressin, but not oxytocin, is released in response to water deprivation in conscious goats. J. Endocr. 110, 335–340 (1986)
[17] Sheldrick, E.L., Flint, A.P.F.: Circulating concentrations of oxytocin during the estrous cycle and early pregnancy in sheep. Prostaglandins 22, 631–636 (1981)
[18] Noden, P., Oxender, W., Hafs, H.: The cycle of oestrus, ovulation and plasma levels of hormones in the mares. J. Reprod Fert. 23, 189–192 (1975)

[19] Meinecke, B., Gips, H., Meinecke-Tillmann, S.: Progestagen, androgen and oestrogen levels in plasma and ovarian follicular fluid during the oestrous cycle of the mare. Anim. Reprod Sci. 12, 255–265 (1987)

[20] Madill, S., Troedsson, M.H.T., Alexander, S.L., Shand, N., Santschi, E.M., Irvine, C.H.G.: Simutaneous recording of pituitary oxytocin secretion and myometrial activity in estrous mares exposed to various breeding stimuli. J. Reprod Fert. 56(suppl.), 351–361 (2000)

[21] Troedsson, M.H.T., Liu, I.K.M., Ing, M., Pascoe, J.: Smooth muscle electrical activity in the oviduct, and the effect of oxytocin, prostaglandin F_2□ and prostaglandin E_2 on the myometrium and the oviduct of the cycling mare. Biol. Reprod Mono. 1, 475–488 (1996)

[22] Ko, J.C.H., Lock, T.F., Davis, J.L., Smith, R.: Spontaneous and oxytocin-induced uterine motility in cyclic and postpartum mares. Theriogenology 32, 643–651 (1989)

[23] Gutjahr, S., Paccamonti, D.L., Pycock, J.F., Taverne, M.A.M., Dieleman, S.J., van der Weijden, G.C.: Effect of dose and day of treatment on uterine response to oxytocin in mares. Theriogenology 54, 447–456 (2000)

[24] Behrendt-Adam, C.Y., Adams, M.H., Simpson, K.S., McDowell, K.J.: Effects of steroids on endometrial oxytocin mRNA production. J. Reprod Fert. Suppl. 56, 297–304 (2000)

[25] Paccamonti, D., Pycock, J.F., Taverne, M., Bevers, G.C., van der Weyden, G.C., Gutjahr, S., Schams, D., Blouin, D.: PGFM response to oxogenous oxytocin and determination of the half-life oxytocin in nonpregnant mare. Equine. Vet. J. 31, 285–288 (1999)

[26] Nikolakopoulos, E.: Factors affecting uterine contractile activity in the mare. University of Edinburgh, PhD thesis (1999)

[27] Fuchs, A.R., Periyasamy, S., Alexandrova, M., Soloff, M.S.: Correlation between oxytocin receptor concentration and responsiveness to oxytocin in pregnant rat myometrium: Effects of ovarian steroids. Endocrinology 113, 742–749 (1983)

[28] Soloff, M.S., Fernstrom, M., Periyasamy, S., Soloff, S., Baldwin, S., Wieder, M.: Regulation of oxytocin receptor concentration in rat uterine explants by estrogen and progesterone. Can. J. Biochem. Cell Biol. 61, 625–630 (1983)

[29] Maggi, M., Fantoni, G., Peri, A., Giannini, S., Brandi, M., Orlando, C., Serio, M.: Steroid modulation of oxytocin/vasopressin receptors in the uterus. J. Steroid Biochem. Molecular Biol. 40, 481–491 (1991)

[30] Watson, E.D., BjØrksten, T., Buckingham, J., Nikolakopoulos, E.: Immunolocalization of oxytocin and neurophysin in the uterus of the mare. J. Reprod Fert. 56(suppl.), 289–296 (2000)

[31] Jones, D.M., Fielden, E.D., Carr, D.: Some physiological and pharmacological factors affecting uterine motility as measured by electromyography in the mare. J. Reprod Fert. 44(suppl.), 357–368 (1991)

[32] Pycock, J.F., Newcombe, J.R.: The relationship between intraluminal uterine fluid, endometritis, and pregnancy rate in the mare. Equine practice 18, 19–22 (1996)

[33] Reilas, T., Katila, T., Makela, O., Huhtinen, M., Koskinen, E.: Intrauterine fluid accumulation in oestrous mares. Acta Vet Scand 38, 69–78 (1997)

[34] Adams, G.P., Kastelic, J.P., Bergfelt, D.R., Ginther, O.: Effect of uterine inflammation and ultrasonically-detected uterine pathology on fertility in the mare. J. Reprod Fert. 35(suppl.), 445–454 (1987)

[35] Squires, E.L., Bournes, C.K., Rowley, H.S., McKinnon, A.O., Pickett, B.W., Shideler, R.K.: Effect of uterine fluid and volume of semen extender on fertility. In: 35th Ann. Conv. Am. Assoc. Equine Pract. Boston, pp. 25–30 (1989)
[36] Cadario, M.E., Thatcher, W.W., Fuchs, A.R., Archbald, L., Thatcher, M.J., LeBlanc, M.M.: Relationship between plasma keto-13,14-dihydro PGF_2□ (PGFM) and uterine clearance of radiocolloid after oxytocin administration in reproductively normal mares and those with delayed in uterine clearance. Proc. Soc. Theriogenology, Baltimore 172 (1998)
[37] Guyton, A., Hall, J.: The autonomic nervous system. In: Textbook of Medical Physiology, pp. 777–778. WB Saunders Co., Philadelphia (1996)
[38] Shew, R.L., Papka, R.E., McNeill, D.L.: Substance P and calcitonin gene-related peptide immunoreactivity in nerves of the rat uterus: localization, colocalization and effects on uterine contractility. Peptides 12, 593–600 (1991)
[39] Shew, R.L., Papka, R.E., McNeill, D.L.: Galanin and calcitonin gene-related peptide immunoreactivity in nerves of the rat uterus: localization, colocalization, and effects on uterine contractility. Peptides 13, 273–279 (1992)
[40] Garfield, R.E.: Control of myometrial function in preterm versus term labor. Clin Obstet Gynecol. 27, 572–591 (1984)
[41] Cadario, M.E., Thatcher, M.J.D., LeBlanc, M.M.: Relationship between prostaglandin and uterine clearance of radiocolloid in the mare. Biol. Reprod VI(Monography 1), 495–500 (1995)

Author Index

The 35th Anniversary of the KSEAUK Foundation

KSEAUK

THE KOREAN SCIENTISTS AND ENGINEERS ASSOCIATION IN THE UK

재영한인과학기술자협회

The Korean Scientists and Engineers Association in UK (KSEAUK) is a non-profit professional organization.

Since it was established in 1974, the KSEAUK has been contributed to networking among Korean scientists and engineers in the UK and the development of science and technology for our homeland.

The main objectives of the KSEAUK are to:

- Promote the application of science and technology for the general welfare of society
- Foster international cooperation especially between the UK and Korea
- Help Korean Scientists and Engineers developing their full career potential

KSEAUK has seven 'Special Interest Group' (SIGs) which consist of SIG-P (Professional), SIG-T (Telecommunication), SIG-B (Bio), SIG-E (Energy and Environment), SIG-W (Women), SIG-MD (Medical and Dental), and SIG-Y (Young Generation).

The representative activities of KSEAUK are

- General Conference, Workshops, and Symposium
- Publication of News Letters and Webzines
- Web Services
- Young Generation Forum
- Recruitment Shows
- Annual Math Competition

- Technical Evaluation and Consultation
- Short-term Programs for the General Public

Webpage of KSEAUK – http://www.kseauk.org

The number of KSEAUK members is 1069 (counted in July, 2009), which makes KSEAUK the largest Korean scientists and engineering association in Europe. KSEAUK had its efforts publicly recognized for the contribution to promotion of science and technology by the government of Republic of Korea on 21/04/2009 (Science Day) in the form of a Medal of honour (Mr Jinil Kim, Former KSEAUK President, 2002-2008). This is the first we have seen since the establishments of the European science associations and this is also in recognition of our efforts in Europe and is only beneficial for further development of the KSEAUK and Europe's Korean Science and Technology community as a whole.

Jin-il Kim, former president of KSEAUK is awarded by the President of Republic of Korea for his contribution to the promotion of Korea Science and Technology

2009 KSEAUK Staffs

	KSEAUK President Dr. Joung Hwan Lee Ph.D., DIC (The University of Sheffield)		Vice President (Internal Affairs) Mr. Soohwan Kim MSc. (Computer Associates)
	Vice President (External Affairs) Dr. Jung-Sik Kim BA, MEng, Ph.D. (Loughborough University)		Public Secretary (External) Mr. Namshik Han LLB, BEng, MEng (The University of Manchester)
	Treasurer Mr. Sungwook Park (The University of Sheffield)		Academics Secretary Dr. Sungjoo Lee Ph.D. (Ajou University)
	Document Secretary Dr. Minsoo Kim Ph.D. (The University of Sheffield)		Information Secretary Miss Jawon Song (University of Cambridge)
	Public Secretary (Internal) Mr. Seongjin (Leon) Baek (University of Cambridge)		Human Resource Secretary Dr. Yongduck Kang BSc, MSc, Ph.D. (University of Nottingham)
	SIG-Environment Leader Dr. Jae-min Kim Ph.D. (University of Strathclyde)		SIG-Woman scientist Leader Dr. Sung-Eun Bae MSc., D.V.M., Ph.D. (University of Surrey)
	SIG-Bio Leader Dr. Byung-Wook Yun Ph.D. (University of Edinburgh)		SIG-Youth Group Leader Miss Anne Kim MSc. (King's College London)

<table>
<tr><td></td><td>SIG-Telecom Leader

Dr. Habin Lee Ph.D.
(Brunel University)</td><td></td><td>General Secretary

Miss Rira Kim BSc
(Imperial College London)</td></tr>
<tr><td rowspan="2"></td><td rowspan="2">Auditor

Dr. Ik Soo Kim
BEng, Ph.D., CEng
MIMechE
(Doosan Babcock Energy)</td><td colspan="2">SIG-Professional Leader
Mr. Joseph Lee (Accenture)</td></tr>
<tr><td colspan="2">SIG-Medical Leader
Mr. Chang-Ho Choi (Imperial College London)</td></tr>
</table>

List of KSEAUK Presidents

	Duration	Name		Duration	Name
1	1974.11.19	J.M. Kim	16	1990	J.R. Park
2	1975/76	W.S. Shin	17	1991	S.G. Seo
3	1976/77	W.S. Shin	18	1992	S.G. Seo
4	1977/78	W.S. Shin	19	1993	M.J. Kim
5	1978/79	W.S. Shin	20	1994	W.S. Shin
6	1979/80	S.Y. Jung	21	1995	W.S. Shin
7	1980/81	W.D. Lee	22	1996	W.S. Shin
8	1981/82	J.K Kim	23	1997–1998	I.Y. Kim
9	1982/83	W.S. Shin	24	1999–2000	J.J. Jeong
10	1983/84	S.S. Kim	25	2001–2002.6	S.I. Nam
11	1984/85	U.W. Lee	26	2002.6–2002.12	J.I. Kim
12	1986	U.W. Lee	27	2003–2004	J.I. Kim
13	1987	H.S. Yoo	28	2005–2006	J.I. Kim
14	1988	Y.J. Hong	29	2007–2008	J.I. Kim
15	1989	H.J. Park	30	2009.1 –	J.H. Lee

Message from EKC2009 Co-chair

Honorable President Seok Joon-weon of the Korean-German Scientists and Engineers Association, President Han Man-wook of Korean-Australian Scientists and Engineers Association, President Jun Chang-hoon of the Korean-French Scientists and Engineers Association, and in particular, President Lee Joung-hwan of the Korean-British Scientists and Engineers Association, the host of EKC 2009. I would like to express my deepest gratitude and appreciation to you all for organizing this significant meeting.

As co-chairman of EKC 2009 as well as President of the Korean Federation of Science and Technology Societies, I would like to extend warmhearted welcome to each and every one of scientists and engineers here today who had to travel all the way from Korea and the Europe for the conference.

Distinguished Korean scientists and engineers!

This year marks the 35th anniversary of the Korean-British Scientists and Engineers Association. It has hosted the conference twice a year in the UK and defined itself as a solid organization with 1,000 members from 6 regions. I highly appreciate all staffs of the KSEAUK serving as bridgehead for promoting cooperation in science and technology between Korea and the UK.

The primary mission of the Korean Federation of Science and Technology Societies is to serve as a focal point of mobilizing Korean scientists and engineers across the globe. In this regard, followed by the first conference held in Heidelberg last year, EKC-2009 bears great significance in exchanging latest information on science and technology and discussing efficient, effective ways to improve cooperation.

Now is the time for Europe to promote exchange based on human network with Korea and introduce our excellent science and technology, and I hope the conference today will provide an opportunity for further understanding of science and technology between Korea and Europe. In so doing, I hope each of you here today plays a role of promoting cooperation between Korea and the EU.

In order to achieve cooperation and progress in science and cooperation among Korean race, KOFST will make greater efforts with each European associations of science and technology. Please always remember KOFST is to help you every step of the way.

The festival for Korean and American scientists and engineers is now officially started. The EU-Korea conference on science and technology is a meaningful event, and I hope every one of you with great capability share good ideas of mutual interest and exchange expertise.

I would like to conclude by thanking President Joung Hwan Lee and concerned European Scientists and Engineers Association for organizing the 2nd conference once again.

Lastly, I wish your participation in this meeting to be meaningful and fruitful.

Thank you very much.

Dr. Ki-Jun Lee
President of The Korean Federation of Science
and Technology Societies (KOFST)

Message from the Korean Ambassador

Good morning. Ladies and gentlemen, I am Chun Young-Woo, the Ambassador of the Republic of Korea to the UK.

I am delighted to be a part of the Europe-Korea Conference 2009 (EKC2009) in the UK. I welcome the Verein Koreanischer Naturwissenschaftler und Ingenieare in der BRD e.V. (VeKNI), the Association des Scientifiques Coreens en France (ASCoF), the Korean Scientists and Engineers Association in Austria (KOSEAA), Europe's Korean Compatriot scientists and engineers and the many scientists and engineers from the Korea Association of Industry, Academy and Research Institute.

I also extend a warm welcome to the many foreign scientists and engineers who join us for today's conference.

The theme for EKC2009 is: "Science and Technology Closer to Humanity: Greenness". It will provide the latest science and technology news and opportunities to network with experts in various fields. The variety of symposiums and forums on subjects such as 'Green Technology', 'ICT', 'Bio Technology', 'Energy and Environment' and 'European Young Generation' offer the ideal environment for all our issues such as climate change.

I am sure this conference will be a very productive event as this year also marks he 35th Anniversary of establishment of the Korean Scientists' Association in the UK.

We can expect to have a synergetic effect from the networking of scientists and engineers between the UK and Korea as well as between Koreans and foreigners, who are experts of such technologies as ET-IT-BT, is likely to produce a synergetic effect.

I know many Ph.D.students are here for the conference and I hope you can deepen your understanding of other fields which may overlap your area of expertise. This is also a great opportunity for you to deepen your understanding of your own field and get inspired for new ideas through networking at today's conference.

Finally, I would like to thank KSEAUK for their efforts for hosting a successful EKC2009, and also would like to express much gratitude to KOFST and other sponsors for their support, without which this conference could not have happened.

Thank you.

Mr. Young-Woo Chun
The Ambassador of The Republic of
Korea to UK

Congratulatory Address for the 35th Anniversary of the KSEAUK Foundation

Dr. Ki-Jun Lee, President of the Korean Federation of Science and Technology Societies, Dr. Woo-Sik Kim, Professor at KAIS, Distinguished Guests, Ladies and Gentlemen!

It is my great pleasure to address today's EU-Korea joint conference of Science and Technology, namely pan-EU EKC 2009. I am honored to be here today as we celebrate the 35th Anniversary of the KSEAUK Foundation in Wokefield Park Hotel, Reading, England. As a founder member of KSEAUK in 1974 I am deeply moved by the growth and development of the Association, which has been working well with the counterparts in Germany, France and Austria. In 1974, we started a group of 14 Korean Scientist and Engineers in the UK and now we have about 1,200 registered members within the Association.

Let me take this opportunity to offer my congratulations to everyone involved with the Association. I am confident that the Association will support cooperation among each KSEA in major European countries as well as Korea, and accomplish its aim when talented professionals cover the exchange of information and to work out strategies for the development of European cooperation.

Ladies and Gentlemen!

Now. the world is experiencing the rapid trend of globalization, so that activities of science and technology, information and human resources are flowing more freely beyond their borders. We are now recognizing the importance of establishing cooperative ties among our regional neighbors. Let us make good use of this opportunity.

I expect we will have in-depth discussion on science and technology closer to Humanity-Greenness topics today.

During this conference, I hope we will have constructive discussions on establishing a framework for cooperation and cementing KSEA cooperation with the counterparts in Korea. It is my firm belief that this joint conference and a symposium hosted by KOFST (the Korean Federation of Science and Technology Societies) will provide a good opportunity to facilitate exchanges and cooperation on the area of environmentally friendly technology and energy. I am hopeful that this gathering will further disseminate our consensus on our cooperation in Europe and Korea.

In closing, I would like to express my warm appreciation to KSEAUK to organize this event, in particular Dr Joung Hwan Lee and his team, and wish all of you the best.

Thank you.

Dr. Woo Seung Shin
World Federation of Overseas Korean
Traders Association, President in UK

Plenary Talks

Plenary Talk I : Thursday (6th August, 2009) 10:30 – 11:10

"Global Challenge for Climate Change and Green Growth"

Prof. Chang-Sun Hong (KAIST)

Dr. Chang-Sun Hong is **currently Professor of Aerospace Engineering at Korea Advanced Institute of Science and Technology** (known as KAIST) after serving as a member of *Korean National Assembly* from 2004 to 2008 on leave from KAIST. He served as the Vice Chair and a member of *Science-Technology, Information and Telecommunication Committee of Korean National Assembly.* Before joining legislator service in 2004, he has been a Professor of Aerospace Engineering, Dean of Academic Affairs, Dean of Engineering and the President of KAIST. He received his Ph.D. degree in Engineering Mechanics at the Penn State University in the United States. After two years of research at NASA-Langley Research Center in Hampton, Virginia, he came back to Korea and joined the faculty members of KAIST in 1979. His research interests were composite materials and structures and smart structure using optical fiber sensors until he assumed the administrative job. He served as president of *Korean Society for Aeronautical and Space Sciences* in 1996-97 and president of *Korean Society of Composite Materials* in 1999-2002. He is a founding member of *Korea Academy of Science and Technology* and a member of *National Academy of Engineering of Korea.*

Plenary Talk II : Thursday (6th August, 2009) 11:20 – 12:00
"Global Fossil Fuel Supply and Demand and the Challenge of Climate Change"
Prof. Alexander Kemp (University of Aberdeen)

Dr. Alexander G. Kemp is **currently Professor of Petroleum Economics at the University of Aberdeen.** He is also Director of Aberdeen University Petroleum and Economic Consultants (AUPEC) which provides consultancy service in petroleum economics. For many years he has specialised in his research in petroleum economics with special reference to licensing and taxation issues. He has published over 200 papers and books on this field, including *Petroleum Rent Collection Around the World,* Institute for Research on Public Policy, (Canada), 1988. For many years he has been a consultant on petroleum contracts and legislation to a large number of Governments, the World Bank, the United Nations, oil companies, the European Commission, the UK Know-How Fund and the Commonwealth Secretariat. He was a specialist adviser to the UK House of Commons Select Committee on Energy (1980-1992). In 1993 he was appointed by the Minister of Energy to the UK Government Energy Advisory Panel. he has been appointed Official Historian by the Prime Minister to write the Official History of North Sea Oil and Gas. He was awarded the Alick Buchanan-Smith Memorial Award (1999) and the OBE (2006) for personal achievement and contribution to the Offshore Oil and Gas Industries.

Symposium

Title: Information and Communications Technology (ICT)

ICT1 (6th Aug 2009, 14:00 – 15:20, Canary1),
Session Chair : Dr. Habin Lee (UK)

Paper presentation

1. Dong Han Ham, The state of the art of visual analytics. (UK) – 15+5 min
2. Joon Sang Baek, An inquiry on how collaborative services on the digital platform can contribute to the quality of social commons. (Italy) – 15+5 min
3. Soo Hwan Kim, NIST – SCAP (UK) – 15+5 min
4. Young Saeng Park, Reconfigurable Automation with SOA based on Web Services. (UK) – 15+5 min

ICT2 (6th Aug 2009, 15:40 – 17:00, Canary1),
Session Chairs : Dr. Dong Han Ham (UK), Dr. Habin Lee (UK)

Keynote speech

- Marco Sbarrini, Mobile Technology Evolution (LGE Europe R&D Lab., Italy)

Dr. Marco Sbarrini is currently **the Director of Programs for LG Europe R&D.** His task is to control the overall delivery process from the handset specification down to the acceptance, to supervise the production chain and maintain relationships with chipset vendors. He was awarded the second prize at the World Graph Drawing Symposium 98 (1998), the LG gold medal as best R&D center (2007), and the LG award as Global Innovation Leader (2008)

Paper presentation

1. Ha Yoon Song, Clustering of Mobile Ground Nodes for HAP MBS Placement. (Austria) – 15+5 min

2. Inwha Kim, The cultural impact on the design of web based information systems. (UK) – 15+5 min

ICT3 (7th Aug 2009, 11:10 – 12:30, Forest2),
Session Chairs : Mr. Soowhan Kim (UK), Dr. Dong Han Ham(UK)

Paper presentation

1. Young Hoon Chun, Tunable Bandstop Resonator and Filter on Si-Substrate with PST Thin Film by Sol-Gel Deposition. (UK) – 15+5 min
2. Eun Jung Kim, Model Index regression: application to non-life insurance. (France) – at others, 15+5 min
3. Gakhee Lee, How to use mathematical models to forecast the KOSPI 200 Index futures price: In combination with computing program. (Austria) – 15+5 min
4. Karim Al-Yafi, Habin Lee, Centralized versus Market based Task Allocation in the Presence of Uncertainty. UK) – 15+5 min

Poster Presentation (6th Aug 2009, 12:45 – 14:00)

1. Donghoon Kim, Face Components Detection using SURF Descriptors and SVMs (Ireland)
2. Jong Gun Lee, Predicting the Macroscopic Lifetime of the Threads of Discussion Forums with User Comment Information (France)
3. Chang-Hun Lee, Twist angle measurement in liquid crystal (LC) cells (UK)

Title: Green Technology (GT)

GT Papers 1 (6th Aug 2009, 08:00 – 09:20, Canary2),
Session Chairs : Dr. Jung-Sik Kim (UK), Mr. Seung-ho Lee (France)

Keynote speech

- Gook-Sup Song, Human response in thermal environment(Korea)- 20+5 min

Dr. Gook-Sup Song is a **Professor at Bucheon University (1981-Present), and the vice president of Korean Society of Living Environmental System.** He received PhD in Architectural Engineering at Chung-Ang University. He has investigated human responses in various building environments, and one of experts in the field of the under floor heating system; ONDOL.

Paper presentation

1. Who-Seung Lee, Frontiers in animal behavior strategy and application in energy, economy, and environment (UK) – 7+ 3 min
2. Jung-Rae Kim, Application of bioelectrochemical processes for electricity generation and sustainable wastewater treatment(UK)-7+3 min
3. Myung-Joo Kang , Sustainable and Competitive Product Service System Solutions(Austria) – 7+3 min
4. Dowon Kim, Inter-firm Collaboration for Low Carbon Society(UK) -7+3 min

Green Policy & Strategy Forum (6th Aug 2009, 14:00 – 16:15, Canary2),
Session Chairs : Prof. Eui-chan Jeon (Korea), Dr. Jae Min Kim (UK)

Keynote speech: Policy and strategy for green growth

1. Dr. HeungJin Choi (Director General, Presidential Committee on Green Growth), Government strategy of green growth in Korea

   Dr. Heung-Jin Choi is **the Director-General of Climate Change Policy at Presidential Committee on Green Growth.** He has worked for Korea Ministry of Environment since 1987. He earned his PhD in Civil and Environmental Engineering at the University of Delaware, USA.

2. Dr. Walter Wehrmeyer (Professor, Centre for Environmental Strategy at University of Surrey), Business strategy for sustainable economic growth in EU

   Dr. Walter Wehrmeyer is **Reader and deputy director of Centre for Environmental Strategy in University of Surrey**. He is also editor of Greener Management International, sits on the Editorial Board of Environment, Development and Sustainability, and is Co-editor, Case Studies of the Journal of Industrial Ecology. His current interest cover organisational approaches to innovation and sustainable development, participatory approaches to decision-making, and
   corporate social responsibility, in particular, the role of culture and ethics in defining and forming business strategies.

3. Mr. Yul Choi (President, Korea Green Foundation), Engaging citizen in tackling climate change: global NGOs activities

   Mr. Yul Choi is **the President of Korea Green Foundation.** He has been worked for Korean Anti-Pollution Organization. He also worked for The Korean Federation for Environment Movement. He received National Recognition in Commemoration of World Environmental Day in 1997 and Goldman Environmental Prize in 1995.

Green Technologies Forum (6th Aug 2009, 16:30 – 18:45, Canary2),
Session Chairs : Dr. Doo Bong Chang (Germany), In Sun Lee (UK)

Keynote speech: Technologies for green growth

1. Dr Andrew Grant (Senior Lecturer, Dept. of Mechanical Engineering at University of Strathclyde), Renewable Technologies in EU: 2nd generation tidal turbine

   Andrew Grant is **a Senior lecturer at University of Strathclyde.** He has studied on renewable energy conversion systems dates from the late 1970's. He has contributed the development of small novel devices for the utilisation of energy from ocean waves and marine currents. Subsequently diversified into wind turbine aerodynamics, and the integration of renewable energy conversion systems into networks at local and regional levels.

2. Mr. Bill Dunster (Principal Director, Zedfactory ltd.), Towards Zero carbon Buildings

   Mr. Bill Dunster is **an award winning eco architect.** He designed BedZED, and since 1999, practice has completed number of award winning zero carbon/ low carbonbuilding projects. He has also taught at the Architectural Association and Kingston University and regularly speaks at a range of seminars and conferences all over the world.

3. Dr. Yeonmook Nah (Dan-Kook University), Green IT Strategy in Korea

   Dr. Yeonmook Nah is **a Professor at Dankook University (1993-Present).** He earned PhD in Computer Engineering at Seoul National University. His interests are databases, multimedia information retrieval, internet search engines, distributed data management, location-based service, sustainable consumtion & production and green IT.

GT Papers 2 (7th Aug 2009, 08:00 – 09:30, Canary2),
Session Chairs : Mr. Dowon Kim (UK), Dr. Dong-Seon Kim (Austria)

Paper presentation

1. Sungjoo Lee, Technology co-evolution analysis in the energy sector(Korea), 7+3 min
2. Seung-Ho Lee, Green Architecture(France), 7+3 min
3. Soo-Young Bae, Wireless monitoring system for hybrid power generation system(Korea), 7+3 min
4. Ji-Yeon Kim, Research on the application method for renewable energy complex system for school buildings(Korea)-7+3 min
5. Ji-Young Lee, Optimisation methodology of low carbon mixed energy systems using the bees algorithm(UK)-7+3 min

6. Lae-Hyun Kim, Development of district heat demand forecasting model using neural networks(Korea)-7+3 min
7. Ui-Sik Kim, Optimal operation system of the Integrated district heating system with multiple regional branches(Korea)-7+3 min

Poster Presentation (7th Aug 2009, 12:10 – 12:30)

1. Hyo-Soon Park, An analysis on the plan to save demand for energy and introduce a renewable energy system in Innovation cities(Korea)
2. Jae Min Kim, Low-Carbon new city energy planning using hourly demand/supply profiles matching analysis(UK)
3. Dong-Soo Jeong, Korean creative green vehicles for the future global market(korea)-7+3 min

Title: Bio Technology (BT)

BT1 (6th Aug 2009, 14:00 – 15:20, Sky5),
Session Chair : Dr. Byung-Wook Yun (UK)

Paper presentation

1. Jung Wook Sim, Study of cell signalling in microfluidics (UK) – 15+3 min
2. Chang-Hoon Nam, Analysis of immunoglobulin kappa variable regions expressed by high and low antibody responder pre-immune Biozzi mice (Germany) – 15+3 min
3. Soohyun Kim, Sex, smell and human diseases (UK) – 15+3 min

BT2 (7th Aug 2009, 09:30 – 10:50, Sky5),
Session Chair : Dr. Hun Mok (Ireland)

Keynote speech

- Sang-Soo Kwak (Environmental Biotechnology Research Center, KRIBB) Industrial transgenic plants for sustainable development on marginal lands (Korea) – 35+5 min

  Dr. Sang-Soo Kwak is **a Principal Researcher at KRIBB.** He earned his PhD in Plant Biochemistry at The University of Tokyo. His main research topics are study of antioxidation mechanism in plant cells and its application to develop transgenic plants with enhanced traits. He has been in charge of the Korea-China collaboration project on combating desertification using biotechnology since 2004.

Paper presentation

1. Haseong Kim, Steady-state analysis of gene regulatory networks via G-networks: Detecting abnormal behaviors of genes in budding yeast cell cycle gene regulatory networks (UK) –15+3 min
2. Namshik Han, A novel strategy for modeling and visualizing of transcription factor networks (UK) – 15+3 min

BT3 (7th Aug 2009, 11:10 – 12:30, Sky5),
Session Chairs : Dr. Sang-Soo Kwak (Korea), Prof. Soohyun Kim (UK)

Paper presentation

1. Choi, Young Pyo, An approach to understanding human prion disease from a biochemical viewpoint (UK) – 15+3 min
2. Hanna Sheu, HIV prevention in schools by empowerment (Austria) – 15+3 min
3. Byung-Wook Yun, S-nitrosothiols regulate hypersensitive cell death during the plant disease resistance response (UK) – 15+3 min

Poster Presentation (7th Aug 2009, 12:30 – 14:00)

1. Sung Eun Bae, Defining changes in expression of mineralocorticoid receptor (MR), glucocorticoid receptor (GR) and Bcl-2 family in human hippocampus following transient global cerebral ischaemia (UK)
2. Jongho Sun, Green fluorescent protein as a reporter for spatial and temporal gene expression in Streptomyces coelicolor A3(2) (UK)
3. Ji-Hoon Yoo, Lack of genotype effect on D1, D2 and dopamine transporter binding in triple μ-, δ-, and κ-opioid receptor knockout mice of three different genetic backgrounds (UK)
4. Jongwoon Kim, Comparative study of risk assessment approaches based on different methods for deriving PNEC and DNEL of chemical mixtures (Germany) – 15+3 min

Title: Mechatronics and Mechanical Engineering (MME)

MME 1 (14:00 – 15:20, 6th Aug, Sky2)

Session Chair : Dr. Jung-Sik Kim (UK)

Keynote speech

- Changduk Kong, Past, Present and Future Contribution on Research and Development of Aero-Propulsion Systems in Korea-Springer – 25+5 min

Prof. Chang-Duk Kong is **a Professor in the Department of Aeospace Engineering at Chosun University.** He gained a PhD at Osaka Prefecture University. He has served as an vice president of the Korea society for aeronautics and space sciences (KSAS) .His main research topics are the structure of aircraft and a gas turbine

Paper presentation

1. Kyoungwoo Park, Numerical Analysis of aerodynamic characteristics for Wing-in-Ground Effect Craft with DUP device – 15+5 min
2. Chul-Ho Kim, Optimum Aerodynamic Design of Express Bus for the Reduction of Fuel Consumption and Global Warming Gas – 15+5 min

MME 2 (09:30 – 10:50, 7th Aug, Canary1)
Session Chair : Dr. Dong-Seon Kim (Austria)

Paper presentation

1. Man-Wook Han, Navigation of a mobile robot – 15+5 min
2. Young Saeng Park, Reconfigurable Automation with SOA based on Web Services-15+5min
3. Sungwoo Choi, -free control of automotive engine and brake for Stop-and-Go Model – 15+5 min
4. Dong Hwa Kim, Artificial Intelligence, Computational Intelligence, Emotional Technology, Robot – 15+5 min.

MME 3 (11:10 – 12:30, 7th Aug, Canary1)
Session Chair : Dr. Man-Wook Han (Austria)

Paper presentation

1. Jung-Sik Kim, Fabrication of Three-Dimensional Magnetic Microcomponents – 15+5 min
2. Dong Seon Kim, Solar air conditioning in Mediterranean climate – 15+5 min
3. Won-Kwang Park, Level-set method for reconstructing thin electromagnetic inclusions – 15+5 min
4. Byeong Sam,Kim, Computational Analysis of Non Contact Transportation System for Wafer Hand – 15+5 min

Poster presentation

1. Wan-Doo Kim, Nature-inspired Superhydrophobic Surface
2. Yong Duck Kang, Energy saving technology
3. Lee Kangseok, Development of the Motorized seat Belt and Belt Tension Estimation

Title: Pink Technology (PT)

PT1 (6th Aug 2009, 15:40 – 17:00, Sky2),
Session Chair : Dr. Jung-Sik Kim (UK)

Keynote speech

- Dr Alma Hodzic, Turning Polymers into Structural Materials, (A.Hodzic@sheffield.ac.uk, Aerospace and Mechanical Engineering, University of Sheffield, UK) – 30 min presentation + 10 min discussion

Dr. Alma Hodzic is **a Senior Lecturer in the Department of Mechanical Engineering at the University of Sheffield.** Her research includes design of better interfaces between all types of reinforcing fibres and corresponding matrices, with utilization of new research tools in nanotechnology. Dr. Hodzic manages Composite Systems Innovation Centre and she is actively involved with editorial and review panels in 15 international scientific journals. She also serves as Intreader on the panel of Australian Research Council and as External Examiner for Universities of Ulster and Salford.

Paper presentation

1. Jang-Hui Han, Generation and Application of High Brightness Electron Beams. (UK) – 15+5 min
2. Jun-Yun Kang, Phase analysis of advanced high strength steels by electron backscatter diffraction. (France) – 15+5 min

PT2 (7th Aug 2009, 09:30 – 10:50, Sky2),
Session Chair : Dr. Jang-Hui Han (UK)

Paper presentation

1. Hyeok Kim, Molecular weight dependence of polymeric insulators on the leakage current in organic thin film transistors. (France) – 15+5 min
2. Jung-Sik Kim, Constrained sintering of SOFC electrolytes. (UK) – 15+5 min
3. Sungjune Jung, Collision of non-Newtonian impinging jets and its application for drop-on-demand inkjet printing. (UK) – 15+5 min
4. Sung Taek Oh, Introduction of Biological Fuel Cell technologies for renewable energy. (UK) – 15+5 min

PT3 (7th Aug 2009, 11:10 – 12:30, Sky2),
Session Chair : Dr. Sungjoo Lee (Korea)

Paper presentation

1. Jinwon Seo, Technopark in the Utopian Theory's point of view. (France) – 15+5 min
2. Jiwon Yoon, Methodology for analysis of the research trend in Korea. (UK) – 15+5 min
3. Chanhoon Chung, Search for New Physics with Positrons in Space. (Germany) – 15+5 min
4. Jeong Jin Hong, Enhanced Predictive Modelling Using Multi-Block PLS. (UK) – 15+5 min

Poster Presentation (7th Aug 2009, 12:45 – 14:00)

1. Yun Ki Byeun, Synthesis of InN and InMnN nanowires by halide vapor phase epitaxy. (Germany)
2. Il Jung, High efficient donor–acceptor ruthenium complex for dye-sensitized solar cell applications. (Switzerland)
3. Jehyun Lee, Contribution of the curved surfaces on magnetization behavior. (Austria)
4. Chul Woo Hyun, Museum of Accident, (Austria)
5. Boung Shik Yoon, Geotechnical Centrifuge Technology and Its Applications (UK)

Title: The 2nd Women in Science, Engineering and Technology in Europe (WiSETiE) Forum

The mission of the WiSETiE (Women in Science, Engineering and Technology in Europe) Forum is to establish and strengthen the network of Korean women in SET-related fields in Europe, and to facilitate communication and cooperation between our members to promote their career development and success. This is our second Forum and our first forum was held in Heidelberg, Germany in 2008 at EKC2008. We welcome all female & male delegates to our forum.

WiSETiE Networking (5th Aug 2009, 21:30 – late , Mansion house Mortimer Room)

Everybody is welcome especially if you are a women scientist!

2nd WiSETiE Forum (6th Aug 2009, 15:40 – 17:00, Sky5)
Session Chair : Dr. Sung- Eun Bae (UK)

- **Opening ceremony :** Dr. Sun-Hwa Han (Director of KISTI)

- **Introduction of WiSETiE :** Dr. Myung-Joo Kang & Ms. Young-Min Kim (WiSETiE OC)

- **Plenary Speaker**
    - Mrs. Sung-Ja Chang, Love, Marriage, Work and Success, 21Century Superwoman, (Namseoul University, Korea) – 15 min presentation + 5 min discussion

    Dr. Sung-Ja Chang is a Professor at Namseoul University. She obtained her MA in Mass Communications at the University of Washington, USA. She was the President of the Korean Institute for Gender Equality Promotion and Education, and the Chief Executive of Women's Policy at the Ministry of Gender Equality.

- **Invited Speaker (Mentoring)**
    - Dr. Hye-Shuk Seo, My life story – 15 mins presentation +5 min discussion

    Dr. Hye-Shuk Seo (KOSEAA). She gained her MD degree at the University of Vienna. She worked as a GP (general practitioner) running her own clinic until recently and is now retired. After retirement, she continues to consult and organise educational seminars and advice for second generation youth. She has been a very active member of KOSEAA since its very beginning.

- **Question & Answer time**

- **Closing Comments**

Title: The 3rd European Young Generation Forum (EYGF)

This year marks the third holding of the EYGF (European Young Generation Forum). The EYGF operates with the aim of acting as a medium where students can establish new networks, as well as introducing students to stimulating concepts in science and its development in this age. This will ultimately serve to form the foundations of a working partnership between outstanding individuals and to aid the advancement of science in Korea.

EGYF Networking (5th AUG 2009, 21:30 – late, Mansion house)

We cordially invite our members to come along and participate at the EYGF social. The social is aimed to introduce members to new acquaintances and thus develop the web of network among our European members. You are also welcome to join the other SIG Networking Bars on the second evening of the conference (SIG-W, SIG-T, SIG-B and SIG-E).
By doing so, you will have the opportunity to engage with professionals and experts dedicated to your field of interest and further develop your interest and knowledge of the field.

3rd EYG Forum
(6th AUG 2009, 14:00~15:20, 15:40~16:30 and 7th AUG 2009, 14:00 - 15:40, Forest 2), *Session Chair : Miss Anne Kim (UK)*

- Co-chaired by :
• Miss Su Yeon Lee of ENST Bretagne, France
• Miss Silvia Lee, Austria
• Mr. Bong Gu Kang of RWTH Aachen, Germany

- Welcome Speech
Miss Anne Kim (SIG-Y Leader)

- Plenary Speech
• Dr. Sun-Hwa Hahn (KISTI)

Dr. Sun-Hwa Hahn is the chief of Department of Policy Research in Korea Institute of Science and Technology Information (KISTI). She obtained her PhD in computer science from KAIST in 1997. She directs research on future strategy, mid and long-term development planning, and provision of policy support to government planning on science and technology.

- Team presentation

- Teambuilding: "Anne-chimedes' dilemma"

Title: ULTRA Programme

ULTRA programme provides a meeting place between the worldwide Korean scholars residing in foreign countries and domestic Korean scholars to find out ways contributing to the development of Korean science. (Website: http://www.ultra.or.kr)

Topics: Environment, energy, and low carbon for sustainable green economic growth

Ultra Programme (9:30 – 12:30, 7th Aug 2009, Canary2)

Speaker: Dr. Jaemin Kim, ESRU, University of Strathclyde
Title: Low Carbon Network: Strategy and Operation

Ultra Programme (14:00 – 17:00, 7th Aug 2009, Canary2)
Speaker: Dr Jung Ha Yoo, Max-Planck-Institute of Plasma PhysicsMaterials Research Division
Title: Issues of Materials and Structure Design for Fusion Reactor Technology

Title: KOFST Programme (Executive Meeting)

Round table meeting with Korea Science and Technology Leaders

Expected Participants
- From Korea: government officials, presidents of universities, presidents of governmental research institutes, directors of industrial research centres
- From EU: related Korean scholars

KOFST Programme (14:00 – 16:00, 6th Aug 2009, Courtyard 2)

Abstracts

Plenary Talks

"Global Challenge for Climate Change and Green Growth"
Professor Chang-Sun Hong (KAIST)

The climate is changing. The earth is warming up, and there is now overwhelming scientific consensus that it is happening, and human-induced. The United Nations Intergovernmental Panel on Climate Change (IPCC) released a report in 2007 that said global warming was "very likely" - meaning an at least 90 per cent certainty - caused by human activity. It's been almost seventeen years since the United

Nations Framework Convention on Climate Change entered into force and more than ten years since its Kyoto Protocol was adopted. Many governments around the world have signed the Kyoto Protocol, an agreement between countries to reduce the amount of greenhouse gases their countries produce. South Korea was classified as a developing country, during the first commitment period (2008-2012), and thus shouldered no obligation for reduction of greenhouse gases. However, in the second commitment period (2012-2016), Korea has to fulfill the obligation befitting its status as an advanced developing country.

In December 2009, the final post-Kyoto negotiations will be held in Copenhagen, Denmark. Korea has been heavily dependent on energy-consuming industries such as steel, petro-chemistry and automobiles. Energy efficiency in manufacturing industries will be crucial as recognized by comparing the energy consumption with the Japanese counterpart. Korea has not been known as a leader in environmental technology. Its industries have lagged behind Japan and Germany in implementing energy efficient manufacturing technologies. In August 2008, the President of Korea said: I want to put forward 'Low Carbon, Green Growth' as the core of the Republic's new vision. Green growth refers to sustainable growth which helps reduce greenhouse gas emission and environmental pollution. Korea has launched campaigns to increase global awareness of the importance of cooperation in promoting green growth in the international community. It is also a new national development paradigm that creates new growth engines and jobs with green technology and clean energy.

Every economy takes the view that research in this green technology should be boosted and clear environmental performance targets should be adopted. There is quite some controversy on the topic of climate change and its solution due to different views and interpretations. In order to gain some insight in this field, the general questions are addressed and sound policy for each country should be implemented to get it done. Each country has different situation for different renewable energy. We should not make same mistakes which the front runner made in the race of the renewable energy.

"Global Fossil Fuel Supply and Demand and the Challenge of Climate Change"

<u>Professor Alex Kemp (University of Aberdeen)</u>

There have always been noteworthy uncertainties surrounding both the supply and demand for petroleum (and coal), whether on a worldwide or country specific context. Hence predicting oil and gas prices is fraught with difficulties. The degree of unpredictability of oil/gas prices has probably increased in recent years. This presentation will examine the longer term prospects for fossil fuel supply and demand worldwide with individual country examples, highlighting the sources of the uncertainties. Attention will be given to the effects of climate change policies on both demand and supply. These policies are themselves very uncertain both in their design and in their effects. The consequences of different policies adopted by governments towards climate change on the markets for fossil fuels will be highlighted.

ICT Symposium – Keynote Speech

"Mobile Technology Evolution"

Marco Sbarrini (Director of Programs for LG Europe R&D) : marcosbarrini@lge.com

Agenda:
1. Mobile Market Trend
2. Key Driving Technology
3. LG technology closer to Greenness
4. LG MC R&D Introduction

Short summary
In last last year the economic downturn lead to a deep industry transformation.

The mobile market is going in a direction where top 3 vendors play more and more an important role.

In this sense also their responsibilities are getting higher in terms of environment.

Key Success Factors cannot be just based on "cost cutting' as a 'Consumer oriented technology" taking care of health, ecology and customer satisfaction is becoming a must.

Presentation will go through a general introduction of key driving technologies (from LTE to Open OS, HW and Services) and will complete with a high level description

of LG activities in Green Product Development Strategy.

The presentation will finish with a short introduction of LG MC R&D.

ICT Symposium – Oral

"The state of the art of visual analytics"

Dong-Han Ham (Middlesex University) : donghan.ham@gmail.com

One of the critical issues challenging human decision makers in the information age is how to find out data relevant to their tasks and how to derive meaningful information from these data. As a new discipline for dealing with this issue, visual analytics has recently emerged. Visual analytics is defined as the science of analytical reasoning facilitated by interactive visual interfaces. It is a multidisciplinary subject that is related to data mining, information visualization, knowledge science, decision science, and so on. Four important areas within visual analytics are: (1) analytical reasoning techniques, (2) visual representations and interaction techniques, (3) data representations and transformations, and (4) production, presentations, and dissemination of the analysis results. This paper reviews the state of the art of visual analytics and suggests that the problems of visual analytics should be considered in the context of human-computer interaction and joint cognitive systems. Based on the review results, this paper proposes a conceptual framework for organizing research problems studied so far and suggesting viable future research directions. The author hopes that this paper will provide comprehensive information about visual analytics and be a good source for researchers who are interested in this new discipline.

"An inquiry on how collaborative services on the digital platform can contribute to the quality of social commons"

Joon Sang Baek (Italy) : joonsbaek@gmail.com

In this paper, an ongoing PhD research on collaborative services on the digital platform and their impact on sustainable local development has been introduced. The notion of collaborative services was introduced and its relation with collaborative production in the digital sphere was discussed. 24 cases of collaborative services on the digital platform has been collected and analyzed to understand how social innovation at the grassroots level can combine with digital technologies and contribute to local development. In the light analysis, basic information of the cases was collected and their contribution to local development was assessed focusing on the tangible effects of the solutions on economic, environmental and social aspects of sustainability. The future work will be to analyze the cases in depth using qualitative research methods such as interview and ethnography. In the in-depth research, the contribution of collaborative services on the quality of social commons of local communities will be analyzed. Finally, the role of digital technologies in this process will be identified. The result of this research is expected to be a reference for designers who design collaborative services for the local development. It can also be a guide for the local government and institutions who are willing to adopt design for sustainability in their policies. Finally, it will hopefully enlighten people their potentials to innovate society through their creativity and collaboration.

"NIST-SCAP"

Soo Hwan, Kim (CA, Senior research engineer) : windsor.kim@gmail.com

Under the direction of OMB (Office of Management and Budget) and in collaboration with DHS, DISA, NSA, USAF, and Microsoft, NIST has set and provided the desktop security standards and make it mandate for the federal government agencies now.Therfore, federal goverment dsktop boxes should be compliant with this new security standards.I want to introduce how CA (Computer Associates) develops and integrate this SCAP (Secure contents application protocol) capable solutions into current CA desktop & server management product line from the development point of view and also address how desktop machine security configurations managed in efficiency.

"Reconfigurable Automation with SOA based on Web Services"

Young Saeng Park (Loughborough University, Research Associate) : y.park@lboro.ac.uk

In order to remain competitive in the global marketplace such as demands of product variety, frequent product change, less production cost and even energy savings, Service Oriented Architecture (SOA) based on Web Services is starting to be considered as the technology in the manufacturing environment that will enable an increase agile and collaborative manufacturing capabilities. A main strength of SOA is the ability to compose composite applications consisting of services that exist within separate domains and organisations. Prior to SOA, this level of application flexibility was stifled by vendor specific communications and the need to present data to legacy system interfaces. SOA offers a step change in approach as Web Service protocols present open standards based communication, and common standards based interfaces. Depending on process requirement in a shop floor, SOA enabled devices are diversely composed to provide production process and the system will be easily reconfigurable by changing or adding one or more services. This research as a part of European IST FP6 project, Service-Oriented Cross-layer infRAstructure for Distributed smart Embedded deviceS (SOCRADES), explores the application of SOA in shop floor automation using Device Profile for Web Services (DPWS) which allows SOA on a device. Orchestration framework, which is supported by the engineering tools developed at Loughborough University, has been introduced to effectively control and easily reconfigure SOA services on devices. This also enables the adoption of collaborative technology for all the levels in manufacturing automation from sensors and actuators to enterprise applications. The orchestrator has been implemented and demonstrated with the prototype test rig in terms of control and dynamic management of services on devices. To demonstrate the integration with high level applications, SAP MII (Manufacturing Integration and Intelligence) is directly integrated with the orchestrator. With the readily updated data, it is fully aware of the changes happening on the test rig and thus able to successfully drive proactive decision-making measures for production cost, time and even energy savings.

"Clustering of Mobile Ground Nodes for HAP MBS Placement"

Ha Yoon Song (Vienna University of Technology, Guestprofessor) : song@ict.tuwien.ac.at

HAP (High Altitude Platform) is a stationary aerial platform positioned in the stratosphere between 17Km and 22Km height and it act as an MBS (Mobile Base Station) of wireless network. HAP based Network has advantages of both satellite and terrestrial communication systems. In this research we study the efficient deployment of HAP MBS that can provide communication for users by clustering of mobile ground nodes. Mobile ground nodes have their own mobility and thus the characteristics of mobility must be considered in order to keep the clusters of mobile nodes as stable as possible. Among the various clustering algorithms, several candidates are thoroughly studied and we use an EM (Expectation Maximization) since we conclude the characteristics of EM clustering algorithm will cope with the stability problem of mobile node clusters with high mobility of ground nodes. For the verification, an island with naval area is selected since it contains traffics of ground vehicles, persons, horse as well as naval vehicles. We evaluate mobile ground node clustering by EM algorithm considering distance and movement speed of nodes based on RWP (Random Waypoint) node mobility with population density.

"The cultural impact on the design of Web based Information Systems"
Inhwa Kim (Brunel University) : inhwakim@gmail.com

The international and multicultural nature of the World Wide Web enables people from different geographical backgrounds to communicate with each other and share information easily. Although the Internet is a global medium, its creators and users have different backgrounds and belong to different cultures, thus it poses many communications& challenges and often leads to misunderstandings and problems. Various studies have described cultural influences and the need for research on the role of culture in interface design and websites. Although current research attempts to identify design elements affected due to cultural differences, the classification of cultural markers and user perception has yet to be addressed in detail. This is because the approaches targeted web sites designed by professional designers or developers rather than actual users themselves. Professional designers are usually trained based on universal web design standards most of which rarely consider the cultural factors in website design. This paper aims to investigate the impact of culture on the design of web based information systems (WIS) from end user&s perspective. A web-based blog is selected as a target WIS since it is entirely under users& control and hence believed to reflect their set of values and preferences that stem from their cultural background. 200 personal blogs from <a href="http://blogger.com" target="_blank">blogger.com</a> are collected in the United Kingdom and South Korea to see any cultural difference in the design and use of the blogging system. Content analysis is employed to compare the design elements used in blogs and statistical test is employed to measure the significance of the differences. Finally the findings are explained in terms of cultural impact and some implications are suggested for the design of web based information systems.

"Tunable Bandstop Resonator and Filter on Si-Substrate with PST Thin Film by Sol-Gel Deposition"
Young-Hoon Chun (University of Birmingham, Research Fellow) : young-hoon@ieee.org

Recent wireless communication systems like cognitive radios require smart radio transmitters and receivers which enable to adjust their operating frequency band and bandwidth. Another requirement can be a fast communication which needs a wide bandwidth or a high frequency band. Ultra-wideband (UWB) systems are proper example of them. As recognized, the frequency, as a resource, is valuable and limited, so the frequency spectrum is always crowded for several purposes, which means there are full of un-wanted signals. Bandstop filters (BSF) or notch filters are useful to avoid unwanted signals especially for wideband applications. Moreover, for the requirements of recent wireless communications, tunable microwave filters become more critical. Among various tunable devices, ferroelectric materials have recently become more attractive for the development of electronically tunable microwave circuits for frequency agile applications as they show low loss and good tunability at microwave frequency. Lead strontium titanate (PST) thin film is a candidate for this intention. In this paper, we introduce a fabrication of PST thin film and its characteristics at the first section. A discussion of design and modeling of parallel plated varactors using the PST thin film is presented. According to the modeled values, tunable resonator and BSF with slotted ground or electromagnetic bandgap structure (EBG) on a CPW substrate is designed by means of a novel tuning method for PST varactor to tune the resonant frequency.Fabricated resonators and 3-pole bandstop filters are measured at the center frequency of 4.23 and 7.1 GHz and its maximum rejection is more than 13.8 dB at the stop band while the insertion loss at the pass band is less than 3 dB.

"Model Index regression: application to non-life insurance"

Eun Jung Kim (CREST (LS), LSTA, Ph.D.student) : ejk_kim@yahoo.fr

Generalized linear models are universally used in pricing vehicle insurance. The usedmodels assume a known distribution for the response variable Y (occurence of claims,claim frequency and claim costs) and assume that it is related to a vector of regressors Xby a relationship : E[Y I X = x] = g(x `b) ( & ` & signifies the scalar product) for a specified real function g. We extend this approach by using a semiparametric modelization with the only assumption that the existence of :b1,&,bk (k >= 1), g : E[Y I X = x]=g(x`b1,&,x`bk)where g is not specified. By usinging the method rMAVE (refined Minimum AverageVariance Estimation) to determine k, bi and g, we motivate this approach by simulationargument and an application to classical real data in non-life insurance and compare theobtained results with those given by a generalized linear model.

"How to use Mathematical Models to forecast the KOSPI 200 InDEX Futures Price"

Gakhee Lee (fin4cast, Austria) : gakhee.lee@fin4cast.com

The central theme of financial analysis goes around the expectation or forecasting of financial asset prices. Basically, there used to be two types of analyses of financial instruments: fundamental or technical. Recently a new method, so called "quantitative analysis", has been developed to meet an alternative investment strategy. Quantitative analysis could be looked upon as an integration of the above-mentioned two analyses. Long-term time series data are put into use to make forecasting models. Mathematical Models to forecast the Futures price should be capable of describing the causal relationships between Futures price movement and its most influencing factors. The most crucial part of building forecasting models lies in analyzing the input data with respect to the price movement of Futures to be predicted. Based on the statistical and economical criteria, the most fitting combination of inputs is chosen. The forecasting models are scheduled to be calculated once a week. Finally, the models are evaluated by statistical parameters like the hit rate and R square and their forward performance. The weekly forecasts of models are then available for investors.

"Centralized versus Market based Task Allocation in the presence of Uncertainty"

Habin Lee, Karim Al-Yafi (Brunel University) : habin.lee@brunel.ac.uk

While various types of approaches to solving task allocation problems have been proposed, they are classified into two major categories: centralized and distributed approaches. Centralized approach is characterized by a central controller generates schedules based on collected data. On the other hand, distributed approaches distribute the decision of the centralized controller to distributed actors or resources. Decisions are made by local knowledge owned actors and the system coordinates any conflicts among the decisions of the individual actors. Market based approach is one of the distributed approaches in which the conflicts among local decisions (i.e., multiple workers want to execute the same task) are coordinated via a market mechanism (or price). Due to the major differences on the two approaches, there have been researches on the comparison of the two approaches to find the optimal match between TAMs and their target application domains. The comparisons have been focused on two aspects: efficiency (cost reduction) and effectiveness (performance optimization). However, the comparison existing researches on the comparison of the two different methods have been focused on stable working environments like in-office business processes or resource allocation within buildings. The fieldforce scheduling problem is characterized with higher uncertainty mainly due to the mobility of the resources (field-forces) and the travelling cost by the resources. In mobile working environment, travelling cost plays a significant role in the total cost of the workforce coordination. The exceptions such as traffic jam and road work sometimes lead to reallocation of the tasks to other field-forces, and this results in a sunk cost which should also be considered in the total cost. Furthermore, the quality of service (QoS) which is affected by the travelling time by field-forces is one of the major factors that should be considered in the comparison of the two approaches. The QoS mainly affected by the completion of the task in due time. The comparison of TAMs in mobile working environment requires considering many various and dynamic factors that affect the efficiency and performance of alternative methods.This paper aims to propose a multi-agent based simulator to compare centralized and market based approaches to mobile task allocation problems (MTAPs). While the two approaches have been used and compared for static task allocation problems in the literature, the uncertainty and dynamism imposed in the mobile task allocation problem require new approaches to the comparison of the two approaches as the traditional mathematical modeling approaches have limitations to incorporate the uncertainly, complexity and dynamism of MTAPs. The fundamental features and structure of MTAPs are described and a simulation model of the simulator is described. Then the basic models following the two approaches are introduced. In particular, four different bidding strategy of the market based approach is detailed and the basic result of the simulation is presented. Finally the future research direction is proposed to conclude the paper.

ICT Symposium – Poster

"Face Components Detection using SURF Descriptors and SVMs"

Donghoon Kim (Ireland) : donghook@cs.tcd.ie

We present a feature-based method to classify salient points as belonging to objects in the face or background classes. We use SURF local descriptors (Speeded Up Robust Features) to generate feature vectors and use SVMs (Support Vector Machines) as classifiers. Our system consists of a two-layer hierarchy of SVMs classifiers. On the first layer, a single classifier checks whether feature vectors are from face images or not. On the second layer, component labeling is operated using each component classifier of eye, mouth, and nose. This approach has the advantage about operating time because windows scanning procedure is not needed. Finally, this system performs the procedure to apply geometrical constraints to labeled descriptors. We show experimentally the efficiency of our approach.

"Predicting the Macroscopic Lifetime of the Threads of Discussion Forums with User Comment Information"

Jong Gun Lee (University Pierre and Marie Curie, Ph.D.student) : jonggunLee@gmail.com

In this poster we propose a way to predict the lifetime of the threads of discussion forums in the macroscopic level. To predict the lifetime, we apply Cox proportional hazard regression model, which is one of non-parametric survival analysis methods, and use online user comment information as risk factors of our Cox proportional hazard model. In order to validate our approach we collect two dataset from two different online discussion forums, Dpreview and Myspace, and present our approach can predict the macroscopic lifetime of Dpreview and Myspace threads only with the first 6 days and 24 hours comments information, respectively.

"Twist angle measurement in liquid crystal (LC) cells"

Chang-Hun Lee (UK) : lch0329@empal.com

High contrast ratio (CR) is becoming more important for a high definition LCD TV. An even higher CR than 1000:1 is continuously being pursued as a distinctive quality. Reducing the light leakage of the black state is the key technology to improve CR. The residual twist angle of the cell induced by the manufacturing process can introduce light leakage, but so far not so much effort has been focused on this subject. In this presentation, we suggest an accurate and simple way to measure the small twist angle of homogeneously aligned cells.

GT Symposium – Keynote Speech

"Human Responses in Thermal Environment"

Gook-Sup Song (Bucheon University) :

The human body responds psychologically and physiologically to the thermal environment to which it is exposed. A lot of researches have been conducted to get a comfortable environment considering psychological responses, and consequently energy consumption has been increased to maintain more comfortable indoor air quality. A comfortable environment does not necessarily make for a healthy environment. The indoor air quality links to obesity, diabetes, neurodevelopmental disorders and semen quality. They mostly considered subjective thermal sensation and productivity without deep consideration for the effect on human health derived from biological and physiological responses. Mild thermal exposure increases tissue microcirculation and consequently produces an appropriate supply of nutrients and oxygen to tissue cells; thus, it can modulate the proliferation of tissue cells. Excessive exposure to heat leads to cell aging, lesions, and elastosis because of oxidative damage, and is harmful to spermatogenesis. A cold environment may induce a decrease in blood flow, leading to a decrease of the supply of nutrients and oxygen due to ischemia. A comfortable temperature theory leads the sick building syndrome. To reduce energy consumption and to get healthy environment in a building, biological and physiological responses should be considered in the future green development.

"Green IT Strategies of Korea"

Yunmook Nah (Dankook University) : ymnah@dankook.ac.kr

The unstable oil price caused by the rapid increase of energy consumptions in the surrounding countries, such as China and India, and the climate change issues caused by Kyoto Protocol and Bali Roadmap are pushing Korean government to consider the green growth as one of the highest priority national agenda. From last year, almost all of Korean ministries, such as Ministry of Knowledge and Economy (MKE), Korea Communications Commission (KCC) and Ministry of Public Administration and Security (MOPAS) have started to build long term plans related with green growth and green IT.

In January 2009, the Information Communication Industry Policy Office of MKE announced "Green IT: IT Industry Strategy for green growth," which focused on greening of IT and building sound basis for the green growth by using IT. The Green Growth Commission was initiated directly under the president office on February 16 and the Green Growth Basic Act was finalized on February 25. The Green Growth Commission announced "Green IT National Strategy" by combining and reorganizing various plans of MKE, KCC and MOPAS, on May 13, 2009. This strategy includes 3 projects related with Green of IT and 6 projects related with Green by IT.

In this talk, we introduce the green IT concepts of Korea, national green IT strategies of Korea and major activities related with green IT in Korea.

GT Symposium – Oral

"Frontiers in Animal Behavior Strategy and Application in Energy, Economy, and Environment"

Who Seung Lee (University of Glasgow, Ph.D. student) : whoseung@gmail.com

In natural, animal behavior is always decided &strategically&. The strategic choice is important because animals can survive or be died by the choice. So, the selected-behavior could induce advantage and disadvantage in life. The study of animal behavior is to find the optimal choice by trade-off between benefits and costs in animal behavior. The strategic selection and trade-off is also important keywords in other fields & such as economy, energy technology and environmental science. In the development of strategies or new solutions, it is important to know which choice is fit. To give a brief idea, here, I demonstrated three case-studies to suggest the possibility of applying of strategies in animal behavior: 1) the optimal working time of heating and cooling systems using the incubation strategy in birds, 2) analyzing economic growth or development pattern using animal growth strategy, and 3) the optimal electric power supply using food distribution strategy in ants. Although this talk is the first step to approach the application of animal behavior in other fields, I suggest that animal behavior strategy would be applied to find the optimal solution.

"Application of bioelectrochemical processes for electricity generation and sustainable wastewater treatment"

Jung Rae Kim (University of Glamorgan) : jkim@glam.ac.uk

Microbial fuel cells (MFCs) are an emerging technology which converts biodegradable organic matter to electrical energy using a biofilm on the electrode as the biocatalyst. It has recently been shown that waste-to-energy technology based on MFC can treat organic contaminant in domestic or industrial wastewater and simultaneously produce electricity up to 1.6 kW/m3 based on reactor volume. Bioelectrochemical systems may reduce the energy consumption for wastewater treatment by replacing energy intensive aeration of present treatment systems, while generate electrical energy from waste. In addition, the biomass production in MFCs has been reported to be 10-50& of conventional wastewater treatment, leading to reduce environmental impact and disposal costs.Various electrochemically active bacteria metabolize biodegradable organic compounds then discharge electrons to an extracellular electron acceptor for bacterial respiration. These bacteria also transfer electrons to electrodes by direct electron transfer, electron mediators or shuttles, and electrically conductive nanowires. Investigation of bacterial electron transport mechanisms may improve understanding of the biomaterial involved and metabolic pathways as well as improving power from MFCs. Biofuel cell systems require interdisciplinary research ranging from electrochemistry, microbiology, material science and surface chemistry to engineering such as reactor design, operation and modelling. Collaboration within each study and integration of systems might increase the performance and feasibility of MFC for sustainable energy.

"Sustainable and Competitive Product Service System Solutions"

Myung-Joo Kang (Vienna University of Technology) : kang@grat.at

Product Service Systems (PSS) are known as system solutions that directly fulfill client demands with a combined set of products and services. Since the PSS strategy allows companies to gain profits by providing highly customised and contextualised offers of products, labour, and knowledge, producing and selling greater volume of products is not necessarily their major concern. Many client demands can be met by less material consumption if non-material resources such as information, knowledge, experience, know-how, time, images, and stories are effectively utilised. As a decoupling agent of economic value creation and material consumption, and as a source of new employment for manpower-based services, PSS are regarded to be one of the most promising business strategies for sustainable development. However, PSS solutions need to be systematically designed to consider environmental, social, and economic aspects profoundly, in order to manifest their potential for sustainable development. With the intention to support systematic PSS design, a number of methodologies were suggested in recent years. Yet, reflective application of these instruments for actual PSS solution development has not been widely recognised in the business world. Therefore this study aims to help companies understand the PSS concept in a more practical and concrete way, to verify and improve some of the design methods for PSS, and to find out critical factors for market success. To reach the goals, three real-life PSS development case studies were conducted in collaboration with large and small companies in different business sectors and regions. Strategic advice and practical toolset encourage business organisations to develop and implement sustainable and competitive PSS solutions in the market.

"Inter-firm Collaboration for Low Carbon Society"

Dowon Kim (UK) : dowon.kim@uea.ac.uk

The world is now faced with the crisis of \'triple crunch\' and many countries have introduced greening approaches to address the crisis. The Korean government also recently claimed \'Low carbon green growth policy\', but the policy seems to have a bias toward new green products rather than low carbon society. Although green products will certainly contribute to economic growth as well as carbon reduction in Korea, the green products need long time to be developed and applied in practice, while the action to reduce carbon emission is urgent. In particular, Korea has high industrial intensity with approximately 500 industrial estates and the carbon emission in Korea mainly comes from industrial activities such as energy intensive manufacturing processes rather than residential or transportation area, while the green products have limitations to address the main production processes of industrial sector. Therefore more endeavours seem necessary to reduce carbon in the existing industries in Korea.

"Technology co-evolution analysis in the energy sector"

Sungjoo Lee (Ajou University, Professor) : sungjoo@ajou.ac.kr

Renewable and sustainable energy technology is indispensable for the future. This paper investigates the interactive nature of relationships between energy technologies in terms of their innovations and diffusion. For the purpose, USPTO database was used and patent citation analysis and Lotka-Volterra equations were applied to the patent data, resulting in two types of network. These networks were employed to investigate the interaction patterns among energy technologies, allowing co-evolution process across them to be analysed. Research findings will provide numerous implications for policy-making and strategic planning for energy technology development.

"Green Architecture"

Seung-Ho Lee (Cabinet d'Architecture, Seung-Ho LEE Architecte DPLG (Owner)) : archi.shl@free.fr

Green ArchitectureToday, the environment is becoming the main subject in lots of science disciplines and the industrial development due to the global warming. This paper presents the analysis of the tendency of Green Architecture in France on the threes axes:- Regulations and Approach for the Sustainable Architecture (Certificate and Standard)- Renewable Materials (Green Materials)- Sustainable Technology Strategies (Equipments)The definition of &Green Architecture& will be cited in the introduction and the question of the interdisciplinary for technological development in &Green Architecture& will be raised up in the conclusion.

"Optimization of well turbine output through the analysis of self-starting experiment"

Soo-Young Bae (Korea Maritime University) : sybae@hhu.ac.kr

This paper presents control system for optimum output power of Wells turbine which is used in wave power generation system of the marine facilities. Power generation characteristics of Wells turbine is not efficient at low velocity of turbine blade,Output power varies according to the velocity of turbine blade under the same condition of input wave frequency and height.Velocity of turbine blade changes with electric load of generator.Wave power charging control system using buck converter for Wells turbine is developed considering characteristic of generator. Experimental test is conducted to prove its performance and efficiency.

"A Research on the Application Method of a New and Renewable Energy Complex System for School"

Ji-Yeon Kim (Korea Institute of Energy Research, Researcher) : jykim@kier.re.kr

With the revision of the &New Energy and Renewable Energy Development, Utilization and Diffusion Promotion Act& (proclaimed on Mar. 14, 2008), the Act expands on its coverage of new/renewable energy facility installation subject buildings including school facilities from &new building& to &enlarge/rebuilding&. As the majority of newly built schools are in dimensions of over 3,000 ㎡, the mandatory new and renewable energy systems should be installed in these schools. The new and renewable energy system that should be installed in school buildings is confined to solar beam, solar heat and ground heat, among others. Even if new and renewable energy facility equipment with good performance would be introduced along with its accompanying policy, achieving acceptable savings in energy may not be accomplished or system building may not be easy, unless an optimum capacity standard on the introduction size of new and renewable energy for educational facilities or optimum system building and economic feasibility would be analyzed and evaluated. In this regard, the research aims to study a new, optimum and renewable energy application method that can cover the minimum energy and operation costs within a range of school budgets. By deriving the optimum application method, it is expected to maximize the cooling/heating and water heating energy saving efficiencies for educational facilities. Therefore, this research carried out a study on the new/renewable energy utilization technique diffusion expansion method and the optimum method. As a result, with the definition of a mid-sized school as one with dimensions ranging from 7,000m² to 12,000m² and class number ranging from 16 to 35, we carried out a simulation in a mid-sized school with 24 classes and dimensions of 10,000 m². The first optimum plan was introduced with the multi-type geothermal heat pump 80HP + solar heat collector 164 m² + highly efficient electronic cooling/heating device (EHP) 214HP. On the other hand, the second optimum plan was induced as the multi-type geothermal heat pump 136HP + + highly efficient electronic cooling/heating device (EHP) 158HP. In conclusion, the energy saving effect would be significant if a solar heat collector is designed by considering water heating load of a school and installing a multi-type geothermal heat pump with remaining new and renewable energy installation liability-expenses. Moreover, the initial investment cost is expected to be considerably reduced if multi-type geothermal heat pumps are installed with the remaining new and renewable energy installation liability expenses while highly efficient electronic cooling/heating devices (EHP) are put in charge of the remaining cooling/heating loads.

"Multi-objective Environmental/Economic Power Dispatch using the Bees Algorithm with Pareto optimality"

Ji Young Lee (Cardiff University, Ph.D.student) : leej5@cardiff.ac.uk

This paper presents the use of the Bees Algorithm for the Environmental/Economic (Power) Dispatch (EED) problem which is formulated as a nonlinear constrained multi-objective optimisation problem. In this problem, both fuel cost and nitrogen oxides emission are to be simultaneously minimised. Simulation results presented for the standard IEEE 30-bus system using the Bees Algorithm with Pareto optimality are compared to those obtained using other approaches. The comparison shows the superiority of the Bees Algorithm and confirms its suitability for solving the multi-objective EED problem.

"Development of District Heat Demand Forecasting Model Using Neural Networks"

Lae Hyun Kim (Seoul National University of Technology) : lhkim@snut.ac.kr

Efficient use of energy and heightened the awareness of the power source as a multi-map of the distribution of district heating systems are being expanded. Therefore, the optimal economic operation of district heating systems are required. Analyze data of heat supply in the past and predict heat demand are basic steps in the optimal economic operation system.The purpose of this study is to develop heat demand forecasting model by analysing the patterns of heat supply in district heating for integrated energy supply system. We reviewed the characteristic, structure, and kinds of the artificial neural network, and the major algorithms used in neural network. The heat demand forecasting model was developed by applying the heat supply data of &G& branches of Korea District Heating Corporation using MATLAB Neural Network. The pattern of heat supply considering the outdoor air temperature was analysed using the heat supply data of the year 2007 obtained from the G branches of Korea District Heating Corporation. The suitable type of data for developing heat demand forecasting model was chosen and applied to the neural network. The input parameters for training were outdoor air temperature and the amount of heat supply of the year 2007 reflecting past trends. The amount of heat supply of the year 2007 and 2008 were simulated using the network model established through the training process. The comparison between forecasted data and actual data showed that the R_2007\'s of G office was 0.99605 and the R_2008 \'s was 0.99573, respectively. And apply for the Korea Meteorological Administration&s forecast data, heat demands were forecast for the week. The heat demand forecasting model using neural network model was practically implemented in order to plan the integrated energy management system of district heating. This forecasting model could be contributed to realize the district energy management system for the optimal operation.

"Optimal Operation System of the Integrated District Heating System with Multiple Regional Branches"

Ui Sik Kim (Hanyang University) : uisik119@gmail.com

This paper presents an optimal production and distribution management for structural and operational optimization of the integrated district heating system (DHS) with multiple regional branches. A DHS consists of energy suppliers and consumers, district heating pipelines network and heat storage facilities in the covered region. The integrated DHS considered in this paper is Korea District Heating Company (KDHC) Seoul Area, which consists of 12 regional DHS branches and 12 regional consumers. In the optimal management system, production of heat and electric power, regional heat demand, electric power bidding and sales, transport and storage of heat at each regional DHS are taken into account. The optimal management system is formulated as a mixed integer linear programming (MILP) where the objectives is to maximize the profits, so that to minimize the overall cost, of the integrated DHS while satisfying the operation constraints of heat units and networks as well as fulfilling heating demands from consumers. In this system, heat demand is estimated based on the environment temperature using neural network model, integrated to the optimal management system. Piecewise linear formulation of the production cost function and stairwise formulation of the start-up cost function are used to compute nonlinear cost function approximately. Evaluation of the operation cost is based on weekly operations at each district heat branches of KDHC, covers vast area of Seoul and Gyeonggi-do in Korea. Numerical simulations show the increase of energy efficiency due to the introduction of the present optimal management system.

GT Symposium – Poster

"An analysis on the plan to save demand for energy and introduce a renewable energy system in Innovat"

Hyo-Soon Park (Korea Institute of Energy Research) : hspark@kier.re.kr

An analysis on the plan to save demand for energy and introduce a renewable energy system in Innovation CityInnovation city was promoted, firstly, to make a local city specialize in future-oriented activities that make it highly innovative where public organizations, industry, academia and research centers cooperate closely and where exists a high-end government environment for dwelling, education, medication, and culture. Secondly, it was aimed at building a foundation for regional independence. The method for establishing regional independence should be planned in consideration not only of economic aspects but also life quality enhancement and a city&s sustainable development. Therefore, to build an environment-friendly and sustainable city, first of all a well-balanced urban planning would be required and a foundation for energy independence should be composed. Moreover, energy and environmental problems should be actively tackled by designing energy-efficient buildings. The research has analyzed the application adequacy, accompanied by comparisons and evaluations, of energy demands and new/renewable energy system introducing plans for the multifunctional administrative city and for each innovation city. As a result, the reduction plans showed a maximum of 12.30 percent of regional deviations, and no city satisfied 5 percent of new and renewable energy supply. However, the innovation city (Naju) of Gwangju and Southern Jeonra Province were planned through an external research to satisfy the targeted 5 percent of national new and renewable energy introduction. The energy utilization plan serves as important data for the regional district unit plan. Therefore, the energy utilization plan should be established and executed for sustainable development and eco-friendly city construction in consideration of regional circumstances and economic funding acquisition plans.

"Low-Carbon new city energy planning using hourly demand/supply profiles matching analysis"

Jaemin Kim (University of Strathclyde) : min@esru.strath.ac.uk

When adopting zero and low carbon demand measures and supply systems including Renewable Energy (RE) technologies in a city, identifying demand/supply match and the best combination of the mixed energy systems is a challenging issue in terms of energy efficiency and carbon emission reduction. RE-based supply profiles are intermittent and depend on local weather. The city energy planning with more explicit data is a challenging task in predicting energy demand and setting up carbon emission reduction strategy. This paper presents an approach for the city-scale energy planning on the basis of hourly-based demand/supply profiles. Statistical data and simulated data are adopted to predict hourly-based energy demand profiles of buildings and non-buildings demand in a new city. A case study of the approach is implemented within a construction project of a new city in Korea.

"Korean Creative Green Vehicles for the Future Global Market"

Dong-Soo Jeong (KOSEF, National R&D Project Bureau, Director General) : dsjeong@kosef.re.kr

Types of green vehicles include vehicles that function fully or partly on alternative energy sources other than fossil fuel, e.g. HEV, FCEV, etc. Another alternative is the use of alternative fuel composition in conventional fossil fuel-based vehicles, making them function partially on renewable energy sources.In order to compete in the future world automotive market, Korean creative green vehicles should be concentrated to develop.

BT Symposium – Keynote Speech

"Industrial Transgenic Plants for Sustainable Development on Marginal Lands"

Sang-Soo Kwak (Korea Research Institute of Bioscience and Biotechnology, Principal Researcher) : sskwak@kribb.re.kr

Who will support the world population of 9 billion people in 2050 ? The food self-support rate on the basis of cereals in Korea is approximately 27&, which will threaten the food security. The dramatic increase in population accompanied by rapid industrialization in developing countries has caused imbalances in the supply of food and energy. To cope with these global crises over food and energy supplies as well as environmental problems, it is urgently required to develop new crop varieties to be grown on marginal lands including desertification areas for sustainable development. In this respect, we are developing several transgenic crops such as sweetpotato (Ipomoea batatas), potato (Solanum tuberosum), tall fescue (Festuca arundinacea), alfalfa (Medicago sativa) expressing genes involved in multiple stress tolerance under the control of an oxidative stress-inducible peroxidase (SWPA2) promoter. Sweetpotato (Ipomoea batatas (L.) Lam.) ranks seventh in annual production among food crops in the world. It is also an alternative source of bio-energy and industrial materials such as natural antioxidants. Moreover, it does not require large amounts of fertilizers and pesticides and is rather tolerant to some environmental stresses. The nonprofit Center for Science in the Public Interest (CSPI) designated sweetpotato as one of ten super foods for better health (2007). Recently, USDA (2008) reported that sweetpotato can yield 2~3 times as much fuel ethanol as field corn, approaching the amount that sugarcane can produce in Maryland and Alabama. It would be worthwhile to start pilot programs to study growing sweetpotato for ethanol production, especially on marginal lands. Taken together, the sweetpotato can be developed to produce valuable industrial materials including bio-energy on marginal lands by molecular breeding. In this respect, we focus on the metabolic engineering for sweetpotato as an industrial plant on marginal lands to contribute in solving global environment, energy and food problems. For this purpose, we are developing industrial platforms such as stress-inducible gene expression system, root-specific gene expression system, intragenic vector system and metabolic engineering of antioxidants. In the presentation, the results and prospects of our studies will be introduced. In addition, the strategies for the development of industrial transgenic crops to combat desertification in China will be introduced in terms of global collaboration.

BT Symposium – Oral

"Study of cell signalling in microfluidics"

Jung-uk Shim (UK)

Quorum sensing (QS) or cell signalling in bacteria is involved in diverse biological functions including virulence, biofilm formation, ecological competence and bioluminescence. Biofilm formed by QS has been widely studied, however, the response of bacterial cells to their microenviroment is not completely understood. Thus, control of this microenvironment through compartmentalization of cells is useful to further investigate the QS phenomenon. Microdroplet provides such a platform for compartmentalization of cell and precise control of signalling molecules concentration.

We have developed a microfluidic system to investigate the expression level of a protein, Green Fluorescence Protein (RFP), depending on the concentration of auto-inducer in single cell resolution. Pico-liter sized aqueous droplets were formed in oil, stored in wells, and monitored. The time-course measurement onto significant number of individual cell requires indexing them to identify where cells are located. The storage well built in the device provides a physical address as droplets are securely docked after storage. Also, the droplet shrinking due to water diffusion into the PDMS matrix that jeopardize quantitative measurements of droplet contents was resolved by integrating a reservoir underneath the wells to continuously supply water to the stored droplets to keep the water content constant over long periods.

"Analysis of Immunoglobulin Kappa Variable Regions Expressed by High and Low Antibody Responder Pre-immune Biozzi Mice"

Chang-Hoon Nam (KIST-Europe) : chang@kist-europe.de

Fifty V regions expressed by B lymphocytes from high (H) or low (L) antibody responder naïve Biozzi mice were sequenced and analyzed according to the IMGT Ig domain system and NCBI nomenclature. 38% L V belong to family 9, whereas H V show a more heterogeneous distribution, with three dominant families (4/5, 9 and 21). When each V sequence was compared to V germline genes, H V exhibited a lower alignment score than L V, indicating that they are subjected to more somatic mutations. 642 point mutations were found in the H V sequences as compared to 201 in the L sequences. The number of mutations found in CDR1 and

CDR2 was six times higher in H mice. A base per base comparison of 8 H and 12 L V sequences with the related germline gene, cw9, showed that, in all H V sequences, short stretches of DNA containing nucleotide substitutions could be found in other family 9 germline genes, by contrast, to no V in L mice, suggesting a somatic gene conversion-like mechanism. Thus, qualitative differences in the natural V repertoire expression exist between adult H and L Biozzi mice.

"Sex, smell and human diseases"

Soo-Hyun Kim (St. George's Medical School, University of London, Senior Lecturer) : skim@sgul.ac.uk

Cells respond to the environmental stimuli and modify their behaviour accordingly. Growth factors and cytokines mediate these extracellular signalling events, thus play key roles in normal human development as well as disease processes. Defects of a relatively unknown protein called anosmin-1 cause Kallmann Syndrome, an inherited human disorder resulting from abnormal development of neurons involved in smell and sex hormone-releasing functions. We recently discovered that anosmin-1 functions through interactions with major cellular signalling pathways involving fibroblast growth factor receptor 1 (FGFR1). Mis-regulation of FGFR1 underlies a large number of pathological conditions. Our recent work produced some evidence that anosmin-1 may be capable of abnormally activating the FGFR1 pathways, thus understanding the exact mechanisms by which anosmin-1 brings its effect may discover possible therapeutic targets. In this project we are planning to investigate the detail molecular mechanisms of anosmin-1-dependent cellular events and how anosmin-1 links FGFR1 and other cellular signalling pathways. This is of particular interest since it is currently unknown how these pathways crosstalk with each other. We will use a variety of biochemical, molecular and cell biology techniques to examine the interactions of anosmin-1 with different molecules, and their functional effects.

"Steady-state Analysis of Gene Regulatory Networks via G-Networks: Detecting Abnormal Behaviors of Genes in Budding Yeast Cell Cycle Gene Regulatory Networks"

Haseong Kim (Imperial College London, Ph.D.student) :
haseong.kim08@imperial.ac.uk

Steady-states of gene regulatory networks (GRNs) can provide crucial evidences of detecting disease-causing genes. However, monitoring the dynamics of GRNs is not easy because biological data do not properly reflect the dynamical behavior of living organisms. Also most of the network methods only provide the inferring algorithms of a GRN structure which is a snap-shot of the given dataset. Probabilistic Boolean Network (PBN) is the most widely known algorithm for its ability of monitoring dynamics of GRNs. However, the PBN is only available with a small set of genes because of the computational time and space limitations.We have modeled the GRNs using G-network which is a queuing network with positive and negative signals. Based on the analytical studies of the G-network, we estimate the two important model parameters, the steady-state probability and income rate of a queue which represents the number of mRNA. The steady-state probability represents the probability that a gene is expressed in steady-state. The income rate of a queue is the transcription initiation rate of a gene. These two model parameters are estimated by using the expectation maximization method. We analyze the simulation dataset which is generated by stochastic gene expressions model and detect the abnormally behaved genes that show different expressions. The proposed method is also applied to two microarray datasets of the budding yeast cell cycle. We compare the dynamics of their GRNs and our method properly finds that the CLB genes are differently expressed in the treatment dataset where the CLB genes are knocked out. In this study, G-networks are utilized to monitor the dynamics of GRNs. The proposed methods have efficiently found the steady-state of GRNs in both simulation and yeast cell cycle datasets. Recently, due to the development of high-throughput technology, the amount of microarray gene expression data greatly increases. Our method enables us to analyze the microarray datasets by monitoring the behavior of their GRNs under various conditions. Also the candidate genes that are affected by given conditions such as diseases and environments can be determined by the proposed method.

"A Novel Strategy for Modeling and Visulaizing of Transcription Factor Networks"

Namshik Han (The University of Manchester) :
namshik.han@cs.manchester.ac.uk

Transcription factor (TF) proteins bind to a promoter element to control the transfer of genetic information from a DNA coding sequence to mRNA. Thus, a comprehensive analysis of transcription factor&s activities is an essential step for discovering the mechanism of regulation of the cell. We suggest a novel strategy for modeling and visualizing of TF network, which utilizes the TF-gene connectivity dataset and gene expression dataset. The computational pipeline infers a set of the strength and the sign of gene-specific TF activities (TFAs) and the level of TFs& concentrations (TFCs); identifies statistically significant up- and down-regulation by utilizing TFAs and TFCs; outputs significant results in a format that makes it easy to visualize trends and patterns. We validated our pipeline by comparing its prediction with a known result from a published biological study that reported novel functions of p38& mitogen-activated protein kinases (MAPK) in postnatal development and tumorigenesis. The comparison was successful and our results predict new biological hypotheses on a genome-wide scale. Our strategy thus provides a concrete framework for discovering the transcriptional regulatory networks.

"An approach to understanding human prion disease from a biochemical viewpoint"

Young Pyo Choi (UK) : s0682740@sms.ed.ac.uk

Transmissible spongiform encephalopathies (TSEs) or prion diseases are a group of fatal neurodegenerative diseases affecting humans and many animal species. Despite their rare occurrence, prion diseases have recently drawn considerable attention worldwide because of the appearance of a new form of human prion disease caused by interspecies transmission of bovine spongiform encephalopathy (BSE) through the consumption of contaminated animal tissue. Prion diseases are characterized by the conversion of the host encoded prion protein (PrPC) into a disease-associated isoform (PrPSc), which is thought to be the main component of the infectious agent. PrPSc has been traditionally distinguished from PrPC by its biochemical properties such as partial resistance to proteolysis and detergent-insolubility. In the absence of a foreign nucleic acid genome associated with the prion diseases, the strain phenomenon in prion diseases has been explained by the conformational difference of PrPSc (according to the prion hypothesis) and thus efforts to provide molecular definitions of prion strains have focused on biochemical characterization of PrPSc. In human prion diseases, much evidence supporting conformation-based strain diversity has been gathered by Western blot analysis of PK-resistant PrPSc. Our research, however, indicates this kind of analysis may provide an incomplete description of PrPSc in CJD brains, since considerable complexity in PrPSc has been identified both between different forms of human prion disease and within single cases of individual disease entities.

"HIV Prevention in Schools by Empowerment"

Hanna Sheu (Austria) : hanna.sheu@mistelbach.lknoe.at

Research BackgroundAccording to recent WHO data, the number of HIV positive patients in Austria is rising. Experts assume that the increasing number of HIV infections is due to the decrease of awareness, particularly among teenagers. Because of the duration of latency period (5-15a)1, it is necessary to start the primary prevention during adolescence.Study ObjectivesThis study aims at inducing effective learning processes with potential risk groups by providing them resources and competence. The project constitutes action research, which shall sensitize teenagers to the HIV/ AIDS issue.MethodsThe project is designed as interventional cohort study. It applies an interactive form of clarifying facts about HIV/AIDS with psychologically motivating techniques of creativity. Schools are the best option to approach adolescents and are therefore chosen as place for conducting the study targeting pupils aged 14 to 17 years. About 200 pupils in this age range are involved in this study conducted in Austria.

"Comparative study of risk assessment approaches based on different methods for deriving PNEC and DNEL of chemical mixtures"

Jongwoon Kim (KIST Europe Forschungsgesellschaft mbH) : jwkim@kist-europe.de

Most living organisms are actually exposed to the chemical mixtures rather than individual substances. But current chemical risk assessment frequently focuses on the single chemical substances. The latest EU chemical regulation, REACH (Registration, Evaluation, Authorization and Restriction of Chemicals) entered into force in June 2007, puts less focus on the chemical mixtures.EU currently presented some methods for deriving PNEC (Predicted No Effect Concentration) and DNEL (Derived No Effect Level) for risk assessment of mixtures (coating materials). However, the differences were estimated by 1.01 - 5.41 folds in our former study on comparing the two methods (the key critical component method and the composite reciprocal DNEL &PNEC method) for derivation of PNEC and DNEL of mixtures.The objective of this study is to compare the results of risk assessment approaches based on different methods for deriving PNEC and DNEL of chemical mixtures, according to the EU technical guidance notes.The exposure assessments in risk assessment were carried out on the basis of the different results from the former study. It is needed to develop advanced risk assessment methodology and models based on available individual toxicity and physicochemical data for various substances in mixtures.Keywords: mixture toxicity, chemical risk assessment, DNEL, PNEC, exposure assessment

"S-nitrosothiols regulate hypersensitive cell death during the plant disease resistance response"

Byung-Wook Yun (The University of Edinburgh, Senior Research Fellow) : bwyun@ed.ac.uk

Nitric oxide (NO) is a fundamental regulator of both eukaryotic and prokaryotic redox signalling networks and its biology impinges upon a plethora of key physiological processes. In animals, NO-based signals were historically thought to be transduced via the activation of guanylate cyclase (GC), following the binding of NO to the cognate heam group and the subsequent modulation of cGMP-dependent kinases, gated channels and phosphodiesterases. However, S-nitrosylation, the addition of a NO moiety to a reactive cysteine thiol, to form a nitrosothiol (SNO), has recently emerged as an alternative mechanism to convey NO bioactivity and as key regulators of apoptosis in animals. For example, S-nitrosylation of nuclear factor κB (NF-κB) or c-Jun-N-terminal kinase (JNK) either promotes or inhibits cell death, respectively. In spite of this, plant genomes do not possess GCs with heam-binding motifs. Thus, S-nitrosylation is likely to be a predominant conduit of NO-related signals in this kingdom. Despite its potential significance, however, the roles of this post-translational modification in plant biology remain largely unexplored. Recently, we provided the first genetic evidence consistent with a key role for S-nitrosylation in plants. Arabidopsis thaliana S-nitrosoglutathione reductase 1 (AtGSNOR1), was demonstrated to control cellular SNO levels in both naÏve and pathogen challenged plants. Furthermore we have identified SNOs as pivotal regulators of plant hypersensitive cell death (HCD). Collectively, our data implies that the kinetics of NO and ROI accumulation are inversely controlled through changes in SNO formation and turnover, which ultimately modulates the rate and magnitude of HCD development. Thus, AtGSNOR1 functions as a binary control switch at the centre of HR regulation.In the presentation, the molecular mechanism underlying cell death via S-nitrosylation of certain protein responsible for ROI accumulation and prospects of our studies will be addressed.

BT Symposium – Poster

"Defining changes in expression of mineralocorticoid receptor (MR), glucocorticoid receptor (GR) and Bcl-2 family in human hippocampus following transient global cerebral ischaemia"

Sung Eun Bae (University of Surrey, Post-Doc Researcher) : s.bae@surrey.ac.uk

Mineralocorticoid receptor (MR) is a ligand-activated transcription factor belonging to the steroid hormone family of receptors. We have previously shown that neuronal stress increases MR expression, and that this is associated with neuroprotection both in primary neuronal culture *in vitro* and *in vivo* in a rodent models of cerebral ischaemia. Here we tested the hypothesis that MR expression is upregulated in human brain following transient global ischaemia, and also investigated the expression of glucocorticoid receptor (GR) and apoptosis related Bcl-2 family members, Bcl-2 (anti-apoptotic) and Bax (pro-apoptotic) in human hippocampus. Our results showed that MR-mRNA and GR-mRNA was present in all regions of the hippocampus in both control and ischaemic group. *In situ* hybridisation qualitative studies showed that MR-mRNA expression was significantly higher ($P<0.01$) in all hippocampal region examined in cases compared with controls, but there was no significant difference in the expression of GR-mRNA. Also immunoreactivity for Bcl-2 was more intense in cases compared to controls. Bax-immunoreactivity was also more intense in cases compared to controls.

In this study, we have shown that MR-mRNA expression is upregulated, but not GR-mRNA, in human hippocampus following transient global ischaemia. MR is known to contribute to regulation of select cell survival and cell death related genes by regulating Bcl-2 and this may form a part of the neuroprotective mechanism of MR. Currently studies are on the way to colocalize MR and Bcl-2 and quantify their expression level. Given the neuroprotective effects of increased MR expression observed in primary cell culture and animal models of cerebral ischaemia, these changes are consistent with similar protective mechanisms operating in human brain. Neuroprotective strategies based on the upregulation of MR based on findings in cell culture and animal models of stroke, therefore, have relevance to human pathophysiology

"Green fluorescent protein as a reporter for spatial and temporal gene expression in Streptomyces coelicolor A3(2)"

Jongho Sun (John Innes Centre, Project Scientist) : sun@bbsrc.ac.uk

The enhanced green fluorescent protein (EGFP) gene is a modified version of the green fluorescent protein gene of the jellyfish Aequorea victoria with a codon usage that corresponds well to that found in many GC-rich streptomycete genes. Here the use of EGFP as a reporter for the analysis of spatially and temporally regulated gene expression in Streptomyces coelicolor A3(2) is demonstrated. The EGFP gene was inserted into plasmids that can replicate in Escherichia coli, greatly facilitating the construction of EGFP gene fusions. The plasmids can be transferred readily to S. coelicolor by conjugation, whereupon two of them (pIJ8630 and pIJ8660) integrate at the chromosomal attachment site for the temperate phage C31. These vectors were used to analyse the spatial and temporal expression of sigF, which encodes a sigma factor required for spore maturation, and of redD, a pathway-specific regulatory gene for the production of undecylprodigiosin, one of the four antibiotics made by S. coelicolor. While transcription of sigF appeared to be confined to developing and mature spore chains, transcription of redD occurred only in ageing substrate mycelium. A further plasmid derivative (pIJ8668) was made that lacks the phage C31 attachment site, allowing the EGFP gene to be fused transcriptionally to genes of interest at their native chromosomal locations.

"Lack of genotype effect on D1, D2 and dopamine transporter binding in triple μ-, δ-, and κ-opioid receptor knockout mice of three different genetic backgrounds"

Ji-Hoon Yoo (University of Surrey) : j.yoo@surrey.ac.uk

We investigated D1, D2 receptor and dopamine transporter (DAT) binding levels in mice lacking all three opioid receptors and wild-type (WT) mice on three different genetic backgrounds. Quantitative autoradiography was used to determine the level of radioligand binding to the D1 and D2 receptors and DAT labelled with [^{3}H]SCH23390, [^{3}H]raclopride, and [^{3}H]mazindol, respectively in triple opioid receptor knockout (KO) and WT maintained on C57BL/6 (B6) and 129/SvEvTac (129) as well as C57BL/6 × 129/SvPas (B6 × 129) strains. No significant genotype effect was observed in D1, D2 and DAT binding in any regions analyzed in any of the strains studied, suggesting that a lack of all three opioid receptors does not influence D1, D2 receptor and DAT expression, irrespective of their genetic strain background. However, strain differences were observed in D1 binding between the three strains of mice studied. Lower levels of D1 binding were observed in the substantia nigra of B6 × 129 WT mice compared to the 129 WT mice and in the olfactory tubercle of B6 × 129 WT compared to B6 WT and 129 WT mice. Lower levels of D1 binding were observed in the caudate putamen of B6 × 129 KO mice compared to 129 KO mice. In contrast, no significant strain differences were observed in D2 and DAT binding between the three strains of mice in any regions analyzed. Overall, these results indicate a lack of modulation of the dopaminergic system by the deletion of all three opioid receptors regardless of different background strains.

MME Symposium – Keynote Speech

"Current Status on Research and Development of Air-Breathing Propulsion Systems in Korea"

Changduk Kong (Chosun University, Professor) : cdgong@chosun.ac.kr

The government of Republic of Korea has recently allocated large amount of research and development funds to both science and engineering projects, for instances, information technology which is currently one of the top world suppliers, transportation sector focusing on engineering research and development of advanced maritime transportation and high speed train, aerospace engineering, etc. Among them, aerospace engineering becomes an importance area for the next generation economic and strategic growth promotion.

This paper hence aims to reflect past, present and future contribution on the research and development of air-breathing propulsion systems which have been carried out by several major aero propulsion related research institutes, companies, universities and societies in Korea. Research institutions, companies and universities for aerospace engineering R&D in the Republic of Korea are structured in a more or less departmental manner where upcoming research projects are designated to be carried out by certain specialized research centers to handle the given project. However it is rational to classify them according to the research areas, activities performed and according to their viable long-term future anticipated projects.

MME Symposium – Oral

"Numerical Analysis of aerodynamic characteristics for Wing-in-Ground Effect Craft with DUP device"
Kyoungwoo Park (Hoseo University, Professor) : kpark@hoseo.edu

A WIG (wing-in-ground effect) craft has the advantages of increasing lift and decreasing drag when it flies close to the water or ground surface by utilizing an air cushion of relatively high pressurized air between airfoil and ground. DUP(direct undesirable pressurization) device in WIG craft can considerably reduce a take-off speed and then minimize the hump drag. In the present work, aerodynamic characteristics and the static height stability of the WIG craft with DUP device are numerically investigated by solving a full Navier-Stokes equation with turbulent model.

"Overview of Research in Austria"
Man-Wook Han (KOSEAA) : han@koseaa.net

In this paper the current research efforts in Austria will be described. The research in Austria is divides in two categories: Research in Universities and Research in Non-University organizationsAustria invests in Renewable Energy, Recycling, Biotechnology etc. In this paper the each area will be described in more detail

"Model-free control of automotive engine and brake for Stop-and-Go"
Sungwoo Choi (CAOR- Centre de Robotique, Ph.D.student) : seongwoo.choi@gmail.com

In this paper, we propose a model-free control approach to deal with automotive longitudinal control in the case of Stop-and-Go scenario. Stop-and-Go attempts to place automatically the vehicle at safe distances in urban environments. In this case, accelerating and braking maneuvers are very demanding in short time intervals, but the corresponding torques are often difficult to obtain because of the uncertainty of dynamics, not only at the chassis level but also at the engine/brake model. In order to compensate the losses related to the neglected dynamics of model, a model-free control approach will be used at the engine/brake level. Model simulation results are presented to validate our method.

"Solar air conditioning in Mediterranean climate"
Dong Seon Kim (Arsenal research, Senior Scientist) : dong-seon.kim@arsenal.ac.at

The purpose of this study is to evaluate performance of various absorption chillers for solar cooling in Mediterranean climate. Various solar absorption cooling systems have been simulated with meteorological data for two Italian cities and system efficiency and specific collector cost were estimated for each of the systems. The results show that certain types of absorption chillers are advantageous for low investment cost.

"Fabrication of Three-Dimensional Magnetic Microcomponents"
Jung-Sik Kim (Imperial College London) : j.kim@kseauk.org

This paper presents a new fabrication process for producing magnetic microcomponents. The process is based on micro powder injection moulding technology, while the micro moulds are produced using microelectromechanical systems (MEMS) technology. The process involves (1) fabrication of polydimethylsiloxane (PDMS) micro moulds from SU-8 masters, which are produced using UV photolithography process; (2) mould filling with magnetic powder and demoulding; and (3) sintering of moulded components in vacuum. The proposed process has been used to sinter Nd-Fe-B microcomponents successfully without using binder system. This research proposes a new approach to fabricate 3-D magnetic microcomponents to meet the needs in applications where magnetic field required.Keywords: Magnetic microcomponents, PIM, MEMS, PDMS micro moulds, SU-8 moulds, Nd-Fe-B, hydrogen decrepitation

"Level-set method for reconstructing thin electromagnetic inclusions"
Won-Kwang Park (L'Ecole Superieure de'Electricite) : park@lss.supelec.fr

A challenging inverse problem in the field of nondestructive testing is to retrieve the shape of thin electromagnetic inhomogeneities (crack) from boundary measurement for the scattering of time-harmonic electromagnetic waves.

Recent promising techniques which use a level set representation of shapes for solving inverse scattering problems have been successfully developed. Starting out from the original binary approach of Santosa for solving the shape reconstruction problem, more recent generalizations have been developed, such as for example using color or vector level sets and various techniques for incorporating regularization into the shape inverse problem using level sets. So far, all these techniques are mostly focusing on finding volumetric objects or regions with known properties.

In this paper, we consider a technique of electromagnetic imaging (at a single, non-zero frequency) which uses the level set evolution method for reconstructing a thin electromagnetic inhomogeneity with dielectric or magnetic contrast with respect to an homogeneous embedding medium. Emphasis is on proof of concept, the scattering problem being so far based on a two-dimensional model. To do so, two level set functions are employed; the first one describes location and shape, and the other one connectivity and length. Speeds of evolution of level set functions are calculated via Fréchet derivatives. Several numerical experiments on noiseless and noisy data as well illustrate how the proposed method behaves.

"Computational Analysis of Non Contact Transportation System for Wafer Hand"

Byeong Sam,Kim (Hoseo University, Professor) : kbs@hoseo.edu

The high-integration trend in semiconductor production processes inevitably has brought the increase of contact or contaminants generated by shock are prevented during wafer handling. This wafer hand device with atmospheric pressure or chemical vapor deposition equipment for transporting and transferring wafers, robot hands carry these longer and wider wafers can also be easily handled. As a contact handling system composed of several problems in increased potential for fracture. A non contact handling system is required to solve this problem. We propose a new non contact transportation of air bearing system with combining air suction and blowout. The numerical analysis and experimental is, therefore, should be performed to obtain compared to results achieved with non contact solutions. This wafer non contact handler shows its strength in maintaining high cleanliness levels for semiconductor production processes.

MME Symposium – Poster

"Nature-inspired Superhydrophobic Surface"

Wan-Doo Kim (Korea Institute of Machinery and Materials) : wdkim@kimm.re.kr

In this talk, we report the fabrication results of transparent and superhydrophobic glass surface which is inspired by nature surface such as lotus leaf, rose petal, taro leaf, wing of butterfly and so on. The amazing applications of superhydrophobic surface cause the interest of scientists and engineers in study on the various artificial superhydrophobic surface. They can apply the environmentally friendly surface like self-cleaning high building windows, car mirrors and windows, goggles, etc. There are two factors which control the surface wettability ; surface composition and roughness. For the transparent and superhydrophobic glass surface, the nanostructured glass surface was prepared using the colloidal lithography and reactive ion etching to increase the surface roughness. The diameter of the nanostructure is in the range of less than 100nm for the transparency. The deposition of FOTS self-assembled monolayers was followed using the vapor deposition method. These fabrication methods results in the transparent and superhydrophobic glass surface showing the contact angle of 145°±5°. This contact angle values originated the chemical composition of SAMs coated glass outmost layer having the low surface energy and the small contact areas between the glass nanostructure topside and water drop. Several geometrical factors are also considered to reduce the sliding angle of transparent and superhydrophobic surface.

"Direct intervention of hairpin vortex in boundary layer to reduce friction drag"

Yong Duck Kang (University of Nottingham, Postdoctoral Researcher) : eaxydk@nottingham.ac.uk

Experiment studies have been carried out to confirm the effectiveness of a strategy controlling the hairpin vortices in the outer region of boundary layers. Many researches in friction drag reduction mainly focusing on the near wall region to target near-wall structures such as sweep events, ejection events and low-speed streaks. Once quasi-streamwise vortices grow into outer region of the turbulent boundary layer, they stretch their legs from the wall to form a hairpin vortex. As the head of hairpin vortex reaches to the edge of boundary layer, vortex core size of the legs is reduced and vorticity is increased. This top-down process of hairpin vortex, where the large-scale turbulence structures influence the small scale near-wall turbulence structures, can contribute to the friction drag in turbulent boundary layers. To interrupt this process, one can intervene in the process of forming and developing hairpin vortices, which could lead to a reduction in energy production and wall-skin friction. In order to verify this concept, a flow visualisation and hot-film measurement were carried out in a water channel. Here, hairpin vortex was generated beneath the test plate in a laminar boundary layer. The results demonstrated that a jet can destroy the neck of hairpin vortex by disconnecting the link between the head and legs of the hairpin vortex. Based on this study, an experiment was carried out in a fully developed turbulent boundary layer. When hairpin vortex passed a detection probe, it was targeted by a jet to counteract with the up-wash motion in the head of hairpin vortex. The upwash motion was stopped by the jet issued from a nozzle and then pushed down toward the wall. This led to a reduction in velocity near the wall, thereby reducing in the friction velocity. The legs of a hairpin vortex seem to be disconnected from the head, reducing the vorticity around them. It is believed that the reduced vorticity of hairpin legs leads to a weakening of the sweep events in bringing high momentum toward the wall. The results show that about 5& reductions in spanwise averaged coefficient of friction coefficient (C_f) is achieved.

"Development of the Motorized seat Belt and Belt Tension Estimation"

Kangseok Lee (Austria) : ks1549@gmail.com

This paper describes of the verification of a relationship between operation of the Motorized Seat Belt (MSB) and the tension by the MSB through the model. Recently, increasing of interest to car safety, the research to improve the traditional seat belt to the MSB is progressed. The MSB is one of the representative of Active Safety System which can protect passengers before car accidents. The MSB can warn or call the passenger\'s attention to a car accident by generating different tension which correspond to the situations. Besides, the MSB can offer more comfortable driving environment by some operation for accommodation. For a realization of these operations, to recognize the relationship between MSB operations and the tension by the MSB is important. The MSB can offer more comfortable and safe driving environment, and helpful for development of the MSB system by recognize the relationship correctly.

PT Symposium – Keynote Speech

"Turning Polymers into Structural Materials"

Alma Hodzic (The University of Sheffield, Professor) : A.Hodzic@sheffield.ac.uk

Polymers offer unique solutions in modern materials design for engineering applications due to their variety and choice of properties, both in thermoplastic and thermosetting families. Recent advances in sustainable materials design set new directions for the use of polymeric materials in engineering applications, and enabled 'commodity' materials to be effectively reinforced and used in structural applications. The aspects such as recyclability, life cycle analysis, biodegradability and low-cost manufacturing have become increasingly important in governments' agendas, and such initiatives drove development of new products based on either selection of re-usable materials, or recycled polymers from industrial and municipal waste. Optimisation of reused polymer blends is a challenge both for manufacturing processes and design solutions due to the unpredictability of recycled compounds. Nanotechnology offers solutions that reduce batch variability and provide more consistent properties, whilst effective fibre reinforcement with appropriate selection of coupling agents and additives increases the mechanical properties several times over that of the base polymer compound. Such solutions turn unusable waste into resources, and offer low-cost alternatives to the standard engineering materials whilst preserving the same degree of performance.

At the other end of the spectrum, the advanced structural materials such as glass and carbon fibre composites, have recently evolved into more functional systems which allow higher impact resistance, higher electrical and thermal conductivity, blast dissipation, self-sensing and self healing among other properties. The aerospace industry has driven the use of composites and fibre metal laminates into aircraft structures, with Boeing 787 currently using composites in 50% of its structure. The next stage is to develop a self-monitoring and self-repairing system that will effectively reduce the need for standard maintenance and increase safety of multi-functional structures.

Nanotechnology plays here even a bigger part, as the effective dispersion of nanoparticles and orientation of nano-carbon structures in polymeric materials have recently shown improvement in thermo-electrical and mechanical properties without compromising the overall performance of these advanced structural materials. Tailoring materials on a nanometer level is geared towards not only better structural performance; it also offers new joining and fastening solutions that will reduce the number of parts and further improve the weight saving in composite structures.

The next era of materials technology will see a sound selection of sustainable technologies for low-performance structural applications and multi-functional advanced systems for high-performance structural solutions. Both evolutionary paths are geared towards reduction of fuel and energy use required in production and service of such applications.

PT Symposium – Oral

"Generation and Application of High Brightness Electron Beams"

Jang-Hui Han (Diamond Light Source, Senior Researcher) : jang-hui.han@diamond.ac.uk

High brightness electron beams are useful for many applications like high energy accelerator, medical therapy and scientific experiment. In this presentation, we will review the state of the art technology for electron beam generation and the characteristics of such beams. Some examples of application will be discussed as well.

"Phase analysis of advanced high strength steels by electron backscatter diffraction"

Jun-Yun Kang (Universite Paris 13, Postdoctoral Researcher) : kjy04@snu.ac.kr

There have been increasing demands for the weight reduction of automobiles during recent a few decades. Weight reduction means higher fuel efficiency and less gas emission, so is of great importance in economics and environmental issues. Steels which have long been the major material for constructing auto body are now facing a new wave, high strength for gauge reduction. The resulting new products, so called advanced high strength steels (AHSS) are usually composed of multi-microconstituents or microscopic phases. Quantitative evaluation of the phase volume fraction is therefore necessary to estimate the properties of these steels. Electron backscatter diffraction (EBSD) which has provided many new possibilities in microscopy for material characterization can be a proper choice for this purpose. In this study, a new analytic method was developed and applied to the phase separation and volume fraction estimation in two representative AHSS steels, dual phase (DP) and transformation induced plasticity (TRIP) steel.

"Molecular weight dependence of polymeric insulators on the leakage current in organic thin film transistors"

Hyeok Kim (Ecole Polytechnique/Univ.Paris7) : fomalhout@gmail.com

Organic thin film transistors which have cost advantage to fabricate flexible devices have been an issue over the decades. However, no research has been performed about molecular weight of polymeric insulators in organic thin film transistors. Moreover, a lot of research has not been carried out about the leakage current, that is one of the most important characteristics of organic thin film transistors. In this paper, the change of leakage current in organic thin film transistors has been analyzed physically as molecular weight of polymeric insulator varied. Especially, when binary mixture of polymers which had two molecular weights was used as the gate insulator, the leakage current decreased over ten times linearly in organic thin film transistors than when polymer which had a unique molecular weight was used as gate insulator. It is established the reason that hydroxyl group density of polymeric insulators is distinguished according to the change of molecular weight when the polymeric insulators are thermally crosslinked.

"Constrained sintering of SOFC electrolytes"

Jung-Sik Kim (Imperial College London) : j.kim@kseauk.org

In some planar solid oxide fuel cells (SOFCs) a thin film electrolyte is sintered onto a rigid substrate which constrains the film. Here we describe measurement of the constrained densification kinetics and sintering stress for zirconia films on fully dense zirconia substrates. 10 μm thick 3YSZ green layers were screen printed on 150 or 300 μm YSZ substrates. The stress induced by sintering (to 1300 and 1350°C) was deduced from monitoring the bending displacement of strips of bi-layers using a long-distance microscope and found to be 2.1 0.2 MPa. The sintering kinetics of the constrained films were measured, using a sensitive optical dilatometer, and compared with the sintering kinetics of a free-standing film and bulk powder compacts. The effect of the constraint is to slow the sintering rate by an amount equivalent to a reduction of approximately 200 K in temperature and limit the final density to approximately 93&. The sintering stress is sufficient to induce relatively large defects in the constrained film which probably originate from local regions of lower than average packing density in the green film.

"Collision of non-Newtonian impinging jets and its application for drop-on-demand inkjet printing"

Sungjune Jung (University of Cambridge, Ph.D.student) : sjj37@cam.ac.uk

Experimental observations and analysis are presented for the formation and atomisation of the fluid sheet created by obliquely colliding non-Newtonian jets. A series of solutions of monodisperse polystyrene (PS) dissolved in diethylphthalate (DEP) and polydisperse polyethylene oxide (PEO) in a glycerin-water mixture were tested to investigate the viscoelastic effect on the impinging jets ejected from needles with an internal diameter of 0.85 mm. The oblique collision of two continuous liquid jets leads to the formation of a thin oval liquid sheet bounded by a thicker rim which disintegrates into ligaments and droplets. Under certain conditions the flow structure exhibits a remarkably symmetrical &fishbone& pattern composed of a regular succession of longitudinal ligaments and droplets. We discovered for the first time that the series of fluid regimes for viscoelastic fluids differs from those for a Newtonian fluid, and that they also show different behaviours depending on their viscoelasticity. In order to investigate viscoelastic effects on the resulting fluid, we focus on the periodic atomisation, the so-called fishbone pattern. Single-flash illumination reveals that the presence of low concentrations of polymers significantly affects the evolution of the sheet and the fragmentation, the shape of the sheet and the ligament, and the drop size. Maximum Fishbone Angle has proved to be a useful tool to distinguish viscoelastic regimes in terms of diluteness, exhibiting a power-law scaling with the reduced concentration (c/c*). Finally, we investigate the possibility of using this method to predict the jettability of the ink in a piezo inkjet printhead. Jetting experiments with the same fluids in a Xaar piezoelectric drop-on-demand print head were also used to characterise their jetting performance in terms of the maximum ligament length, a crucial parameter in determining the printability of the fluid. There are close similarities between the ligament collapse behaviour in both experiments. Good correlation was found between the maximum fishbone angle and the maximum ligament length in the jetting experiments, which suggests that a test based on oblique impinging jets may be useful in the development of fluids for inkjet printing.

"Introduction of Biological Fuel Cell technologies for renewable energy"

Sung Taek Oh (University of Glasgow, EPSRC advanced research fellow) : sungtaek25@hotmail.com

A microbial fuel cell (MFC) is a device to use for bio electrochemical energy production. The MFC typically consists of two components (an anode and a cathode separated by an ion conducting membrane). Biological catalyst (protein or living cell) at the anode oxidises fuel and transfers electrons through external wire to the cathode. Protons migrate across the membrane to the cathode where they combine with oxygen and electrons to form water. The device may provide new opportunities for the renewable energy in not only nano medical industries for battery but also renewable power plant such as waste water/swage treatment plants. Here, the purpose of this paper is to provide an overview of microbial fuel cell technology and review the current MFC applications. The research trends of the new technology may help many researchers making a bridge between current nano biotechnology, information and environmental technology.

"Technopark in the Utopian Theory's point of view"

Jinwon Seo (France) : tradarch@gmail.com

1. Could we talk about the Utopian Theory?

The word 'utopia' takes a negative meaning in the normal linguistic use such as 'no realizable', 'only in the dream'. But we can not deny also its positive effect like 'will to be', 'perspective historic' mostly found in the prospective urbanism projects especially in the nineteenth and twentieth centuries. We are trying to convert this negativity of the 'utopia' to the positive 'devenir' in the purpose of the Technopark.

2. Technopark in the hyperdisciplinary process

In the development of the Technopole (some kind of Technopark in French), the multi(pluri)disciplinary, interdisciplinary and transdisciplinary approaches took the important place in the system educative and productive. Now, we propose the hyperdisciplinary producing approach in the prospective point of view about the Technopark especially will be constructed in Korea.

"Search for New Physics with Positrons in Space"

Chanhoon Chung (RWTH Aachen University, PostDoc) : chchung@physik.rwth-aachen.de

Today the universe consists of 4.6& of ordinary matter, 23.3& of dark matter and 72.1& of dark energy. The dark matter is generally assumed be stable, non-relativistic and only weakly interacting. But we do not know what the dark matter is made of and how it is distributed within our Galaxy.In general, the cosmic anti-particles are expected as secondary products of interactions of the primary cosmic-rays (CRs) with the interstellar medium during propagation. While the measurements of CR positrons have become more precise, the results still do not match with the pure secondary origins.The comparison of the expected background of positrons with experimental data have been performed using CR propagation models. A phenomenological study based on the beyond Standard Model is carried out and shows a better interpretation of CR fluxes including heavy dark matter annihilations in the galactic halo or center. However, future experiments have to provide measurements of CRs up to TeV energy scale with higher precision.The Alpha Magentric Spectrometer (AMS-02) will be the major particle physics experiment on the International Space Station (ISS) and measure high energy CRs with unprecedented accuracy for at least three years in space.The mission will extend our knowledge on primary CR origins, acceleration and propagation mechanisms. Especially, the precise measurement of the positron flux may be the most promising tool for the detection of the dark matter in space, since the predicted positron flux is less sensitive to the astrophysical parameters responsible for the propagation and the dark matter halo profile.The fully assembled AMS-02 detector will be forseen high energy beam test at CERN in Switzerland. Afterwards a space qualification test using a large space simulator at the ESA-ESTEC in Netherlands this year.Then it will be delivered to NASA-KSC to prepare for the launch with a space shuttle \'Discovery\' at end of this year. A launch date for AMS delivery (STS-134) to ISS is currently targeted on September 16, 2010.

"Enhanced Predictive Modelling Using Multi-Block PLS"

Jeong Jin Hong (Newcastle University, Ph.D.student) : j.j.hong@ncl.ac.uk

To analyse data measured from different process stages/phases or different types of measurement is difficult using the conventional statistical methods such as PCA and PLS which all data go into one matrix as it is hard to find out the relationships among the variables or stages within one big matrix. Multi-block methods enable the data to be separated into sub-blocks by appropriate knowledge of the systems. For example, different types of measured data, different process stage or phase. Sub-blocks can be analysed individually. Therefore, it can provide information about within sub-block as well as between sub-blocks. And it is useful to understand the process better. However, many literatures stated that there is no advantage of using the multi-block methods for predictive modelling only as prediction performances of PLS model and multi-block PLS model show same. Thus, multi-block methods usually used for process fault detection and diagnosis. Hypothesis that prediction performance can be improved if appropriate variable selection in each sub-block is obtained is applied into this research. To obtain the appropriate variable selection, model parameters from MBPLS models is used as high magnitude of model parameters can represent high contributions to the processes. In other words, variables in each sub-block showing higher contributions according to model parameters are selected to build models. Data is produced using penicillin fed-batch simulator, and data is divided into two sub-blocks by different process modes, batch mode and fed-batch mode. 10 cases having sets of training and validation data with different batches are used to build model. Each training and validation data set is used to build a model and to select appropriate number of latent variables. Data is not selected as training and validation data is used as unseen testing data to evaluate the models. Models are built by PLS, MBPLS and MBPLS with selected variables using proposed method. All cases using proposed variable selection method show improved prediction performances on validation data and 8 cases out of 10 cases show better results on testing data than those from conventional PLS and MBPLS models.

PT Symposium – Poster

"Synthesis of InN and InMnN nanowires by halide vapor phase epitaxy"
Yun-Ki Byeun (Institute fuer Gesteinshuettenkunde in RWTH-Aachen) : byeun@ghi.rwth-aachen.de

Indum nitride (InN) is an interesting and potentially important semiconducting material with superior electronic transport properties. Compared to other group-III nitrides, InN possesses the lowest effective mass, the highest mobility, and the highest saturation velocity. Therefore it is very suitable for high speed and high fresquency electronic device applications. Rescently, this material has received much attention due to the discovery of its narrow direction bandgap value is correct, (Al,Ga,In)N-based lighting emitting devices can be operated in the wavelength range from near-infrared to violet(0.7~6.2eV). Spite of having those advantages, high quality InN has not been achieved mainly due to the lack of a lattice mismatch and restricted growth conditions owing to low dissociation temperature of InN itself. So we investigate about growth method of InN and InMnN having high quality crystalline, as well as characteristic of grown 1-D InN nanostructure by HVPE.

"High efficient donor–acceptor ruthenium complex for dye-sensitized solar cell applications"
Il Jung (EPFL, Post-Doctor) : il.jung@epfl.ch

A highly efficient heteroleptic ruthenium (II) complex cis-di(thiocyanato)(4,4\'-<WBR>dicarboxylic acid-2,2\'-bipyridine)(4,4\'-<WBR>di-(2-(4-ditolylamine phenyl)ethenyl)-2,2\'-<WBR>bipyridine) ruthenium (II) (IJ-1) was synthesized and characterized, which when anchored on nanocrystalline TiO2 films exhibited high power conversion efficiency, 10.3%, and incident photon to electron conversion efficiency, 87%

"Contribution of the curved surfaces on magnetization behavior"
Jehyun Lee (Vienna University of Technology) : jehyun.lee@tuwien.ac.at

Micromagnetic simulation is a powerful tool to investigate the magnetization behavior of the magnetic recording media. The main parameters of micromagnetic simulation are the crystalline anisotropy KU, saturation magnetization MS, and the shape of the particle that determines the exchange coupling and magnetostatic interactions with its neighbor particles. The importance of the shape has been pointed out throughout the development of magnetic materials as well as those of KU and MS. Nevertheless, the shape effect of the grain was not studied deeply due to the difficulty in modeling of realistic curved surfaces. In this study we have investigated the effect of the convex and concave surfaces in magnetic recording media, aided by the Surface Evolver; an open source program that calculates the equilibrium shape under given surface tensions. With given grains with curved surfaces, the contributions of the intergranular exchange and magnetostatic interactions are studied. For intergranular exchange study, an analytic equation is derived by modifying the Stoner-Wohlfarth equation, and its results are proved by finite element micromagnetic simulation. Besides, the study of magnetostatic interaction resulted in an opposite result to the previous simulation, but agrees to the experiments.

"사고박물관"
Chul woo Hyun : ch@moserarchitekten.at

60년대,70년대 한국은 `경제성장 우선주의`라는 환각상태에 도취돼 허울좋은 성장에 성장을 거듭해 왔던게 사실이다. 그에 댓가라도 치르듯 `재난 왕국`이라는 오명을 씌우게 될 만큼 부끄러운 인재(人災)들로 한국 근현대사를 가득 채우고 말았다. 소중한 생명을 앗아간 수많은 대형참사들을 기억하기위해 널리 알리고 그 고인들과 유가족들을 위로하며, 일선 공무원으로서 국민의 안전의식향상을 위한 작은 건축적 움직임으로써 시작한 이 프로젝트는 금세기 한국 대형참사 1위로 5백 2명의 무고한 희생을 치른 전 삼풍백화점자리를 계획대지로 선택하여 대지선정에서부터 사고를 기념의 의미를 부여하였다. 역사적으로 조선시대 상궁과 현관들의 무덤이기고 했던 이 터에서 미스터리하게도 대형붕괴사고를 낳았고 그 후 이 지역은 정부로 의해 특별재해지역으로 선포되기도 하고 삼풍기념공원으로 남겨두자는 수 많은 국민들의 염원이 있었음에도 불구하고 주거시설로 전혀 적합하지 않은 땅에 거대한 주상복합건축물이 용도변경의 의해 완공되었다. 이런 사실들을 토대로 남겨진 그대로를 비워둠으로써 채움의 의미를 갖는 컨셉을 두고 `지하사고박물관`에 접근했다. 정적인 장소로써의 박물관에만 집중하던 중 개인의 선택에 의해 관람을 목적으로 하는 박물관기능 이외에 유동인구의 동선을 자연스럽게 유도하여 동적인 장소로써의 기능 또한 추가, 의식적으로든 무의식적으로든 이 곳을 지나며 사고박물관을 인식하게 되는 장치로 계획대지 좌측으로 근접해 지나는 지하철 3호선을 이용한 새로운 지하철역을 택했고

아파트단지들과 관공서들이 밀집된 지역성을 가만해 지하주차장의 기능도 두었다. 이런 세가능 기능은 제각기 커튼월구조로 분리되어있지만 위로는 공원으로써 보행가능한 하나의 지붕구조가 아래로 지하 중앙홀을 중심으로 안으로 둘러싸고 있다. 이렇게 동일한 천장아래 분리된 기능을 가진 내부에 지름 2.8m, 2m, 1m 인 서로 다른 크기의 천창을 통해 본질적 의미에서 영혼이 잠든 곳에 낮에는 새로운 다양한 생명의 빛을 드리우고 밤에는 재생된 삶의 빛이 어둠의 세상을 밝힐것이다. 남북 쌍방으로 경사진 본래의 대지조건을 최대한 이용해 그 자체로 만들어내는 경사를 남북 주출입구의 축으로 삼았으며 3호선 지하철 노선을 대지 안으로 끌어들이면서 전체적인 평면계획이 그 연장선상에서 진행되었다. 내부 만남의 장소에 큰 고목을 두고 박물관 입구 전면에 국화꽃 진열대를 마련해 일반 지하철역과는 다른 추모의 장소로써 오가는 사람들의 머리속에 자연히 새겨질것이다. 외부로는 높이 30cm, 50cm, 70cm 로 마감재료를 목재로 한 벤치를 뒤 짙은 희색에 콘크리트의 우울함을 중화시킨다. 대지면적 15,397 m2 전체가 비워채 남겨진 공원으로써 각종 추모행사들을 가능하게 할 것이며 하나의 거대한 기념비으로써의 그 역할을 해낼것이다.

"Geotechnical Centrifuge Technology and Its Applications"

Boung Shik Yoon (The University of Nottingham) : ybs1024@hotmail.com

Physical and model testing plays an important role in the research of geotechnical engineering. Earth structures (e.g. retaining walls, embankment and cutting slopes, and tunnel etc.) are typically several metres in size, so that &full-scale& tests are rarely undertaken, and tests at reduced scale are considerably more attractive. However, the behaviour of &soil& shows extreme dependency on the &ambient& stress, which is generally high (e.g. a few metres below the ground surface), so that reduced scale testing at normal gravity does not enable to duplicate realistic stress regime. Geotechnical centrifuge modelling allows full-scale stresses to be reproduced in a reduced (in this case 1/50th) scale model by applying a correspondingly increased centrifugal &gravity& force, which effectively increases the self-weight and therefore size of the model. Geotechnical centrifuges may be the most efficient and robust tool both in physical testing and in practical design.This paper will review the fundamentals of geotechnical centrifuge and present current centrifuge modelling technique with its applications to various geotechnical problems.

GPSR Compliance

The European Union's (EU) General Product Safety Regulation (GPSR) is a set of rules that requires consumer products to be safe and our obligations to ensure this.

If you have any concerns about our products, you can contact us on ProductSafety@springernature.com

In case Publisher is established outside the EU, the EU authorized representative is:

Springer Nature Customer Service Center GmbH
Europaplatz 3
69115 Heidelberg, Germany

Batch number: 10755284

Printed by Printforce, the Netherlands